CLIMATE CHANGE AND MICROBES

Impacts and Vulnerability

CLIMATE CHANGE AND MICROBES

Impacts and Vulnerability

Edited by
Javid A. Parray, PhD
Suhaib A. Bandh, PhD
Nowsheen Shameem, PhD

AAP APPLE ACADEMIC PRESS

First edition published 2022

Apple Academic Press Inc.
1265 Goldenrod Circle, NE,
Palm Bay, FL 32905 USA

4164 Lakeshore Road, Burlington,
ON, L7L 1A4 Canada

CRC Press
6000 Broken Sound Parkway NW,
Suite 300, Boca Raton, FL 33487-2742 USA

2 Park Square, Milton Park,
Abingdon, Oxon, OX14 4RN UK

Apple Academic Press exclusively co-publishes with CRC Press, an imprint of Taylor & Francis Group, LLC

Library and Archives Canada Cataloguing in Publication

Title: Climate change and microbes : impacts and vulnerability / edited by Javid A. Parray, PhD, Suhaib A. Bandh, PhD, Nowsheen Shameem, PhD.

Names: Parray, Javid Ahmad, editor. | Bandh, Suhaib A., editor. | Shameem, Nowsheen, editor.

Description: First edition. | Includes bibliographical references and index.

Identifiers: Canadiana (print) 20210370149 | Canadiana (ebook) 20210370203 | ISBN 9781774637210 (hardcover) | ISBN 9781774637968 (softcover) | ISBN 9781003189725 (ebook)

Subjects: LCSH: Microbial ecology. | LCSH: Climatic changes. | LCSH: Microorganisms. | LCSH: Plants—Microbiology.

Classification: LCC QR100 .C53 2022 | DDC 579/.17—dc23

Library of Congress Cataloging-in-Publication Data

CIP data on file with US Library of Congress

ISBN: 978-1-77463-721-0 (hbk)
ISBN: 978-1-77463-796-8 (pbk)
ISBN: 978-1-00318-972-5 (ebk)

About the Editors

Javid A. Parray, PhD

Javid A. Parray, PhD, is an Assistant Professor in the Higher Education Department at Government Degree College Eidgah, Srinagar, India, where he teaches the subject of environmental science. He has published many research articles in reputable refereed international and national journals. He had authored books on different themes, including Approaches to Heavy Metal Tolerance in Plants; Sustainable Agriculture: Biotechniques in Plant Biology; and Sustainable Agriculture: Advances in Plant Metabolome and Microbiome. He was awarded the "Emerging Scientist of Year Award" by the Indian Academy of Environmental Science for the year 2015 in addition to being awarded with an international travel grant for participating in international conferences. Dr. Parray completed a fast-track project entitled "Molecular characterization and metabolic potential of rhizospheric bacteria from Arnebia benthamii across North Western Himalaya" at CORD, University of Kashmir, India. He finished his postdoctorate research associateship on a DBT-funded project entitled "Tissue culture-based network programme on saffron." He earned a PhD in Environmental Science with a specialization in plant microbe interaction on the topic "Evaluating role of Rhizospheric bacteria in saffron culture" from the University of Kashmir, India.

Suhaib A. Bandh, PhD

Suhaib A. Bandh, PhD, is an Assistant Professor in the Higher Education Department of Jammu and Kashmir at Government Degree College, DH Pora, Kulgam, India, where he teaches courses on environmental science and disaster management at the graduate level. He is the President of the Academy of EcoScience in addition to being a life member of the Academy of Plant Sciences, India, and the National Environmental Science Academy. He is a recipient of many awards and has participated in a number of national and international conferences organized by reputed scientific bodies in India and abroad. He has a number of scientific publications to his credit, published in some highly reputed and impacted journals. Dr. Bandh has recently edited and authored a number of books, including *Freshwater Microbiology: Perspectives of Bacterial Dynamics in Lake Ecosystems; Environmental Management: Environmental Issues, Awareness and Abatement,* and *Environmental Perspectives and Issues.* Dr. Bandh has also edited a special issue on "Ecology and Biotechnological Applications of Biofilm" in *BioMed Research International.*

Nowsheen Shameem, PhD

Nowsheen Shameem, PhD, is an Assistant Professor in the Department of Environmental Science at Cluster University Srinagar, India. She has worked as a project associate on the DBT-sanctioned project on "Spawn production for the entrepreneurs of Kashmir Valley" at CORD, University of Kashmir, India. She has also worked as a group leader for drafting a number of research proposals and ideas. She has published many research articles in reputed, refereed international and national journals of Springer, Elsevier, and Hindawi. She has also presented her research work at international and national conferences. Dr. Shameem finished her doctorate through the University of Kashmir on "Phytochemical analysis and nutraceutical value of some wild mushrooms growing in Kashmir Valley" in 2017.

Contents

Contributors

Maajid M. Bandh
Department of Zoology, University of Kashmir, Srinagar, India

Suhaib A. Bandh
Government Degree College, DH Pora, Kulgam, India

Shahnaz Bashir
Sri Pratap College Campus, Cluster University, Srinagar, India

Aashia Beigh
Sri Pratap College, Cluster University Srinagar, India

Basharat A Bhat
Department of Bioresources, University of Kashmir, Srinagar, India

Mohd Yaqub Bhat
Section of Mycology and Plant Pathology, Department of Botany, University of Kashmir, Srinagar, Jammu and Kashmir, India

Zia Ur Rahman Farooqi
Institute of Soil and Environmental Sciences, University of Agriculture, Faisalabad, Pakistan

Aparna B. Gunjal
Department of Microbiology, Dr. D.Y. Patil, Arts, Commerce and Science College, Pimpri, Pune, Maharashtra, India

Muhammad Mahroz Hussain
Institute of Soil and Environmental Sciences, University of Agriculture, Faisalabad, Pakistan

M. G. Ingale
Department of Microbiology, Bajaj College of Science, Wardha, India

S. S. Khandare
Department of Microbiology, Bajaj College of Science, Wardha, India

Aqsa Khursheed
Sri Pratap College, Cluster University Srinagar, India

Ishfaq Majeed
Department of Zoology, University of Kashmir, Srinagar, India

Rakeeb A Mir
Department of Biotechnology, BGSB University, Rajouri, India

Haika Mohi-ud-din
Sri Pratap College, Cluster University Srinagar, India

Waqas Mohy Ud Din
Institute of Soil and Environmental Sciences, University of Agriculture, Faisalabad, Pakistan

Mohazib Nabi
Sri Pratap College Campus, Cluster University, Srinagar, India

Anees Un Nisa
Section of Mycology and Plant Pathology, Department of Botany, University of Kashmir, Srinagar, Jammu and Kashmir, India

Tanzeela Rehman
Sri Pratap College Campus, Cluster University, Srinagar, India

Fahad Rasheed
Department of Forestry & Range Management, University of Agriculture, Faisalabad, Pakistan

Christy BK Sangma
ICAR Research Complex for NEH Region, Nagaland Centre, Jharnapani, Medziphema, Nagaland, India

Sabira Sultana
Gauhati University, Guwahati, Assam, India

Lubna Tariq
Department of Biotechnology, BGSB University, Rajouri, India

Irteza Qayoom
Sri Pratap College, Cluster University Srinagar, India

Abdul Hamid Wani
Section of Mycology and Plant Pathology, Department of Botany, University of Kashmir, Srinagar, Jammu and Kashmir, India

John Mohd War
Section of Mycology and Plant Pathology, Department of Botany, University of Kashmir, Srinagar, Jammu and Kashmir, India

Abbreviations

ACC	1-aminocyclopropane-1-carboxylate
ACO	ACC oxidase
ACS	ACC synthase
AM	arbuscular mycorrhizal
AMF	Arbuscular mycorrhizal fungi
CAC	Codex Alimentarius Commission
CFCs	chlorofluorocarbons
CH_4	methane
CCMs	carbon concentrating mechanisms
CO_2	carbon dioxide
CPDs	cyclobutane pyrimidine dimer
DMS	dimethyl sulfide
DMSP	dimethylsulfoniopropionate
DNA	deoxyribose nucleic acid
DOM	dissolved organic matter
ECM	Ectomycorrhiza
FL	free-living
FUMs	fumonisins
GBS	Guillain-Barre syndrome
GHSs	greenhouse gases
GT	gigatonnes
HABs	harmful algal blooms
HAV	hepatitis A virus
HFCs	hydrofluorocarbons
HGT	horizontal gene transfer
HUS	haemolytic uremic syndrome
IAA	indole acetic acid
INFOSAN	International Food Safety Authorities Network
IOC	Intergovernmental Oceanic Commission
IPCC	Intergovernmental Panel on Climate Change
IR	infrared
ISR	induced systemic resistance
MCP	microbial carbon pump

NH_3	ammonia
N_2O	nitrous oxide
NO	nitric oxide
OA	ocean acidification
OAF	open-air factor
OECD	Organization for Economic Development and Cooperation
OTA	ochratoxin A
OTUs	operational taxonomic units
PA	particle-associated
PAHs	polycyclic aromatic hydrocarbons
PFCs	Perfluorocarbons
PGPF	plant growth-promoting fungi
PGPR	plant growth-promoting rhizobacteria
PLP	pyridoxal-5-phosphate
ppb	parts per billion
ppm	parts per million
ppt	parts per trillion
PSFs	plant-soil feedbacks
RNA	ribose nucleic acid
ROS	reactive oxygen species
SCF	Scientific Committee on Food
SF6	sulfur hexafluoride
SIP	stable isotope detection
SOM	soil organic material
TBARS	thiobarbituric acid-reactive substances
TCT	trichothecenes
TSS	total soluble sugars
UML	upper mixing layer
UV	ultraviolet
UNFCCC	UN Framework Convention on Climate Change
VAM	vesicular-arbuscular mycorrhizal
WHO	World Health Organisation
ZEA	zearalenone

Preface

In the contemporary environmental dialogue, climate change is a big concern. In this discourse, the anthropogenic causes of global warming and the downstream impacts of climate change on flora and fauna have been given a significant part of public interest, study, and funding. The disappearance of macro-species, ecosystems, and habitats has been reasonably well known. However, the debate on the transition of climates, vis-à-vis the possible effect on microbial ecosystems and the contribution by the microbial ecosystems, has mostly been ignored. To propagate more detailed extrapolations of climate change impacts, growing public understanding, and informing policymaking, the microbial composition must be integrated into ecosystem modeling. A "call to arms" addressed in recent researches shows that this acknowledgment has increased, and for this reason, many scientists urge to boost the literacy of societal microbial ecology.

Based on the definition of "colonizing resistance" of the human microbiome, it was presumed that the microbial populations were immune and resistant to invasions. Studies further support the findings that identify disruptions in the soil, marine, engineered, and human-associated habitats. It poured population taxonomic outlines to a new steady-state while preserving somewhat same practical dimensions. In these situations, however, the disruption past and the time taken to achieve this new state of balance for a group are significant. Recent years have seen evidence that neither microbial communities are immune to disruption, nor do they have the capacity to recover completely within a couple of years after a stress event. Several studies have highlighted the possible effects on the functionality of significant compositional changes in gut culture and long-term changes, while conditions are re-established before the disruption. The general result is that microbial habitats and populations are more susceptible than previously assumed to disruptions and stress events.

Functional and compositional changes are the two most critical types of disturbance reactions for microbial ecosystems. Pulse (short-term disruptions), and press (long-term disruptions) are the two most common disruptive forms. Microbial communities' response to disruptions depends on the form, number, duration, and intensity of the disturbances, practical and compositional considerations of the microbial environments, and the levels of practical redundancy. In the wake of ecological and human-induced pressures, a community with a higher degree of functional redundancy also benefits from the stability and vice versa. While a wide range of stress response mechanisms has been developed for unicellular organisms to improve resistance to physico-chemical stress. The secondary result of ecosystem function and composition modification may also be the biological reactions of microorganisms to promote their persistence in presence of environmental stress factors, e.g., ocean acidification causes certain microorganisms to change genetics to encourage cell maintenance than growth. However, the precise extent of how marine life reacts to these dramatic variations in pH is still unclear.

There is a large amount of data regarding microbial communities' high sensitivity to natural or anthropogenic stressors. Due to the lack of a systemic database, these studies are not easy to compare. The Microbiome Stress Project, however, aims to resolve this problem in the global collaborative initiative. Also, after environmental disasters and stress events, natural ecosystems tend to be mostly unexplored. In most studies, manipulation tests under controlled conditions are carried out due to the multiple compounding elements involved in the natural environment. To create a full picture of microbial ecology, population biology, ecological pressures, and other processes need to be understood extra thoroughly and incorporated into ecosystem ecology. Responding to various pressure factors can change the inclusive biospheres' equilibrium and vary dramatically with the resilience and tolerance to different sources of disruption. Thus, each ecosystem needs to be perceived and investigated explicitly for each cause of pressure and appropriate efforts to include it into a broad sense.

CHAPTER 1

Role of Microorganisms as Climate Engineers: Mitigation and Adaptations to Climate Change

MUHAMMAD MAHROZ HUSSAIN[1*], ZIA UR RAHMAN FAROOQI[1], FAHAD RASHEED[2], and WAQAS MOHY UD DIN[1]

[1]*Institute of Soil and Environmental Sciences, University of Agriculture, Faisalabad, Pakistan*

[2]*Department of Forestry & Range Management, University of Agriculture, Faisalabad, Pakistan*

Corresponding author. E-mail: hmahroz@gmail.com

ABSTRACT

As critical components to the carbon and nitrogen cycles, microorganisms participate substantially in the production and utilization of greenhouse gases, comprising carbon dioxide (CO_2), methane (CH_4), nitrous oxide (N_2O), and nitric oxide (NO). Although not often found in the forefront of climate change dialogues, changes, such as increased temperature and atmospheric CO_2 concentrations can significantly impact microbial processes in the soil. They may increase or decrease microbial-mediated greenhouse gas emissions. Microbial practices play a vital part in the global flow of the most important biological greenhouse gases and can react quickly to climate change. The microbial response to climate change will partially regulate the balance between carbon storage and damage in soil under potential temperature and rainfall circumstances. It is not clear whether variations in microbial processes is advantageous to positive or negative response of greenhouse gas emissions. This chapter briefly explained the role of microbes in climate change and the microbial mitigation in climate change. This chapter also highlighted the microbial adaptions to the climate change physiologically and community wise.

1.1 INTRODUCTION

From the first molecular oxygen, which was generated by oceanic cyano-bacteria about 3.5 billion years in the past, to the carbon-producing meth-anogens and the warmth of carbon-abundant bogs, microbial processes react to the most important drivers of climate change (Bartdorff et al., 2008). In general, microorganisms play an essential role in deciding the atmospheric absorptions of greenhouse gases, such as carbon dioxide (CO_2), methane (CH_4), and nitrous oxide (N_2O), which have the utmost influence on radiation propulsion. The history of the Earth is more debat-able about its role in the impending decades and centuries, its important climate feedback and how people use microbial processes to tackle climate change. In times of greenhouse gas flow, the microbial feedback response to climate change can rise or decline the rate of climate change (positive feedback) (Singh et al., 2010).

It is expected that the 21st century will experience one of the fastest climate changes in the history of the Earth and that the bioflow of the most important anthropogenic greenhouse gases is closely linked to the micro-organisms (Averill et al., 2014). Therefore, it has never been so essential to improve our understanding of microbial processes. In terrestrial ecosys-tems, people understand from a physiological and community perspective the reaction of plant communities and commensal microorganisms (e.g., mackerel mushrooms and nitrogen-fixing microbes) to climate change (Bardgett et al., 2008). Though, the reaction of heterotrophic soil micro-bial populations to climate change with global warming and changes in rainfall is unknown. This is an essential element because it affects the type and scope of the response from the terrestrial ecosystems. However, the diversity of microbial communities in the terrestrial environment is so broad and undeveloped that knowledge on the response of microbial populations to climate change is very complex and only a small food web is being built (Morgan, 2002).

Similarly, due to the impact of distinct terrestrial ecosystems (unlike microbial populations, land utilize, other obstacles such as management practices) and different biogeographical designs (spatial and temporal delivery of microbial communities), this situation is worsening (Farooqi et al., 2018). Soil respiration is the primary route for carbon transfer from land to the atmospheric pool (Schlesinger and Andrews, 2000). Soil microorganisms can produce a more significant impact on the formation of organic carbon in the soil than earlier assumed (Potthoff et al., 2008). For instance, mycorrhizal fungi can be the main route by which vegetable carbon penetrates the pond. Mycelium conversion accounts for approximately 60% of organic soil input, and the residual 40% is owing to root turnover and waste (Godbold et al., 2006).

Also, the kind of mycorrhizal fungi can ascertain the carbon content of the soil: Averill et al. (2014) found that the soil nitrogen content of an ecosystem dominated by ectomycorrhizal fungi is 70% higher than the carbon content of an arbuscular mycorrhizal fungal ecosystem. Therefore, the impacts of climate change on the microbial action and physiology of the soil establish the share of the annual carbon stored in the Earth in the long-term storage of organic carbon (Stuart Chapin III et al., 2009). Changes in the composition, frequency, and function of the microbial community have been noted in climate change trials that control temperature, precipitation, CO_2, and their connections (Averill et al., 2014; Godbold et al., 2006; Hawkes et al., 2011; Potthoff et al., 2008; Schlesinger and Andrews, 2000; Stuart Chapin III et al., 2009).

Temperature control in 32 experiments boosted soil respiration by 20% and net nitrogen mineralization by 46% (Rustad et al., 2001). Blankinship et al. (2011) studied 75 experimental climate experiments and observed that heating reduced bacteria, changes in precipitation, fungi, and CO_2 improved the total microbial biomass. However, since there are only two leading researchers in these groups, additional work is required to substantiate the universality of the answers. The processes that regulate the speed of operations in underground ecosystems are complex and offer these systems various, apparently features with great potential for empirical behavior.

1.1.1 ROLE OF MICROBES IN CONTROLLING GREENHOUSE GAS DISCHARGES

An insight into the physiology and dynamic range of microbial populations is essential for a better understanding of the mechanisms for controlling greenhouse gas flow (Allison et al., 2010; Schimel and Gulledge, 1998). The hypothesis is that the structure of microbial communities has nothing to do with significant-size ecosystem models, and the theoretical experience and methods to assess the variety of microbial populations in natural environments to establish relationships with ecosystems. Due to lack of advance modeling and analytical approaches, this issue has received little interest. However, latest innovations in molecular technology and its use to characterize non-cultivable microorganisms are beginning to provide more precise knowledge of microbial management of greenhouse gas emissions (Allison et al., 2010; Singh et al., 2010).

1.1.2   CONTROLLING THE RELEASE OF CO_2 WITH THE HELP OF MICROBES

In the worldwide carbon cycle, the annual CO_2 emissions from smouldering fossil fuels are comparable witho the pure CO_2 that flows into the soil, the sea, and the atmosphere. The current concentration of CO_2 in the atmosphere mainly depends on the balance between photosynthesis and breathing. In the ocean, photosynthesis is mostly done through phytoplankton, but autotrophic and heterotrophic respiration yields most of the carbon absorbed by photosynthesis to the dispersed inorganic carbon pool (Farooqi et al., 2018; Arrigo, 2005). In terrestrial ecosystems, higher plants absorb CO_2 in the environment primarily through net primary production. However, changing their role as symbiotic fungi or pathogens and using nutrients in the soil break down microorganisms. Then breathe heterotrophically and significantly improve net carbon exchange (Van Der Heijden et al., 2008).

Primary land production absorbs about 120 billion tons of carbon annually. It emits about 119 billion tons of carbon, half of which breathes through autotrophic (mainly plants) and the other half through heterotrophic soil microorganisms (Aonofriesei, 2019). Land and sea together form a net carbon sink of approximately 3 billion tons per year, which effectively absorbs 40% of the CO_2 discharges from the current usage of fossil fuels. Changes in land use (significant deforestation in the tropics) add 1–2 billion tons of carbon to the environment every year. Because the soil stores about 2 trillion tons of organic carbon, disruption in agriculture and other land uses in the soil significantly stimulates the rate of organic degradation and the net release of CO_2 into the environment (Smith et al., 2008). For example, microorganisms are more likely to be exposed to burried organic carbon and oxygen, so that ploughing or draining carbon plentiful organic soils is understood to invigorate decay and breathing (Smith, 2008; Farooqi et al., 2018).

Because of this cultivation and confusion, human intervention has lost an estimated 4–90 billion tons of carbon to the ground (Carney et al., 2007). These reactions are mediated by microbial activity, but unlike the production of nitrogen and methane dioxide, the production of CO_2 is due to a variety of microbial practices. However, recent findings challenge this hypothesis by proving that there is a direct link among CO_2 flows and structural and physiological alterations in microbial communities (Allison et al., 2010; Carney et al., 2007).

1.1.3 METHANOGENS ROLE IN CH₄ RELEASE

Microbial control of CH_4 emissions is much more direct worldwide than CH_4 emissions. The production of microbial CH_4 mainly causes natural discharges (around 250 million tons of CH_4 each year). This method is brought out by a group of anaerobic archaea in the gut of marshlands, the sea, rumen, and termites. Discharges from human activities (mainly rice, landfills, fossil fuels, and livestock) (around 320 million tons of CH_4 each year) exceed these natural resources, except some emissions from fossil fuels which is mainly driven by microbes (Solomon et al., 2007). Methanotrophic bacteria are necessary buffers for large amounts of CH_4 that are produced in specific environments. "Low empathy" (involved only at CH_4 concentrations of 40 ppm, also referred to as vegetable CH_4 organisms of type I), vegetable CH_4 organisms that belong mainly to the bacteria of the genus *Propionibacterium* and typically produce greater quantity of CH_4. The Earth polluted it before it escaped into the atmosphere (Reay, 2003). The described "high-empathy" plant CH_4 organisms (involved at CH_4 concentrations below 12 ppm) belong mainly to the bacteria of the genus *Propionibacterium* (also known as type II plant CH_4 organisms) and makeup around 30 million tons of CH_4 in the atmosphere (Reay, 2003).

1.1.4 MICROBIAL ROLE IN NITROUS OXIDE CONTROL

Like CO_2 and CH_4 emissions, global N_2O emissions are primarily based on microbes. Natural and anthropogenic sources come mainly from soil discharge due to microbial nitrification and denitrification (Solomon et al., 2007). For every ton of active nitrogen (mostly manure) that is dumped on the surface of the soil naturally or deliberately, 10–50 kg of N_2O is released (Crutzen et al., 2016). Slight is recognized around the level of microbial management of these practices at the environment intensity. Still, some experiments have been conducted to differentiate the relative impacts of nitrification and denitrification to gain N_2O flow. Most of the N_2O generated by nitrification is the result of the action of autotrophic ammonia (NH_3) oxidizing bacteria, which belong to autotrophic bacteria (β-protein bacteria) (Teske et al., 1994).

Though recent experiments have shown that certain archaea also play an essential role in nitrification, but their comparative role in this process is still under discussion. In conflict, denitrification is a multi-step method in which every step is arbitrated by a group of microbes that have the enzyme required to catalyze this stage (Leininger et al., 2006). Nitrogen dioxide production is usually the result of incomplete denitrification. Denitrifying enzymes (such as nitrate reductase) in every catalytic process are genetically highly conserved, but the denitrifying activity is distributed across the phylogenetic population. Recent studies provide direct evidence that there is a close association between denitrifying bacterial species and soil N_2O release rates (Salles et al., 2009).

1.2   IMPACT OF DIFFERENT MICROBES IN THE MITIGATION OF CLIMATE CHANGE

The manipulation of terrestrial ecosystems offers potentially useful tools for controlling artificial climate change. The following section describes approaches that can be used to prevent soil microbial populations and to mitigate climate change.

1.2.1   REDUCED CARBON DIOXIDE EMISSION THROUGH MODIFIED MICROBIAL POPULATIONS

Today, soils comprise approximately 2000 pg (picogram) of organic carbon, double as much carbon in the environment, and three times as much vegetation (Smith, 2004). Different types of lands (e.g., forest, prairie, arable land) have various carbon storage capacities, suggesting that land use can be managed so that 1 pg of carbon is absorbed into the soil each year (Singh et al., 2010). This potential for soils is very high since complicated biological processes control the effects of the absorption of organic carbon in the soil and changes in the abiotic components, such as water, temperature, land usage, and fortification with nitrogen fertilizers (Reay et al., 2008). Forest soils are incredibly efficient in storing carbon. This is because fungi in the soil are higher than bacteria, which makes carbon binding easier (Castro et al., 2010; Farooqi et al., 2020). Understanding their ecology and function is essential for managing soil microbial communities and increasing carbon sequestration.

This is a challenge due to the failure to distinguish the biodiversity and role of soil microbial populations and the lack of academic standards of microbial ecology, such as species definitions and factors that determine the formation and structure of the community. There are indications that bacteria can be classified according to their carbon mineralization capability and separated into auxotrophic types (categorized by a high growth rate of unstable carbon that prevails in eutrophic environments) (Fierer et al., 2007). It has been indicated that eosinophilic bacteria are oligotrophic bacteria, while *Proteus* and *Actinomycetes* establish plant communities. Controlling land usage (e.g., switching from cultivated land to forestry) and agricultural traditions (e.g., utilizing a small amount of nitrogen for agriculture) can promote the growth of malnourished communities (Farooqi et al., 2020).

However, it is also essential to realize the ecological strategies of other critical microbial taxa. Not all taxa in all categories may be nutritious or oligotrophic, so the class alone may not be a forecaster of soil carbon reduction (Fierer et al., 2007). Therefore, it is essential to use rapidly evolving techniques (such as high data sequencing) to comprehend soil microbial variety well. Besides, new skills, such as *metagenomics, super-transcriptomics, super-proteomics*, and stable isotope detection (sIP) should be used to investigate the physiological functions and functions of different taxa in the ecosystems. This is the only way to use microbial ecology to expect whether a specific soil is a net carbon source or sink. However, it was found that with low or direct sowing, the N_2O emissions of the soil can increase (the denitrification rate increases expected to anaerobic circumstances in the trodden soil). By expanding the content, you can find a compromise that can be compensated for some benefits of reserves (Farooqi et al., 2018). The transformation of cultivated land into endless grassland leads to the accumulation of organic material on the soil surface and increased carbon sequestration (Farooqi et al., 2020). It is also possible to control various functions of plants in degraded or agriculturally improved soils to control the carbon content released into the soil (Steinbeiss et al., 2008).

In some cases, the simultaneous use of nitrogen fertilizers can increase crop yields and carbon storage in soils by suppressing the microbial degradation of persistent organic matter (Schimel and Schaeffer, 2012). Conversely, it is not possible to fully explain how carbon uptake in the soil promotes nitrogen accumulation, as there is an indication of the effect of nitrogen accumulation on carbon storage in the land (Ciais et al., 2014; Erisman et al., 2008). To maximize the long-term carbon sequestration potential in soils, you should better understand the interactions between different climate zones (temperature, water content, and groundwater level) and soils (pH, water content and texture). Biological properties (bacteria, fungi, archaea, soil animals, plants and their consumers) that affect the carbon cycle in the soil are limited (Steinbeiss et al., 2008).

1.3   METHANOGENS MANAGEMENTS AS ALTERNATIVE CLIMATE CHANGE MITIGATION OPTION

By our understanding, methane (most toxic greenhouse gas) emission involved the more complex biochemical microbial processes then CO_2 and N_2O (Reay, 2003). Though, several of the above ambiguities also apply to the management of CH_4 ground traffic. Because atmospheric microorganisms generate nearly all CH_4, it is tentatively possible to monitor most CH_4 releases in the terrestrial ecosystems by controlling the composition and practices of microbial communities (Bartdorff et al., 2008). The natural oxidation of CH_4 by CH_4 nutrient organisms accounts for only 5% of the worldwide atmospheric CH_4 sinks (around 30 million tons each year). However, CH_4-oxidizing bacteria can oxidize 90% of the CH_4 generated in the soil and then release it into the environment. As we all know, the conversion of cultivated land or grassland to forests will run to a substantial reduction in the CH_4 flow (Tate et al., 2007; Chen et al., 2017), and the nature and frequency of the vegetative CH_4 organisms are essential to predict the CH_4 flow.

As current climate models do not take this into account, future studies will have to emphasise on integrating these data and interfaces to enhance the prediction of CH_4 flows in different ecosystems. By altering land usage and managing practices, this information can also be used to reduce CH_4 productions (Smith, 2004). For example, CH_4-oxidizing bacteria have long been essential to absorb some of the CH_4 absorbed during rice growing. Therefore, increasing oxygen consumption in the soil can reduce the number of farming practices that improve the frequency and duration of flooding. Strategies to lessen CH_4 releases from ruminants incorporate improved feed value and direct inhibition of the rumen methanogen communities through the use of antibiotics, vaccines, and alternate electron acceptors (Smith et al., 2008).

1.4 MINIMIZING NITROGEN OXIDES EMISSION THROUGH MICROBIAL POPULATION MANAGEMENT

The primary resource of anthropogenic N_2O emissions is the usage of nitrogen fertilizers in farming. Since a large amount of fertilizer is released in the way of N_2O, it is possible to reduce the nitrogen consumption of microorganisms by applying fertilizers that can reduce N_2O emissions more precisely. Strategies include reducing fertilizer consumption, using fertilizers at the right time (when plants need a high nitrogen content and low leaching rates), and the potential for massive releases or washouts in moist soils where certain forms of nitrogen (such as nitrates) are avoided (Singh et al., 2010). Similarly, improving drainage and managing measures to reduce anaerobic soil circumstances (such as land compaction and excess moisture) will reduce denitrification rates and N_2O emissions. The usage of nitrification inhibitors in fertilizers to reduce nitrate production and the consequent loss of seeping or denitrification is an effective strategy for reducing the N_2O flows generated during agricultural production (Smith, 2008). These and comparable microbial approaches have an excellent ability for lowering greenhouse gas releases from the land usage and farming sectors.

1.5 FEEDBACK RESPONSES AND CLIMATE CHANGE IMPACTS ON DIFFERENT MICROBIAL COMMUNITIES ON GREENHOUSE GAS EMISSIONS

Previous work on climate change and the response concentrated on measuring biogeochemical practices. The data are used to create a projecting climate model. Because the knowledge of microbial responses to climate change is still restricted, we have overlooked the potentially significant impact of terrestrial microbes on climate change, because explaining the process in predictive models is fundamental. To progress our considerate of microbes with climate change and response reactions, we are expanding projecting climate models to provide data about the structure of the microbial community, physiological and biogeographical patterns, and functional connections between microbial and plant communities. There is ample indication that climate change will have a direct and implicit impact on terrestrial microbial populations and how they work. The effects of heightened CO_2 levels on microbial communities are related to their impact on plant metabolism, development and variety, the physical and chemical properties of the soil, such as cascading effects of soil humidity and resource feature (carbon and nitrogen) (Bardgett et al., 2009).

The significant immediate effects of climate change on soil microbes can be affected by fluctuations in temperature and water content. These considerations can alter practices such as greenhouse gas flow in two directions, by changing the physiological conditions of current microbial population and by altering the structure of the microbial population. For instance, most microbes grow at high temperatures and utilize substrates quicker, which can change the frequency of practices such as breathing. However, the management system remains unchanged. In the second case, climate change can promote alterations in the structure of microbial communities, the process rates, and control mechanisms which can change with the physiological changes in new microbial communities (Schimel and Schaeffer, 2012; Schimel and Gulledge, 1998). In severe instances, this can lead to the failure of a particular method (e.g., loss of entire functional groups (e.g., loss of the denitrifier or methanogen)) or previously insignificant processes (e.g., composition and other reasons).

1.5.1   CARBON DIOXIDE

It is usually believed that amplified CO_2 concentration changes the amount and quality of unstable amino acids, plant sugars, and organic acids, which can stimulate the growth and activity of microbes (Bardgett et al., 2009). CO_2 flow can be changed depending on the availability of nutrients (likewise nitrogen) (Diaz et al., 1993). In the long run, increased microbial biomass binds nitrogen in the soil due to increased carbon emissions from roots limits the nitrogen accessible in plants, and generates a negative response that limits imminent plant growth. It is believed that this increases the carbon–nitrogen ratio in the soil and increases fungal dominance and diversity (French et al., 2009). Fungi usually have a higher carbon assimilation efficiency than bacteria (i.e., they absorb more carbon than bacteria). They are metabolized, and the cell wall (phospholipids and peptidoglycans) of fungi mainly consists of carbon polymers (chitin and melanin). This is because bacterial cell membrane carbon polymers (melanin and chitin) are extra resilient to degradation. As a result, mushroom-dominated ecosystems typically have low soil respiration rates and increased sequestration of carbon (Six et al., 2006). However, some studies have shown a significant increase in soil breathing with increasing atmospheric CO_2 levels (Jackson et al., 2009). The underground response to increased CO_2 is typically more significant (Jackson et al., 2009) than the aboveground reaction in the same system.

Conversely, soil microorganisms prefer to use labile carbons rather than complex carbons, which can slow waste decomposition and reduce respiratory CO_2 emissions. Changes of carbon quantity and the quality provided by plants can also affect the response by straight distressing the functioning and structure of soil microbial communities (Balser and Wixon, 2009).

The global average surface temperature is predictable to upsurge by 1.1°C–6.4°C by 2100 (Solomon et al., 2007), which may also affect soil carbon sequestration, possibly by accelerating the activity of heterotrophic microorganisms. The sensitivity of the constant and unstable parts of organic carbon in the soil to temperature changes is believed to differ widely. For example, the increased thawing rate and depth of permanently frozen soils in high latitudes make large amounts of organic carbon reserves (400 Pg or 4 billion tons) in these soils susceptible to the rate of decomposition (Schuur et al., 2009). If the organic carbon input from the ground primary production has no compensating effect, this can lead to a vast and irrepressible positive feedback effect (Zimov et al., 2006). Overall, the rise in temperature is directly related to the upsurge in soil breathing. A worldwide usual temperature rise of 2°C is expected to increase soil carbon release by 10 Pg, primarily due to a 50–52 increase in microbial activity (Solomon et al., 2007).

This is believed because high temperatures rouse the usage of unstable carbon. However, the temperature sensitivity of carbon is unclear due to the diverse and complex structure of carbon. This situation is additionally exacerbated by the role of environmental restraints in the degradation of organic enzymes, including chemical and physical defense alongside enzyme activity, and the effects of scarcity, flooding, and temperature on enzyme activity and oxygenation. Also, these environmental restraints are affected by climate change. It is consequently difficult to predict the result of an increase in temperature on carbon storage. In some cases, high temperatures can lead to lower the concentration of organic carbon in the soil, particularly in temperate ecosystems. Indeed, recent studies have shown that even slight warming (around 1°C) can increase respiratory rate in arctic swamps, especially in underground ecosystems (Dorrepaal et al., 2009).

This shows that organic carbon stored in northern swamps has a large amount of continuous positive feedback to the worldwide climate system, but the process of this reaction is still unknown (Smith and Fang, 2010). An elevated temperature disturbs the composition of the microbial communities because dissimilar microbial groups have different optimal temperature ranges for growth and activity. In a few cases, reducing environmentally friendly microflora can reduce organic carbon loss in the soil (Monson et al., 2006). For instance, elevated temperatures in high latitude ecosystems have changed the fungal community system by reducing up to 50% of bacterial, fungal, and soil respiration (Allison and Treseder, 2008). To further obscure substances, these changes in breathing may be due to changes in the composition and activity of the microbial community or variations in the eminence and extent of organic carbon in the soil (Bradford et al., 2008). There are indications that soil warming has reduced the relative frequency of fungi and changed the structure of the bacterial community of the comparable arctic ecosystems.

Part of the organic carbon in the soil can also be protected by physical and chemical methods to prevent microbial degradation (Rinnan et al., 2007). It is estimated that the loss of carbon caused by climate change is undependable due to too many variables (Kirschbaum, 2006). Soil moisture is another critical factor that determines the structure of terrestrial microbial communities and the amount of organic carbon degradation in the land. This is exaggerated by a 20% upsurge or reduction in precipitation rates predicted by the *Intergovernmental Panel on Climate Change* 2007 (Solomon et al., 2007). The bacterial community responds directly to water content because it needs water for its physiological activity, while the microbial community is indirectly due to changes in soil moisture caused by gas diffusion and oxygenation. Therefore, the effects of changes in precipitation on the response reactions of soil microbes to climate change might be due to the straight impacts on microbial physiology and community structure. Long-term dry circumstances limit the development and degradation of microorganisms and can adversely affect the carbon flow of particular ecosystems. However, soil drying increases the oxygenation of wetlands and swamps and improves the carbon cycle, which creates positive feedback on the CO_2 flow (Freeman et al., 2002). It should be noted that the feedback response triggered by changes in temperature and humidity varies by ecosystem and region. For example, temperature increases have a more significant impact in high-altitude, cold, and temperate areas because microbial growth in these ecosystems is usually imperfect by temperature. Contrariwise, augmented drought circumstances have a more significant impact in the tropics, since they can lead to significant changes in root growth, a significantly reduced microbial biomass or a changed structure of the microbial community (Meier and Leuschner, 2008). The information is then quantified and consolidated at the regional level to provide more reliable forecasts at the global level.

1.5.2   METHANE

From a global climate change perspective, CH_4 is the another most significant anthropogenic greenhouse gas, and microbial use (CH_4 nutrition) is the most extensive land uptake. To be able to predict CH_4 emissions more accurately, we need to understand the CH_4 flow's response to climate change. The reaction of the microbially arbitrated CH_4 flow to variations in climate and atmospheric composition is just as indeterminate as to the response of the CO_2 flow. Current analysis shows that global warming, especially in alpine, can significantly increase net CH_4 emissions from permafrost and wetlands, leading to a significant positive response to global warming (Zhuang et al., 2004). Warming conditions are expected to increase methanogen and methanogen activity without limiting other factors, but whether the reaction will reach equilibrium and how it affects the global CH_4 network, similarly, after warming specific climate systems, net primary productivity, and groundwater depth and soil water content change, thereby promoting methanogenesis and net CH_4 emissions. In other areas, when rainfall and dry soil are used, oxygen utilization (which promotes CH_4 oxidation) and CH_4 (Christensen et al., 2003) net emissions decrease. Numerous studies have shown that an increase in the CO_2 content significantly reduces the microbial uptake of CH_4 in the soil (up to 30%) and increases the CH_4 outflow by (Ineson et al., 1998; Phillips et al., 2001a). However, the mechanism for reducing CH_4 absorption is still unknown. Some studies have shown that plant-related increases in soil moisture (McLain and Ahmann, 2008) reduce CH_4 consumption, while others have found that CH_4 oxidation is reduced and soil moisture is not increased (Phillips et al., 2001b). Increased CO_2 levels can indirectly affect emissions by affecting microbial activity. Increasing the CO_2 content of the soil leads to increased soil moisture in connection with plants, which leads to a lower oxygen content, which increases CH_4 production and depletion.

However, it is always believed that high temperatures and low humidity increase the rate of gas diffusion and microbial oxygen, as well as the exposure of the atmosphere to CH_4, thus increasing the net uptake of CH_4 by terrestrial ecosystems. Reported feedback reactions of the CH_4 flow are also triggered by the changes in the active microbial communities since a high CO_2 content can reduce the methanogenesis of plants by up to 70% (Kolb et al., 2005). Indeed, previous studies have shown that CH_4 flow is associated with changes in nutrient CH_4 biome (Singh et al., 2009) and frequency due to changes in land use (Menyailo et al., 2010). Changes in the CH_4 flow with increased CO_2 values or temperatures can also be associated with changes in the frequency and the community structure of the CH_4 organism in nutrients (Mohanty et al., 2007). The interaction of moisture and CO_2 in the ecosystem is crucial to be able to predict the future CH_4 flow on the ground more precisely.

1.5.3   NITROUS OXIDE

In connection with industrial development and the increased N_2O releases from fossil fuel combustion, the NH_3 emissions associated with strengthening agriculture have increased the amount of active nitrogen released in the last century by a factor of 3–5 (Barnard et al., 2005). An increased supply of active nitrogen in terrestrial ecosystems leads to increased nitrification and denitrification, which increases N_2O production (Guo and Gifford, 2002). Little is known about the nature and degree of the interactions with climate change. In a few studies, it has been reported that denitrification activity decreases with increasing CO_2 (Barnard et al., 2005). Also, some studies report increased N_2O emissions with increasing CO_2 levels, but only if excess mineral nitrogen is only available (Cheng et al., 2006). This can indicate that ecosystems with low active nitrogen consumption reduce N_2O emissions with increasing CO_2 levels. About microorganisms, limited information is available about changes in the N_2O flow. However, it can be observed that as the CO_2 content increases, the community structure of NH_3-oxidizing bacteria changes, and their frequency decreases by 90% (Barnard et al., 2005). Reports of the temperature-dependent nitrifying and denitrifying bacterial communities are also inconsistent. Investigation of earlier studies showed that an elevated temperature is unlikely to have a significant impact on the nitrification or denitrification of enzymes (Barnard et al., 2005). Though, other experiments have shown that the quantity of total N_2O flow allied with nitrification reductions at elevated temperatures. This is associated with structural changes in the NH_3 oxidizing bacterial community (Avrahami et al., 2003). Upcoming investigations on this topic must therefore focus on the denitrification of microbes and the evaluation of denitrification of nitrate in response to changes in active nitrogen and the interaction between climate and atmospheric composition changes.

1.6 ADAPTATION MECHANISM OF MICROBES UNDER DIFFERENT CLIMATE CHANGE RESPONSES

Understanding the underlying mechanisms of microbial response is essential to summarize future bacterial contributions to land microbes in soil carbon cycling and additional ecosystem purposes. When the environment changes due to new disruptions, likewise climate change, variations in functions, likewise soil respiration, the activity of the enzyme, and waste breakdown can be observed (Dorrepaal et al., 2009). However, information is often lacking to identify the mechanism by which these changes occur. The overall functional response of the entire soil is generated by the individual actions of the community of soil microbial diversity. Functions that combine distinct physiology and act with environmental circumstances lead to changes in species classification and gradient composition (Leibold et al., 2004). Microbial community status can also be affected by distribution restrictions and landscape connections (Ehrlén and Eriksson,

2000). Besides, strengthening positive feedback on other stable countries could cover regional classifications and immigrants, which could make the community more dependent on history.

1.7 IMPACT OF MOISTURE CONTENTS ON MICROBIAL POPULATION IN MITIGATING CLIMATE CHANGE

The size and shape of the functional response of microorganisms to soil moisture have a direct impact on the reaction of the ecosystem and how we simulate the response of microbial processes to climate change. Water is the primary regulator of microbial activity in soil (Liu et al., 2009) and limits enzyme activity and soil respiration. Though, the response of microorganisms to moisture can vary widely. Available water can be used as a source to limit microbial activity (immediately or by restricting nutrient access) so that the bacterial response to water can be lined or at least monotonous. The microbial functional response to water has a broad range of consistency, which enables microbially driven carbon processes in the soil to adapt to changes in precipitation conditions and to make direct predictions based on water availability (Singh et al., 2010).

Manzoni et al. (2012) showed that the inferior microbial activity was the limit of consistent, which can be a general limit for the resistance to dispersion and dehydration of solutes. Though, distinct microbial taxa show single physiological retort curves to water, comprising specialized and generalist approaches (Lennon et al., 2012). Consequently, it may be possible to put a lesser limit on the microbial water reaction, but it is expected to change with increasing water content and improving the upper limit. The challenge then is to understand the distribution of the variability of the microbial response to water and to understand whether/how it affects the aggregate function. Microbial water reactions that only occur in ecosystems, biomes, habitats, microbial functional class, cause ancient emergencies that change the response of functions such as breathing to changes like precipitation. For example, arid soils differ fundamentally in the microbial reaction of the soil between dry and inland areas because fungi control them, have a high degradation potential (stubborn substances), and have a healthy separation from the microbial activity of the plants (Collins et al., 2008).

The pulsed rain event can control the transition between another stable state of elevated and little microbial function, and the period of the pulsed event can affect the fast and slow mechanisms of the bacterial community differently. Short-term precipitation events can disproportionately contribute to the carbon cycle in the soil, especially if the microbial reactions are fast and large or if short-term precipitation impulses make older carbon pools available in the soil (Collins et al., 2008). Besides, the microbial response after rewetting can outweigh the significant reduction in process rate during drying, especially when extracellular enzymes remain in dry soil (Schimel and Schaeffer, 2012; Schimel and Gulledge, 1998). However, most microbial impulse response studies are conducted in a single location, and these processes are not fully understood.

1.8   MICROBIAL PHYSIOLOGY

Microorganisms generally have a wide range of physiological abilities, so physiological adaptation to environmental changes can be expected. The general functional response may be due to the physiological breadth of a single taxon within the community but may also represent the physiological diversity between taxa. Functional plasticity has been observed in response to short-term changes in temperature and humidity in the microbial community (Bradford et al., 2010), but not always. Another plastic foam that is widespread in soil microbes can avoid temporary environmental pollution such as drought (Lennon and Jones, 2011). When conditions improve, the rest of the dormant groups quickly predicts functional resilience. Though, the accomplishment of dormant approaches in the face of climate change depends on the determination of resting breeders and the nature of the new atmosphere. For instance, long-lived spores are desired to withstand long-term scarcity.

Despite general malleability, the microbial response to climate change can be limited by regional climate history. Exceeding the historically selected pressure range leads to a more robust response, more extensive changes, or difficulties in predicting. The mutual transplantation of complete soil cores between plant communities supports this view: Waldrop et al. (2006) show only minor changes in the composition of the microbial community and the aggregation function of soil cores that have been transplanted from grassland under oak roofs. It was found that (the environmental conditions are completely inside the range usually initiate in grasslands), but quick changes were observed when oak crowns were transplanted into grassland (outdoor the normal range)—considerate the parameters of microbial physiological plasticity offer possible boundary circumstances.

1.9   MICROBIAL COMMUNITY COMPOSITION

In changing environments, differences in microbial performance can change the composition of the community by changing the relative frequency of existing taxa or expanding from a pool of local species. As the environment changes, certain microbial classes benefit more than others, and their benefits and functions change (Pett-Ridge and Firestone, 2005). Decentralization will also offer new immigration taxa, and the taxonomy of the species should lead to the existence of organisms that are best suited to the local environment. A species classification was observed in bacterial communities (Van der Gucht et al., 2007), but not in protists. The step of sorting and its impact on quality depends on the degree of dispersion, and microorganisms are generally considered to be endless. However, recent studies have shown that microbial spread is restricted and even highly endemic is also supported. Given the local differences in the pools of micro-bial species, the microbial empirical response to trial climate managements may be restricted by available taxa (Stuart Chapin III et al., 2009).

Many climate change experiments have drained small areas in dry, humid environments. If growth is limited, these regions may lack local species pools involving drought-adapted individuals. With positive feedback, some stable states occur if the state of the ecosystem continues (Beisner et al., 2003). In dry grasslands, for example, plants can improve water use efficiency through awning and floor effects. The result is a dramatic pattern of a downfall under strict water regulations (Rietkerk and Van de Koppel, 2008). Similar positive feedback cans a vital role in the response of underground microorganisms to climate change. Positive improvements in existing microbial communities can resist climate change and then change or failure as the degree of change increases. Positive feed-back (sometimes called the avenue effect) eliminates the aggression that adaptive racing brings to the community because it has a small number of founders and cannot endure in low density (Keitt et al., 2001).

The characteristics of the microbial life story usually do not match the effects on the signal path. Pathway effects can be observed when the microbial community needs enough density to achieve synergistic function. Similarly, positive feedback can lead to frequency-dependent competitive asymmetries, so larger traditional populations cannot be substituted by more suitable, low-density invasive species. The result is the great potential for ever-changing historical ecosystems and community structures, which increases temporary resistance, but sudden changes in state ultimately limit elasticity. Drought has observed historical problems with microbial function, but not continuously (Rousk et al., 2013). If positive feedback controls microbial communities in the soil, the historical legacy of past climates will play a significant role in the future change of these communities and the resilience of the functioning of the ecosystem.

1.10   CONCLUSIONS AND PERSPECTIVES

Scientists agree that the Earth's climate is changing and that the average temperature rise of the Earth since the 1900s is mainly due to human activity. However, there is still much uncertainty about future greenhouse gas emission estimates and the response of these emissions to further changes in global climate and climate composition. To overcome this, uncertainty requires a better understanding of terrestrial microbial feedback responses and the management of microbial systems to reduce the potential for climate change. There is a need to improve the mechanistic experience of microorganisms to control greenhouse gas emissions and to coordinate interactions between their various biotic and biological components. This understanding helps eliminate significant uncertainties in predicting the micro response to climate change and enables knowledge in future models of climate change and soil response. It is currently difficult to determine whether changes in climate change processes are due to the effects of climate change on soil microbial communities, changes in soil abiotic factors, or both. It is unclear how microbes respond to climate change and how the climate response affects the entire ecosystem and ecological gradient. Another problem is that most studies focus on greenhouse gas.

- Based on the above information, we justify a range of research topics and require priority research to develop methods for preventing microbial-mediated climate change.
- First, a better understanding of microbial responses to climate change is needed to understand future ecosystem functions.

- Secondly, microbial taxa should be classified according to their function and physiological capacity and this information should be linked to the level of the ecosystem.
- Third, the microcontrol of greenhouse gas emissions and the mechanistic understanding of microbial responses to climate factors (e.g., warming, changing precipitation, increasing CO_2 content) in different ecosystems should be improved.
- Fourth, the inclusion of microbial data (biomass, community, diversity, and activity) in climate models should be established to reduce uncertainty and improve forecasts and forecasts.
- Fifth, a better understanding of the effects of climate change on surface-to-soil interactions and nutrient cycles is required, and the role of these interactions in coordinating ecosystem responses to global change.
- Finally, a framework based on the above five points should be developed to manage potential natural microbial systems to improve carbon sequestration and reduce net greenhouse gas emissions.

KEYWORDS

- **biogeochemical cycles**
- **climate change**
- **soil ecology**

REFERENCES

Allison, S. D.; Treseder, K. K. Warming and Drying Suppress Microbial Activity and Carbon Cycling in Boreal Forest Soils. *Global Change Biol.* **2008,** *14* (12), 2898–2909.

Allison, S. D.; Wallenstein, M. D.; Bradford, M. A. Soil-Carbon Response to Warming Dependent on Microbial Physiology. *Nat. Geosci.* **2010,** *3* (5), 336–340.

Aonofriesei, F Microbial Communities, Microbial Processes, and Global Climate Changes. *Int. Multidisciplinary Sci. GeoConf.erence*: SGEM **2019,** *19* (4.1), 993–1000.

Arrigo, K. R. Marine Microorganisms and Global Nutrient Cycles. Nature **2005,** *437* (7057), 349–355.

Averill, C.; Turner, B. L.; Finzi, A. C. Mycorrhiza-Mediated Competition between Plants and Decomposers Drives Soil Carbon Storage. *Nature* **2014,** *505* (7484), 543–545.

Avrahami, S.; Liesack, W.; Conrad, R. Effects of Temperature and Fertilizer on Activity and Community Structure of Soil Ammonia Oxidizers. *Environ. Microbiol.* **2003,** 5 (8), 691–705.

Balser, T. C.; Wixon, D. L. Investigating Biological Control Over Soil Carbon Temperature Sensitivity. *Global Change Biol.* **2009,** *15* (12), 2935–2949.

Bardgett, R. D.; De Deyn, G. B.; Ostle, N. J. Plant-Soil Interactions and the Carbon Cycle. *J. Ecol.* (Oxf.) 2009, *97* (5), 838–912.

Bardgett, R. D.; Freeman, C.; Ostle, N. J. Microbial Contributions to Climate Change through Carbon Cycle Feedbacks. *ISME J.* **2008,** *2* (8), 805–814.

Barnard, R.; Leadley, P. W.; Hungate, B. A. Global Change, Nitrification, and Denitrification: A Review. *Global Biogeochem. Cycl.* **2005,** *19* (1).

Beisner, B. E.; Haydon, D. T.; Cuddington, K. Alternative Stable States in Ecology. *Front. Ecol. Environ.* **2003,** *1* (7), 376–382.

Blankinship, J. C.; Niklaus, P. A.; Hungate, B. A. A Meta-Analysis of Responses of Soil Biota to Global Change. *Oecologia* **2011,** *165* (3), 553–565.

Bradford, M. A.; Davies, C. A.; Frey, S. D.; Maddox, T. R.; Melillo, J. M.; Mohan, J. E.; Reynolds, J. F.; Treseder, K. K.; Wallenstein, M. D. Thermal Adaptation of Soil Microbial Respiration to Elevated Temperature. *Ecol. Lett.* **2008,** *11* (12), 1316–1327.

Bradford, M. A.; Watts, B. W.; Davies, C. A. Thermal Adaptation of Heterotrophic Soil Respiration in Laboratory Microcosms. *Global Change Biol.* **2010,** *16* (5), 1576–1588.

Carney, K. M.; Hungate, B. A.; Drake, B. G.; Megonigal, J. P. Altered Soil Microbial Community at Elevated CO2 Leads to Loss of Soil Carbon. *Proc. Natl. Acad. Sci.* **2007,** *104* (12), 4990–4995.

Chen, H.; Liu, J.; Zhang, A.; Chen, J.; Cheng, G.; Sun, B.; Pi, X.; Dyck, M.; Si, B.; Zhao, Y. Effects of Straw and Plastic Film Mulching on Greenhouse Gas Emissions in Loess Plateau, China: A Field Study of 2 Consecutive Wheat-Maize Rotation Cycles. *Sci. Total Environ.* **2017,** *579,* 814–824.

Cheng, W.; Yagi, K.; Sakai, H.; Kobayashi, K. Effects of Elevated Atmospheric CO_2 Concentrations on CH_4 and N_2O Emission from Rice Soil: An Experiment in Controlled-Environment Chambers. *Biogeochemistry* **2006,** *77* (3), 351–373.

Christensen, T. R.; Ekberg, A.; Ström, L.; Mastepanov, M.; Panikov, N.; Öquist, M.; Svensson, B. H.; Nykänen, H.; Martikainen, P. J.; Oskarsson, H. Factors Controlling Large Scale Variations in Methane Emissions from Wetlands. *Geophys. Res. Lett.* **2003,** *30* (7).

Collins, S. L.; Sinsabaugh, R. L.; Crenshaw, C.; Green, L.; Porras-Alfaro, A.; Stursova, M.; Zeglin, L. H. Pulse Dynamics and Microbial Processes in Aridland Ecosystems. *J. Ecol.* **2008,** *96* (3), 413–420.

Diaz, S.; Grime, J.; Harris, J.; McPherson, E. Evidence of a Feedback Mechanism Limiting Plant Response to Elevated Carbon Dioxide. *Nature* **1993,** *364* (6438), 616–617.

Dorrepaal, E.; Toet, S.; van Logtestijn, R. S.; Swart, E.; van de Weg, M. J.; Callaghan, T. V.; Aerts, R. Carbon Respiration from Subsurface Peat Accelerated by Climate Warming in the Subarctic. *Nature* 2009, *460* (7255), 616–619.

Ehrlén, J.; Eriksson, O. Dispersal Limitation and Patch Occupancy in Forest Herbs. *Ecology* **2000,** *81* (6), 1667–1674.

Farooqi, Z. U. R.; Sabir, M.; Zeeshan, N.; Hussain, M. M. *Mitigation of Climate Change through Carbon Sequestration in Agricultural Soils., vol 1. Climate Change and Agroforestry Systems: Adaptation and Mitigation Strategies*; CRC-Apple Academic Press, 2020.

Farooqi, Z. U. R.; Sabir, M.; Zeeshan, N.; Naveed, K.; Hussain, M. M. Enhancing Carbon Sequestration Using Organic Amendments and Agricultural Practices. *Carbon Capture, Utilization Sequestration* 2018, 17.

Freeman, C.; Nevison, G.; Kang, H.; Hughes, S.; Reynolds, B.; Hudson, J. Contrasted Effects of Simulated Drought on the Production and Oxidation of Methane in a Mid-Wales Wetland. *Soil Biol. Biochem.* **2002**, *34* (1), 61–67.

French, S.; Levy-Booth, D.; Samarajeewa, A.; Shannon, K.; Smith, J.; Trevors, J. Elevated Temperatures and Carbon Dioxide Concentrations: Effects on Selected Microbial Activities in Temperate Agricultural Soils. *World J. Microbiol. Biotechnol.* **2009**, *25* (11), 1887–1900.

Godbold, D. L.; Hoosbeek, M. R.; Lukac, M.; Cotrufo, M. F.; Janssens, I. A.; Ceulemans, R.; Polle, A.; Velthorst, E. J.; Scarascia-Mugnozza, G.; De Angelis, P. Mycorrhizal Hyphal Turnover as a Dominant Process for Carbon Input into Soil Organic Matter. *Plant Soil* **2006**, *281* (1–2), 15–24.

Guo, L. B.; Gifford, R. Soil Carbon Stocks and Land Use Change: A Meta Analysis. *Global Change Biol.* **2002**, *8* (4), 345–360.

Hawkes, C. V.; Kivlin, S. N.; Rocca, J. D.; Huguet, V.; Thomsen, M. A.; Suttle, K. B. Fungal Community Responses to Precipitation. *Global Change Biol.* **2011**, *17* (4), 1637–1645.

Ineson, P.; Coward, P.; Hartwig, U. Soil Gas Fluxes of N_2O, CH_4 and CO_2 Beneath Lolium Perenne under Elevated CO_2: The Swiss Free Air Carbon Dioxide Enrichment Experiment. *Plant Soil* **1998**, *198* (1), 89–95.

Jackson RB, Cook CW, Pippen JS, Palmer SM Increased Belowground Biomass and Soil CO_2 Fluxes after a Decade of Carbon Dioxide Enrichment in a Warm-Temperate Forest. *Ecology* **2009**, *90* (12), 3352–3366.

Keitt, T. H.; Lewis, M. A.; Holt, R. D. Allee Effects, Invasion Pinning, and Species' Borders. *Am. Nat.* **2001**, *157* (2), 203–216.

Kirschbaum MUF The Temperature Dependence of Organic-Matter Decomposition—Still a Topic of Debate. *Soil Biol. Biochem.* **2006**, *38* (9), 2510–2518.

Kolb, S.; Carbrera, A.; Kammann, C.; Kämpfer, P.; Conrad, R.; Jäckel, U. Quantitative Impact of CO_2 Enriched Atmosphere on Abundances of Methanotrophic Bacteria in a Meadow Soil. *Biol. Fertility Soils* **2005**, *41* (5), 337–342.

Leibold, M. A.; Holyoak, M.; Mouquet, N.; Amarasekare, P.; Chase, J. M.; Hoopes, M. F.; Holt, R. D.; Shurin, J. B.; Law, R.; Tilman, D. The Metacommunity Concept: A Framework for Multi-Scale Community Ecology. *Ecol. Lett.* **2004**, *7* (7), 601–613.

Lennon, J. T.; Aanderud, Z. T.; Lehmkuhl, B.; Schoolmaster, Jr., D. R. Mapping the Niche Space of Soil Microorganisms Using Taxonomy and Traits. *Ecology* **2012**, *93* (8), 1867–1879.

Lennon, J. T.; Jones, S. E. Microbial Seed Banks: The Ecological and Evolutionary Implications of Dormancy. *Nat. Rev. Microbiol.* **2011**, *9* (2), 119–130.

Liu, W.; Zhang, Z.; Wan, S. Predominant Role of Water in Regulating Soil and Microbial Respiration and Their Responses to Climate Change in a Semiarid Grassland. *Global Change Biol.* **2009**, *15* (1), 184–195.

Manzoni, S.; Schimel, J. P.; Porporato, A. Responses of Soil Microbial Communities to Water Stress: Results from a Meta-Analysis. *Ecology* **2012**, *93* (4), 930–938.

McLain, J. E.; Ahmann, D. M. Increased Moisture and Methanogenesis Contribute to Reduced Methane Oxidation in Elevated CO_2 Soils. *Biol. Fertility Soils* **2008**, *44* (4), 623–631.

Meier, I. C.; Leuschner, C. Belowground Drought Response of European Beech: Fine Root Biomass and Carbon Partitioning in 14 Mature Stands across a Precipitation Gradient. *Global Change Biol.* **2008,** *14* (9), 2081–2095.

Menyailo OV, Abraham W-R, Conrad R Tree species affect atmospheric CH4 oxidation without altering community composition of soil methanotrophs. *Soil Biol. Biochem.* **2010,** *42* (1), 101–107.

Mohanty, S. R.; Bodelier, P. L.; Conrad, R. Effect of Temperature on Composition of the Methanotrophic Community in Rice Field and Forest Soil. *FEMS Microbiol. Ecol.* **2007,** *62* (1), 24–31.

Monson, R. K.; Lipson, D. L.; Burns, S. P.; Turnipseed, A. A.; Delany, A. C.; Williams, M. W.; Schmidt, S. K. Winter Forest Soil Respiration Controlled by Climate and Microbial Community Composition. *Nature* **2006,** *439* (7077), 711–714.

Pett-Ridge, J.; Firestone, M. Redox Fluctuation Structures Microbial Communities in a Wet Tropical Soil. *Appl. Environ. Microbiol.* **2005,** *71* (11), 6998–7007.

Phillips, R. L.; Whalen, S. C.; Schlesinger, W. H. Influence of Atmospheric CO_2 Enrichment on Methane Consumption in a Temperate Forest Soil. *Global Change Biol.* **2001a,** *7* (5), 557–563.

Phillips, R. L.; Whalen, S. C.; Schlesinger, W. H. Response of Soil Methanotrophic Activity to Carbon Dioxide Enrichment in a North Carolina Coniferous Forest. *Soil Biol. Biochem.* **2001b,** *33* (6), 793–800.

Potthoff, M.; Dyckmans, J.; Flessa, H.; Beese, F.; Joergensen, R. G. Decomposition of Maize Residues after Nanipulation of Colonization and Its Contribution to the Soil Microbial Biomass. *Biol. Fertility Soils* **2008,** *44* (6), 891–895.

Reay, D Sinking methane. *Biologist (London, England)* 2003, *50* (1), 15–19.

Reay, D. S.; Dentener, F.; Smith, P.; Grace, J.; Feely, R. A. Global Nitrogen Deposition and Carbon Sinks. *Nat. Geosci.* **2008,** *1* (7), 430–437.

Rietkerk, M.; Van de Koppel, J. Regular Pattern Formation in Real Ecosystems. *Trends Ecol. Evol.* **2008,** *23* (3), 169–175.

Rinnan, R.; Michelsen, A.; Bååth, E.; Jonasson, S. Fifteen Years of Climate Change Manipulations Alter Soil Microbial Communities in a Subarctic Heath Ecosystem. *Global Change Biol.* 13 (1), 28–39.

Rousk, J.; Smith, A. R.; Jones, D. L. Investigating the Long-Term Legacy of Drought and Warming on the Soil Microbial Community across Five European Shrubland Ecosystems. *Global Change Biol.* **2013,** *19* (12), 3872–3884.

Rustad, L.; Campbell, J.; Marion, G.; Norby, R.; Mitchell, M.; Hartley, A.; Cornelissen, J.; Gurevitch, J. A Meta-Analysis of the Response of Soil Respiration, Net Nitrogen Mineralization, and Aboveground Plant Growth to Experimental Ecosystem Warming. *Oecologia* **2001,** *126* (4), 543–562.

Salles, J. F.; Poly, F.; Schmid, B.; Roux, X. L. Community Niche Predicts the Functioning of Denitrifying Bacterial Assemblages. *Ecology* **2009,** *90* (12), 3324–3332.

Schimel, J.; Schaeffer S. M. Microbial Control Over Carbon Cycling in Soil. *Front. Microbiol.* **2012,** 3, 348.

Schimel JP, Gulledge J Microbial Community Structure and Global Trace Gases. *Global Change Biol.* **1998,** *4* (7), 745–758.

Schlesinger, W. H.; Andrews, J. A. Soil Respiration and the Global Carbon Cycle. *Biogeochemistry* **2000,** *48* (1), 7–20.

Schuur, E. A.; Vogel, J. G.; Crummer, K. G.; Lee, H.; Sickman, J. O.; Osterkamp, T. The Effect of Permafrost Thaw on Old Carbon Release and Net Carbon Exchange from Tundra. *Nature* **2009,** *459* (7246), 556–559.

Singh, B. K.; Bardgett, R. D.; Smith, P.; Reay, D. S. Microorganisms and Climate Change: Terrestrial Feedbacks and Mitigation Options. *Nat. Rev. Microbiol.* **2010,** *8* (11), 779–790.

Singh, B. K.; Tate, K. R.; Ross, D. J.; Singh, J.; Dando, J.; Thomas, N.; Millard, P.; Murrell, J. C. Soil Methane Oxidation and Methanotroph Responses to Afforestation of Pastures with Pinus Radiata Stands. *Soil Biol. Biochem.* **2009,** *41* (10), 2196–2205.

Six J, Frey S, Thiet R, Batten K Bacterial and Fungal Contributions to Carbon Sequestration in Agroecosystems. *Soil Sci. Soc. Am. J.* **2006,** *70* (2), 555–569.

Smith, P. Land Use Change and Soil Organic Carbon Dynamics. *Nutr. Cycl. Agroecosyst.* **2008,** *81* (2), 169–178.

Smith, P.; Fang, C. A Warm Response by Soils. *Nature* **2010,** *464* (7288), 499–500.

Smith, P.; Martino, D.; Cai, Z.; Gwary, D.; Janzen, H.; Kumar, P.; McCarl, B.; Ogle, S.; O'Mara, F.; Rice, C. Greenhouse Gas Mitigation in Agriculture. *Phil. Trans. R. Soc. B: Biol. Sci.* **2008,** *363* (1492), 789–813.

Solomon, S.; Manning, M.; Marquis, M.; Qin, D. *Climate Change 2007-the Physical Science Basis: Working Group I Contribution to the Fourth Assessment Report of the IPCC,* vol 4; Cambridge University Press, 2007.

Stuart Chapin, III, F.; McFarland, J.; David McGuire, A.;, Euskirchen, E. S.; Ruess, R. W.; Kielland, K. The Changing Global Carbon Cycle: Linking Plant–Soil Carbon Dynamics to Global Consequences. *J. Ecol.* **2009,** *97* (5), 840–850.

Tate, K. R.; Ross, D.; Saggar, S.; Hedley, C.; Dando, J.; Singh, B. K.; Lambie, S. M. Methane Uptake in Soils from Pinus Radiata Plantations, a Reverting Shrubland and Adjacent Pastures: Effects of Land-Use Change, and Soil Texture, Water and Mineral Nitrogen. *Soil Biol. Biochem.* **2007,** *39* (7), 1437–1449.

Teske, A.; Alm, E.; Regan, J.; Toze, S.; Rittmann, B.; Stahl, D. Evolutionary Relationships among Ammonia-and Nitrite-Oxidizing Bacteria. *J. Bacteriol.* **1994,** *176* (21), 6623–6630.

Van der Gucht, K.; Cottenie, K.; Muylaert, K.; Vloemans, N.; Cousin, S.; Declerck, S.; Jeppesen, E.; Conde-Porcuna, J-M.; Schwenk, K.; Zwart, G. The Power of Species Sorting: Local Factors Drive Bacterial Community Composition Over a Wide Range of Spatial Scales. *Proc. Natl. Acad. Sci.* **2007,** *104* (51), 20404–20409.

Van Der Heijden, M. G.; Bardgett, R. D.; Van Straalen, N. M. The Unseen Majority: Soil Microbes as Drivers of Plant Diversity and Productivity in Terrestrial Ecosystems. *Ecol. Lett.* **2008,** *11* (3), 296–310.

Waldrop, M. P.; Zak, D. R.; Blackwood, C. B.; Curtis, C. D.; Tilman, D. Resource Availability Controls Fungal Diversity across a Plant Diversity Gradient. *Ecol. Lett.* **2006,** *9* (10), 1127–1135.

Zhuang, Q.; Melillo, J. M.; Kicklighter, D. W.; Prinn, R. G.; McGuire, A. D.; Steudler, P. A.; Felzer, B. S.; Hu, S. Methane Fluxes between Terrestrial Ecosystems and the Atmosphere at Northern High Latitudes during the Past Century: A Retrospective Analysis with a Process-Based Biogeochemistry Model. *Global Biogeochem. Cycl.* **2004,** *18* (3).

Zimov, S. A.; Schuur, E. A.; Chapin, III F. S. Permafrost and the Global Carbon Budget. *Science(Washington)* 2006, *312* (5780), 1612–1613.

CHAPTER 2

Climate Change-Induced Aggravations in Microbial Populations and Processes: Constraints and Remediations

WAQAS MOHY UD DIN, MUHAMMAD MAHROZ HUSSAIN*, and ZIA UR RAHMAN FAROOQI

Institute of Soil and Environmental Sciences, University of Agriculture, Faisalabad, Pakistan

Corresponding author. E-mail: hmahroz@gmail.com

ABSTRACT

Microbes are drivers of life. They support every life-sustaining activity on the earth right from food crops production through increasing the soil fertility, helping in food processing via fermentation and decomposition of dead animals to save environmental nuisance. Microbes support one of the biggest industries in the world, agricultural, through nutrients cycling, modifying soil properties for better crop production, retaining nutrients for plant supply and fertility, and remediating soil from soil and plant pollutants through bioremediation. Climate change has aggravated or disturbed various processes undertaken by soil microbes, that is, microbial populations, diversity, their strategies, and nutrient cycles by killing through increased temperature and increasing soil salinity and related problems. This chapter is an effort to describe the benefits of microbes in the life comprehensively, effects of climate change on microbial populations and its associated processes, and ultimate impact on the environment.

2.1   INTRODUCTION

Climate change changes the distribution of species, and at the same time, influences the interactions between living organisms (French, 2017; van der Putten et al., 2016). Different natural populations are involved, having other biographical characteristics, such as heat resistance and diffusion capacities. Besides, interactions between the members of the community can be beneficial, pathogenic, or poorly functional, and these relationships can alter due to the harsh environment (Vandenkoornhuyse et al., 2015). A large number of studies have shown that changes in the relationships of different species due to climate change have changed the diversity of living organisms and the functions of the terrestrial ecosystem (Cox et al., 2016; Langley and Hungate, 2014; Urban, 2015), while the soil communities are less (Keesstra et al., 2016).

Living organisms that live on earth interact with each other in a variety of ways and interact with plants to shape and maintain the properties of the ecosystem. The interaction of soil microorganisms with each other and with plants can influence the frequency, diversity, and composition of the landscape pattern of plants and animals (Berg et al., 2010; Bennett et al., 2017). Suppose the total effect of whole soil organisms (such as disease-causing, symbiotic bacteria and degrader) reduces plant's growth and yield. In some cases, the interaction of plants with microorganisms is considered harmful, and if the benefits of soil communities improve plant health (e.g., the interaction of plants and microorganisms is deemed to be positive, and their survival. Given their importance in defining the characteristics of ecosystems, understanding how climate change directly affects soil microbes and plants is a critical research that will address ecosystem functions such as soil (McCormack et al., 2015; Fischer et al., 2014). The annual carbon flow of 120 Gt in and from terrestrial ecosystems exceeds that generated by burning fossil fuels amount of carbon by far (IPCC, 2007).

Therefore, small changes in carbon concentration exchanged between the ecosystem and the atmosphere can have a significant influence on upcoming carbon concentrations in the atmosphere. So far, ecosystem models have mainly been uncertain about the carbon feedback from terrestrial ecosystems to the atmosphere (Classen et al., 2015). As a result, many experimental studies have been focused on making more reliable carbon flow predictions to estimate how many terrestrial ecosystems could store carbon. Both plant and soil biomasses are 2.53 times more than the carbon content of the environment (Monks et al., 2015). Soil can hold a huge amount of carbon and its capability to chelate carbon helps to reduce the rise of carbon dioxide (CO_2)

in the atmosphere. Various elements can influence carbon sequestration, including climate, starting material, soil texture, age, and its structure, soil communities and plants (Jenny, 1941; Monks et al., 2015).

Microorganisms are abundant in the ecosystem, and their existence always influences the whole environment. Observations, which are done by the human influence of microbes on the environment, could be helpful, destructive, or insignificant. The most significant impact of microorganisms on the planet is their ability to recover the main elements of all living systems, primarily carbon, oxygen, and nitrogen (N). Primary production is a photosynthetic organism that absorbs carbon dioxide (CO_2) from the atmosphere and converts it into the organic (cellular) matter. This process, also known as CO_2 fixation, makes up a large part of the organic carbon that can be used for the synthesis of cell materials. Decomposition or biodegradation can cause complex organic materials to decompose into other forms of carbon that other organisms can use. However, certain compounds (such as plastics, insect repellent, Teflon, and pesticides) are prepared synthetically. Some combinations are persistent, while some are decomposing slowly.

The metabolic process of microbial respiration and fermentation ultimately breaks down carbon-containing compound such as CO_2, which is then reverted to the atmosphere for an uninterrupted method of primary production. Nitrogen fixation done by living organisms is a process that only occurs with certain bacteria that eradicate N_2 from the environment and convert it to ammonia (NH_3) for the usage of plants and animals. Fixation of nitrogen can also improve the soil N_2 that has been eliminated by agricultural production. Together with each these advantages, microorganisms have therefore made a significant contribution to maintaining the sustainability of the environment. In this chapter, brief cycling of the carbon, sulfur, nitrogen, and methane has been explained. Also, there are certain factors of microbial role in major environmental factors from formation to destruction that have been unveiled.

2.2 MICROBIAL IMPACTS DUE TO MAJOR ENVIRONMENTAL ISSUES

2.2.1 MICROBIAL CONTRIBUTION IN ATMOSPHERIC FORMATIONS

Due to the pressure caused by dehydration, the atmosphere is a harsh environment for microorganisms. This leads to a limited time for microbial activity. However, some organisms gain confrontation with these extreme conditions through specific mechanisms that can promote the loss of living activity. Spore bearing microbes, fungi, molds, and polyp-forming protozoa have definite means to protect themselves from these stress-inducing grassy environments. Therefore, the capability depends heavily on that environment, the time they devote to the environment, and the sort of microorganisms. However, a lot of other factors also affect the viability of microorganisms (Classen et al., 2015). These factors include temperature, moisture, the content of oxygen and definite ions, ultraviolet radiation, different types of pollutants, and air-related factors.

2.2.2   SOIL FORMATION AND MICROBES

Formation of soil is a gradual process that spans millions of years of physical and chemical weathering and biological processes. Microorganisms play a crucial role in the construction of soil aggregates and soil constancy and give soil fertility and productivity. Soil microorganisms are involved in these processes in many ways. For example, filamentous microorganisms use an extensive mycelium network to assemble clay particles into soil aggregates. Besides, some organisms secrete extracellular polysaccharides or cause clay particle compaction, which promotes aggregation of soil particles. The upper layer of soil is always abundant in indigenous microbes (comprising actinomycetes), algae, mold, and protozoa. Besides, anthropogenic and animal endeavors bring in specific microorganisms into soil in various ways (Vandenkoornhuyse et al., 2015). Anthropogenic activities add bacteria precisely as biodegradation products and bring sludge to agriculture, while the microorganisms introduce microbes via bird drops or defecation.

2.2.3   IMPACT OF TEMPERATURE ON MICROBIAL COMMUNITIES

Temperature is a critical factor that affects microbial activity. Generally, high temperatures cause deactivation due to dehydration and destabilizing proteins, while low temperatures promote extensive endurance rates. At significantly fewer temperatures, certain microorganisms fail in their sustainability due to freezing and the formation of ice crystals on their surfaces (Keesstra et al., 2016).

2.2.4 RELATIVE HUMIDITY

The relative water content present in the air is crucial for the sustainability of microorganisms in the air. Most bacteria having gram-negative properties linked with aerosols can endure longer at lower moisture level, while bacteria having gram-positive properties can take longer at higher moisture level. The capability of microorganisms to survive in aerosols is linked to the superficial biochemistry of the organism. A possible explanation for this fact could be that the structure of the cell membrane lipid bilayer has changed in reaction to the exceptionally low moisture level. Throughout the dehydration process, the double layer of the cell membrane fluctuations from the usual crystal structure to the gel-like system and influences the simple protein configuration, which leads to the deactivation of the cells. Encapsulated virus nucleocapsids (such as cold viruses) have an extended endurance rate in the air when the humidity is lower than 50%, while viruses without nucleocapsids (such as enteroviruses) can be above 50% due to high relative humidity.

2.2.5 OXYGEN, IONS, AND OPEN-AIR FACTOR (OAF)

The combination of oxygen, ions, and open-air factor (OAF) inactivates various types of microorganisms in the air. Some reactive forms of oxygen are generated due to light, mutations cause ultraviolet radiation, or pollution, comprising hydrogen peroxide, superoxide radicals and hydroxide radicals, and DNA damage. Similarly, OAF (a combination of ozone and hydrocarbons) can also inactivate microorganisms by destroying enzymes and nucleic acids. In adding to these considerations, positive ions only trigger physical disintegration, such as the deactivation of cell proteins present on the surface. In contrast, ions negatively charged cause mutually biological and physical damage to DNA.

2.2.6 RADIATION

In general, low-wave radiation, such as ultraviolet and ionizing radiation, is damaging to microorganisms that cause DNA destruction. These rays destroy DNA by creating specific or dual strand break down and altering the structure of bases in nucleic acid. Ultraviolet rays initiate damage through the formation of thymine dimers within the chain, which inhibits biological activities, such as genome replication, transcription, and translation. Various processes, involving the involvement of microorganisms with large atoms, dyes, or carotenoids in the air, higher relative humidity, clouds, etc., can protect microorganisms from these destructive radiations (Leff et al., 2015). However, numerous organisms have developed mechanisms to restore DNA injury caused by ultraviolet rays.

2.3   NUTRIENT CYCLING THROUGH MICROBES

2.3.1   *MICROBES ROLE IN CARBON CYCLE*

Microbes play a significant character in the global carbon cycle and are a vital component of all living organisms. Microbes use carbon for living beings and themselves by extracting carbon from nonliving resources. In aqueous habitats, microorganisms transform carbon in the absence of oxygen and occur in anaerobic areas. Carbon dioxide is the highest universal form of carbon that penetrates the carbon cycle. At the same time, CO_2 is a water-solvable gas that appears in the atmosphere. Photosynthetic algae and plants utilize CO_2 to synthesize carbohydrates through photosynthesis. Besides, chemical autotrophic organisms like bacteria and archaea use CO_2 to produce sugar. This carbon, which is in the type of sugar, is beyond processed during the breathing process by a reaction chain, the tricarboxylic acid cycle, which is then converted into energy. Microorganisms can also utilize carbon under anaerobic environments to generate energy via a procedure called fermentation. Plants are the main elements in terrestrial ecosystems, but symbiotic organisms, such as wild plankton, cyanobacteria, and lichens, also help bind carbon in particular ecosystems.

Nonliving organic substances can be recycled from fungi and bacteria, while saprophytes use carbon-containing substances and generate CO_2 through breathing, thereby providing to the carbon cycle. Higher animals such as herbivores and carnivores also use the intestinal flora in the intestine to digest organic matter and gain energy. This process is called decomposition and ultimately produces inorganic elements, such as CO_2, water, and ammonia. Actinomycetes and proteus can break down solvable organic compounds (Leff et al., 2015). Similarly, bacteria are also involved in the breakdown of poorly degradable carbon-containing compounds, such as lignin, cellulose, and chitin, and use higher amounts of accessible nitrogen to support the making of transport and extracellular enzymes. In contrast, microbes existing in a less nitrogen environment are better able to break down of nitrogen-rich organic compounds. The frequency of α-proteobacteria and bacteroides has a positive effect on carbon mineralization, while acid bacteria are against carbon mineralization (Peiffer et al., 2013).

2.3.2 MICROBES ROLE IN DRIVING NITROGEN CYCLE

Nitrogen is a crucial component in the structure of proteins and nucleic acids. Microorganisms play a key part in the nitrogen cycle in various methods such as fixation of nitrogen, reduction of nitrate, nitrification, and denitrification. Microbial practices limit the production of ecosystems because the accessibility of nitrogen is the restricting factor to produce plant biomass. Both archaea and bacteria can bind nitrogen in the atmosphere by reduction to ammonium. Nitrogenase is an oxygen-sensitive enzyme and can catalyze fixation of nitrogen in a less oxygen environment. Fixation of nitrogen needs energy in the form of ATP (16 moles) per mole of fixed nitrogen.

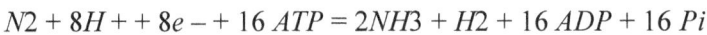

$$N2 + 8H + + 8e - + 16\ ATP = 2NH3 + H2 + 16\ ADP + 16\ Pi$$

Free-living microorganisms (e.g., nitrogen-fixing bacteria, *Burkholderia*, and *Clostridia*) have nitrogen-fixing abilities, and only a few form a symbiosis with plant rhizospheres (such as *Frankia, Mesozoic, Rhizobium*, and *Rhizobia*). Fixation of nitrogen by two- to threefold in the soil where symbiotic rhizobia communities is present. Nitrification consists of two ways: first, the oxidation of ammonia into the nitrite, and second, the conversion of this nitrite into nitrate with the help of microbes. Soil bacteria, such as *Nitrospirillum, Nitrosonas, Crenarchaeal,* or nitros bacteria, are rarely used to oxidize ammonia to nitrite, and then certain bacteria such as nitrifying bacteria oxidize ammonia to nitrite. Nitrification can also change the ionic status where the soil carries positive charge before the process of nitrification, and after that these charges convert from positive to negative with the help of oxidizing ammonia to nitrite and thereby releasing energy. The energy is then absorbed via the nitrifying microorganisms. Denitrification process produces nitrate (NO_3^-), nitrite (NO_2^-), and nitrogen oxide (NO) in reducing form, and then they convert into laughing gas (N_2O), greenhouse gas, or harmless nitrogen (N_2). Because this procedure needs oxygen limitation, most of the oxygen occurs in water-filled zones that give an oxygen-free environment. Denitrification process is a significant part of nitrogen cycle in which nitrogen present in solid form is reverted into soil, water, and then into atmosphere to complete the nitrogen cycle (Graham et al., 2016). Denitrification process is carried out by the chain of microorganisms present in soil such as *Proteus, Actinomycetes,* and *Sclerotinia.* Various types of bacteria lacks one or more of the enzymes participating in the removal of nitrogen (denitrification) and are well known to be deficient denitrifiers, such as lack of nitrous oxide reductase in various fungi and bacteria, which produces N_2O as the end-product. Inadequate denitrification, therefore, leads to greenhouse gas emissions.

2.3.3   SULFUR CYCLE AND ROLE OF MICROBIAL POPULATIONS

Sulfur (S) is a vital element of several vital metabolites, and vitamins occur in two amino acids (methionine and cysteine). Similar to nitrogen and carbon, microorganisms are able to convert sulfur from its best oxidizable type (sulfate or SO_4^-) to its most easily reducible form (sulfide or H_2S). In particular, the sulfur cycle contains some exclusive prokaryotes. Two classes of unrelated prokaryotes help to oxidize H_2S to S and $S°$ to SO_4. In the initial step the oxygen-producing photosynthetic green and purple bacteria oxidize H_2S as an electron supplier for cyclic photophosphorylation. The following steps are carried out by "monochrome sulfur bacteria."

Sulfur oxidizing unicellular are usually thermophilic bacteria located near warm springs (volcanoes) and hot deep-sea caves that are abundant in H_2S. They can also be acidophilic for the reason that they acidify their ecosystem through sulfuric acid production. While SO_4 and S can be treated as electron acceptors in respiration, bacteria help to reduce sulfate to produce H_2S in an oxygen-free environment, like the denitrification process H_2S. The usage of SO_4 as an electron acceptor is a mandatory method that only holds in oxygen-free environments. This procedure leads to anaerobic swamps, H_2S in soils and sediments, which create a special smell. Bacteria and plants consume S as SO_4 for usage and reduction to sulfides. Living organisms can eliminate sulfide groups from proteins as a trace of S through the decay process. These procedures complete the sulfur cycle.

2.3.4 METHANOGENS-DRIVEN METHANE CYCLE

Some microorganisms break down organic compounds through anaerobic or fermentation to organic acids and certain gases such as hydrogen and CO_2. Under severe anaerobic conditions, methanogens can use hydrogen to reduce CO_2 to methane. To achieve the cycle, bacteria help to oxidize methane (such as vegetative methane bacteria) and convert it into water, CO_2, and energy in the presence of oxygen. Other microorganisms contribute to the carbon cycle by breaking down hydrogen sulfide (H_2S) into complexes with carbon through energy generation. Few bacteria such as *T. ferrooxidans* draw power from the oxidation of iron to iron (III), which supports the carbon cycle. There are very few bacteria like succinate, *Clostridium butyrate*, and *Syntrophomonas spp*. When all these species are together (also known as hydrogen transfer between species), carbon is mined anaerobically to produce mass CO_2 and methane.

2.4 INTERACTIONS OF HOST MICROBE WITH PATHOGENIC AND BENEFICIAL MICROBES

The symbiosis of host microorganisms occurs in nearly all living organisms, and such symbiotic bacteria can be beneficial, harmful, or has no effect to the host. For example, innocuous *E. coli* strains frequently discovered in the intestine are a normal component of the intestinal flora and are able to benefit their hosts by supplying vitamin K and preventing disease-causing bacteria from settling in the gut. The interaction between host and microorganism forms a complex network. In contrast, some other strains, such as the *E. coli* O 26 strain, can trigger disease in the host. The collaboration between the host and the microorganism can be useful or harmful but has important environmental implications. If an organism can cause disease even in a host that looks healthy, it is said to be the primary pathogen. However, if it triggers disease only after the host's defenses are compromised, it is said to be a secondary pathogen. Microorganisms that are always correlated with the host are known as flora. These microorganisms have extensive symbiotic relations with their hosts. Certain host-microbe interactions are given below.

Symbiosis: A connection in which two different organisms (symbionts) live in close relationship with one another.

Commensalism: A connection between two species in which one is gained and the other is not altered, neither negatively nor positively.

Mutualism: Equally useful connection between two species.

Parasitism: A connection between two species in which one gain (parasite) from the other (host); it usually includes some damage to the host.

2.5 PLANT-MICROBIAL VALUABLE INTERACTIONS

The valuable interaction between plants and microbes can be divided into three parts. First, microorganisms indirectly support the growth of plants via direct contact with plants or via biological or abiotic considerations that influence the soil resource with minerals. Second, various microorganisms hinder the growth and movement of plant pathogens, thereby promoting plant development. Third, only a few microorganisms generate plant hormones that invigorate plant growth. Besides, certain saprophytic microorganisms create neutral connections with plants with not promoting or damaging them. These microbes decompose organic components to enrich the soil's nutrient content, which affect their efficiency and development. The plant rhizosphere is the most critical soil biological environment in which plants interact with microorganisms. In the rhizosphere, various microorganisms settle around the growing roots, which, depending on the nutritional status of the soil, plant defense mechanisms, soil environment, and the kind of microbial propagation in the rhizosphere, can lead to symbiotic, neutral, or parasitic interactions.

Microbial communities existing in the rhizosphere benefitting plants by endorsing their growth are identified as rhizosphere bacteria (PGR). These PGPRs comprise numerous bacteria such as *Pseudomonas, Bacillus, Azospirillum, Serratia marcescens, Rhizobium, Streptomyces*, and fungi such as *Trichoderma, Coniothyrium*, and *Ampelomyces*. These PGPRs increase fertility of plant, excrete plant hormones, and support plant growth. They are protected from numerous diseases by making antibiotics and provoking plant defense systems. *Bacillus* and *Pseudomonas* have properly explored PGPR and essential bacteria in the root zone. PGPR bacteria have subsequent tasks to play:

PGPR bacteria inhibit the disease-causing microorganisms via secreting low-molecular iron carriers to reduce iron utilization.

PGPR can decrease the movement of disease-causing organisms by initiating plant-induced systemic resistance (ISR) or systemic resistance

(SAR). Signaling particles such as ethylene, jasmonic acid, and salicylic acid stimulate these plant endurance systems.

In addition to its ability to bind nitrogen, PGPR bacteria also increase the production of plant hormones (e.g., auxin, cytokinin, and gibberellin). These plant hormones play a key part in root induction, division, and growth of cells. *Azospirillum sp.* mainly secrete auxin.

Numerous commercial PGPRs help plant growth through a variety of methods (such as biostimulants, bioprotectants, and biofertilizers).

PGPR microbes (e.g., *Spirulina*) also provide plants with nutrients through releasing phosphorus from carbon-containing compounds, and thereby, supporting plant growth. Overlapping chemical-based plant signals drives root colonization by microorganisms. For instance, vegetable flavonoids can be used as chemical attractants for nitrifying bacteria, floating zoospores, and symbiotic fungi.

Through the interaction between microorganisms and surface of the plant body, plants produce signaling molecules in the form of flavonoids in the rhizosphere. These signaling molecules promote the pathogenicity, association, symbiosis, or neutral adaptation of microorganisms to plants. In the symbiosis of legume rhizobia, the coryneform soil bacterium, nitrogen-fixation nodules in the roots of legumes. The bacteria use the bacterial enzyme nitrogenase under nitrogen-poor conditions to reduce about 80% of the chemistry in the atmosphere. Inert nitrogen is converted to ammonia (Nadeem et al., 2014). In this symbiotic relationship, the roots of plants release Nod gene expression initiators; bacteria release Nod factors, while plant roots show ion flow, express nodular protein, and a nodular morphology occur. The plants effectively provide the function of the oxygen-sensitive nitrogenase, which is coded by the bacterial Nif gene and complex carbohydrate, by delivering a microaerobic environment, which supports the metabolism of commensal bacteria within the bacteria.

In return, the bacteria bind nitrogen from the atmosphere so that plants can gather their biological necessities. Other diazobacter, such as rhizosphere fungi, nitrogen-fixing bacteria, and bacteria, particularly *Bacillus*, and *Pseudomonas*, also cooperate with rhizobia, by disturbing the nodule formation and fixation of nitrogen. Interacting microorganisms can benefit from each other's nutrients. There is, therefore, a mutual relationship between nitrogen-fixing bacteria and nitrogen-fixing *Spirulina*. Both cooperate with rhizobia to enhance the growth of the plant. These beneficial impacts are primarily due to improved root growth and increased absorption of water and minerals by the root system (Remigi et al., 2016). The development of rhizobia includes the appearance of specific genes of rhizobia: bacterial genes (Nod genes) and plant genes (Nodulin genes).

In mycorrhizal binding, the fungus colonizes the cells inside the arbuscular mycorrhizal fungi (AMF or AM) or the roots of the host plant extracellularly in the ectomycorrhizal fungi (Jacott et al., 2017). In ectomycorrhizal fungi, the mold does not get into the plant body, but settles in the surface layer of the cell and constructs a Hartig network. The Hartig network links numerous organisms and disease-causing mold and soil microorganisms. In this context, the fungus forms a network structure around the root (hair) to expand entry to soil nutrients. Fungi present in the surface promote the development of tree seedlings and the sprouting of seeds. This interconnected compound offers mushrooms with comparatively constant and directly complex carbohydrates such as glucose and sucrose. Complex carbohydrate is transferred from their resource (usually leaves) to the root tissue and then to the plant's fungal companion.

While the mycelium has a higher absorption capability for minerals and water due to the larger mycelium surface area, which increases the mineral assimilation capacity of the plant, plants having mycorrhizal association are generally more resilient to diseases affected by metal toxicity and soil pathogens. Mycorrhizal fungi, particularly belonging to the class Zygomycetes vesicular-arbuscular mycorrhizal fungus (VAM), play a vital role in the movement of phosphorus in soils with comparatively low accessible phosphorus content, which promotes better growth of grains and legumes.

The basic properties of VAM-forming mushroom types are:

- They all belong to Glomales (Zygote).
- The spores germinating through the plant plasma membrane begin to interact.
- Attachments (attachment points) in the form of mycelium.
- The formation of extracellular hypha systems in apoplasts.
- Formation: penetration in plant cells (intracellular clumps).
- Expand the interaction area.
- Long lifespan: a few days.

The hyphae present in the extracellular surface of mushroom species accumulate nutrients and move them to the mushroom. The combination of legumes and mycorrhizal fungi has a major influence on the growth of roots and sprouts as well as on the absorption of phosphorus, which leads to improved knotting and fixation of nitrogen. Good molds can initiate resistance genes that code for defensin proteins and generate active oxygen through NADH oxidase, thereby protecting plants from pathogenic micro-organisms. Mycorrhizal fungi can increase the yield fourfold.

2.6 RECYCLING OF WASTES AND ITS DETOXIFICATION VIA MICROBES CONTRIBUTION

Numerous types of bacteria and fungi species are capable of the detoxi-fying and degrading compound via several methods; thus, bioremediation practices are widely used.

2.6.1 BIODEGRADATION

Remediation and transformation are the waste management tools through living organisms and is used to eliminate or neutralize harmful waste into less noxious or non-noxious substances. The most frequently used micro-organisms are *Arthrobacter*, *Flavobacterium*, and *Azotobacter*. The treat-ment of waste managing through biotechnological methods includes the utilization of microorganisms to cleanse the air, water, and soil contami-nants at lower temperatures and pressures. Consequently, a small amount of energy is required than with traditional physico-chemical treatment methods. Differing on the type of pollutant monitoring point and satisfac-tory environmental circumstances, bioremediation can be borne away in situ or ex situ (Table 3.1). The processes of biofortification and biostimu-lation promote the breakdown ratio of organic and inorganic chemicals. These technologies have proven to be environmentally friendly and inex-pensive means of controlling pollution that can increase public acceptance and comply with environmental regulations. It is known that heterotrophic microorganisms such as *Pseudomonas, Sphingomonas*, and *Mycobacteria* are associated with hydrocarbon degradation. Pseudomonas is a well-analyzed bacterium that can degrade alkane, monoaromatic hydrocarbon, phenanthrene, and naphthalene under oxygenated environment. Bacteria that break down hydrocarbons dominate the soil contaminated with oil. However, since these elements are assimilated during the biodegradation process, higher concentrations of hydrocarbons can exhaust the accessible nitrogen and phosphorus in this area. Microorganisms (bacteria and fungi) can break down several biodegradable pesticides like atrazine, and its associated derivatives propazine and simazine (Aislabie et al., 2005). At the same time, some other types of pesticide, which is nonbiodegradable, such as DDT (dichlorodiphenyltrichloroethane), remain not easily broken down and are even present in water and soil.

Several fungi species capable of breaking down lignin, such as *Phanerochaete chrysosporium*, can break down a variety of toxins like dioxins and pentachlorophenol. The most outstanding examples are zygotic bacteria that are broken down during the wood treatment process in Whakatane. The biodegradation of several pollutants varies on their physical state and their chemical structure (Guemiza et al., 2017). The biodegradation of pollutants varies on their physical state and their chemical structure. In contrast, different pollutants (such as hydrocarbon) can be easily broken down, while synthetic pollutants (similar with Aldrin and DDT) are nondegradable and remain in the ecosystem. Degradability likewise varies on unique and novel structures and solubility in water, since poorly solvable substances are hard to degrade. Also, pollutants with poor water solubility or low hydrophobicity can easily be bound to clay particles so that microorganisms can easily use them in the soil. These microorganisms present in soil use such pollutants as energy sources and occur in higher concentrations. These pollutants can be toxic to them and cause slowdown biodegradation.

2.7   CLIMATIC CHANGE DIRECTLY IMPACTS ON PLANTS AND SOIL COMMUNITIES

Climate change will change the comparative frequency and role of soil populations due to the different physiological, growth rates, and temperature sensitivity of soil population (Classen et al., 2015; Wang et al., 2016). The immediate impacts of climate change on microbial structure and function have been widely investigated (A'Bear et al., 2014; Chen et al., 2014). For example, warming up to 58°C in temperate forests changed the relative frequency of soil bacteria and improved the ratio of common bacteria to fungi (Classen et al., 2015). Living organisms and communities respond to warming up and other disruptions via resistance caused by the malleability of the microbial characteristics of the strength of the population when it reverts to its original composition after the stress is relieved (Nemergut et al., 2013).

If the functional characteristics of organisms existing in the soil are different, or the steps to control speed limits or fate are controlled, changes in the composition of the microbial community can lead to changes in ecosystem function (Schimel and Schaeffer, 2012). For example, particular microbiota control ecosystem tasks likewise fixation of nitrogen (Urakawa et al., 2014), denitrification (Cayuela et al., 2014), and methane production (Schimel and Schaeffer, 2012). Changes in the relative frequency of organisms that regulate a particular process directly affect the speed of that practice. However, some methods on a relatively large scale (likewise nitrogen mineralization) are more closely related to abiotic considerations (such as humidity and Temperature) than the composition of the microbial community, as several organisms control these practices (Cardinale et al., 2012). Global warming (e.g., warming) immediately changes the respiratory rate of the microbial soil, since soil microbes and their mediated processes are sensitive to temperature. The role of high temperature in microbial metabolic rate has recently got substantial consideration (Classen et al., 2015).

In the absence of changes in the composition of the community, the intrinsic temperature sensitivity of microbial activity is defined as the factor by which microbial activity increases with a temperature rise of 108°C (Q10). Q10 is often used in climate models to illustrate the temperature sensitivity of microorganisms. However, the use of this relationship masks many contacts that affect the temperature sensitivity of microbial processes, likewise decomposition. Consequently, utilizing only Q10 to account for temperature sensitivity in the model can lead to dire projections. Although the degradation of organic matter present in the soil, the soil respiration, and the development of microbial composition usually rise with temperature, these reactions of warming are frequently temporary in field experiments (Burns et al., 2013). It is speculated that the immediate impact of warming on soil communities is due to the depletion of unstable carbon substrates in the soil due to increased microbial activity and the adaptation of the microbial community. The composition changes or limits its biomass to adapt to the changing environment and biological conditions due to compromises availability of substrate (Burns et al., 2013).

Investigational warming may primarily change the structure of the microbial community and the frequency of gram-negative and gram-positive bacteria (Wu et al., 2011), or it might take several years for the warming effect to become visible in the microbial community. Intriguingly, the results of field and laboratory tests frequently contradict each other (Arndt et al., 2013) as well as long-term field tests (Sistla and Schimel, 2013) and short-term laboratory tests (Dungait et al., 2012). Thermal compensation by microbial communities can help the opposite conclusion. These conflicting results have led to controversy over the evidence and mechanism of thermal adaptation (Dungait et al., 2012). The immediate influence of temperature on microbial functioning is complicated and is likely to be facilitated through the transformation, development, and interaction of microbes over time. Changes in temperature are usually associated with alterations in soil moisture, which might explain some conflicting results from trials examining how microbial communities react to climate change.

For instance, the rate of microbial action at higher temperatures can be restricted by dispersion and microbial contact with available substrates (Davidson and Janssens, 2006). Although bacterial communities can react quickly to water rhythms, slower-growing fungal communities can wait behind their reaction (Cregger et al., 2014). Also, scarcity aggravates the difference in temperature sensitivity between populations of bacteria and fungi (Briones et al., 2014). Just as the availability of soil moisture changes slightly (30% less water holding capacity), soil fungus populations can be transferred after one primary member to another while the bacterial population remains unchanged. These models show that the plasticity of fungi is greater than that of bacteria in non-extreme drying and wet cycles (Kaisermann et al., 2015). Adaptation to soil populations with minimal water use efficiency or frequent dry-wet processes can lead to fewer components or functions changing due to shifting water regimes (Tielbörger et al., 2014). At any location, the interaction between microorganisms and background temperature and humidity conditions can alter the composition and role of microorganisms and it changes with climate change.

It is still uncertain: (1) how humidity and temperature and their interactions disturb certain microbial useful groups in the community likewise methanogens; (2) what effects the change in the microbial community has on the breakdown of new and old organic soil; (3) What mechanisms have promoted microeconomic response to the net ecosystem of climate change?

2.8 INFLUENCE OF CLIMATE CHANGE ON MICROBE–MICROBE INTERACTIONS

Microbes have formed a complicated network of connections that always respond to alterations in resources. For example, foraging through mycorrhizal fungi can change the community of autogenous bacteria, thereby changing the transmission of nitrogen from mycorrhiza to plants (Hassani et al., 2018) and the breakdown of organic matter. An increase in temperature leads to an increase in the carbon distribution of the mycorrhizal mycelium, which can change the mycorrhizal association from symbiosis to parasite (Classen et al., 2015). The changes in these connections between mycorrhiza and plants can affect the microbial composition of the soil (Stockmann et al., 2013) and activity (Rashid et al., 2016) to a certain extent in a cascading manner as follows: It can aggravate negative or positive interactions among plants and their associated communities. Different bacterial and fungal interactions in free-living communities can alter ecosystem function and carbon feedback, but little research is available.

2.9 MICROBIAL RESPONSE MECHANISMS TO CLIMATE CHANGE

The insight into the underlying processes of microbial response is crucial to summarizing future microbial impacts from soil microbes to soil carbon cycling and other ecosystem functions. When the environment changes due to new disruptions, for instance, climate change, we can examine alterations in processes, such as enzyme activity, soil respiration, and waste disintegration. Typically, however, the information needed to determine the mechanism by which these changes occur is lacking. The aggregate functional response of the whole soil is generated by the individual activities of the diverse community of soil microorganisms, which means that distinct processes can work concurrently to produce the observed function. At this time, we consider four types of reaction mechanisms that can come into play: physiology, community composition, feedback, and evolution. The characteristics that connect the individual physiology and function with the environmental conditions cause the classification and design of the species to change with the gradient (Tscharntke et al., 2012). The community status can also be affected by distribution restrictions and landscape connection patterns (Hermy and Verheyen, 2007). Furthermore, strengthening positive feedback as a substitute for stable states can replace local classification and immigration and lead to a robust historical dependency on community reactions (Monceau et al., 2014). After all, the difference in characteristics due to evolutionary difference is another method that regulates the community's response to environmental changes, possibly with the least understanding of the contacts with classification, relocation, and positive feedback (Devictor et al., 2010).

2.10   MICROBIAL PHYSIOLOGY

Global carbon storage in the ground is more than three times that of atmospheric carbon, and over-field biomass carbon is four times higher than atmospheric carbon(Allen et al., 2010). According to the foreseeable climate change, the balance among carbon releasing and carbon seques-tration is significant (Allison and Martiny, 2008). During the microbial breakdown of soil organic material (SOM), organic carbon is used to generate energy and stabilize biosynthesis. Several experiments (Frey et al., 2013; Graham et al., 2016) confirmed this hypothesis while using glucose as a substrate at different incubation temperatures. In distinction, experiments with multiple substrates (Frey et al., 2013) and theoretic models (Moyano et al., 2013) have shown that the temperature increases with increasing temperature. Theoretically, this reduction is affected by the different temperature sensitivity of bacterial respiration and substrate absorption (Manzoni et al., 2012). However, our mechanical knowledge of the necessary process is restricted. The changes in substrate productivity with temperature might be due to: (1) biological changes inside the micro-bial population that is dynamically managing the substrate; (2) changes in the structure of the active community; or (3) a blend of the two (Manzoni et al., 2012; von Wintersdorff et al., 2016).

The efficiency of usage of microbial substrates is an important biolog-ical characteristic of the terrestrial C cycle. In the microbial population level, it can be influenced by functional traits and life story approaches (von Wintersdorff et al., 2016). A constructive correlation among the effectiveness and frequency of gram-negative bacteria has recently been observed (Bölscher et al., 2016; Harris et al., 2012), suggesting that they perform an essential role in changing the efficiency of substrate use across the board Community could play. Besides, it is generally believed that fungi use substrates more efficiently than bacteria (e.g., Holland and Coleman, 1987; Ohtonen et al., 1999). This concept has been about for some time, but though some experiments help this view (Bölscher et al., 2016; Kallenbach et al., 2016), others do not (Brant et al., 2006; Thiet et al., 2006). It has been observed that the seasonal dynamics of the composition of the microbial community change with the increasing relative frequency of winter fungi (Buckeridge et al., 2013; Schadt et al., 2003). This transition to comparatively more active fungi at reduces temperatures might explain the increased efficiency observed at lower temperatures. However, few experiments complement immediate measurement of substrate efficacy at all temperatures with population composition studies. Microbes usually have a wide range of physiological abilities, so physiological adaptation to environmental changes can be expected. The overall functional response may be expected to the physiological scope of a single taxon in the population but may also signify the physiological variety between taxon. Functional plasticity has been observed in the response of microbial communities to short-term alterations in temperature and humidity (Buyer et al., 2010; Kuzyakov and Gavrichkova, 2010), though not always (Hawkes and Keitt, 2015). One more plastic form that is common in soil microorganisms can avoid temporary environmental impacts such as drought (Philippot et al., 2013). When the historical selection pressure range is exceeded, the microbial response is more significant, with significant changes or unpredictability. The mutual transplantation of complete soil samples among plant communities supports this view. Kuzyakov and Blagodatskaya (2015) observed that the composition of the microbial communities transplanted from grassland under the oak sky and the aggregation function of the soil samples hardly changed. But rapid changes were observed when the oak core was transplanted into the grass (conditions were out of the usual range). Thus, the insight into the boundaries of microbial physiological plasticity will give boundary restrictions for possible microbial functions in response to climate change.

2.11 MICROBIAL COMMUNITY COMPOSITION

Variations in microbial functioning in different environments can lead to changes in population composition through both changes in the comparative abundance of taxa now present or dispersion from the local species pool. While the climate changes, some microbial taxa will gain more than others, causing changes in supremacy and function (Fierer, 2017). Dispersion will also offer new migrant taxa, and species categorizing should result in the existence of organisms greatest suitable to the regional environment (Tscharntke et al., 2012). Species organizing has been noted in bacterial populations (Peršoh, 2015); however, not in protists (Tedersoo et al., 2014). The degree of ordering versus mass impacts can vary on dispersion, which is often believed to be unrestricted in microbes, though, recent experiments support microbial dispersion limitation (Van Der Heijden et al., 2015) and still a high level of endemism (Nguyen et al., 2016). If there are regional variations in microbial species pools, empirical reactions of microbes in experimental climate influences may be restricted by available taxa. Several climate change trials impose scarcity on small plots implanted in an ambient extreme rainfall area, which may require a local species pool including scarcity-adapted individuals if dispersion is restricted.

2.12   CONCLUSION

If the reaction of plants and soil communities to natural climate fluctuations or at a certain point in time is observed, their interaction can be unpredictable. Anticipated to the temperature sensitivity of the carbon cycle process, minor alterations in temperature can cause massive amounts of soil carbon to be released back into the atmosphere. Vegetable carbon input powerfully conveys the temperature sensitivity of carbon breakdown in the soil. Still, the comparative importance of the directly or unintended impacts of climate change on carbon dynamics in the soil stays unsolved, significantly when changing from one state to another. We believe that the unintended impacts of climate change on plant-facilitated microorganisms can be more challenging than the immediate effects of climate on the structure and function of microbial communities. These effects should be assessed on suitable chronological and spatial scales, preferably in microbial-centered research, in edict to complement the current pattern of plant-centered climate change research. New technical methods will play a key role in microbial-centered research, as our goal is to uncover those groups that are very susceptible to the climate and those groups that respond to changes in the function of microbial communities. Overall, these advances are critical to predicting the essential points of ecosystems, the impact of severe events, and the strength of populations beneath climate change. In summary, if we want to comprehend whether the effects of climate change on microbial-microbial and plant-microbial relations are equivalent to or better than the direct impacts of climate change on the composition and function of ecosystems, we have to regulate, quantify, and expand the best observation methods. Observation, manipulation, and experimental tests along natural gradients such as the sculpting of plant and soil microbial populations and their connections with climate change drivers must be combined to forecast upcoming ecosystem functions.

KEYWORDS

- **soil microbiota**
- **nutrient cycling**
- **carbon sequestration**

REFERENCES

A'Bear, A. D.; Jones, T. H.; Boddy, L. Potential Impacts of Climate Change on Interactions among Saprotrophic Cord-Forming Fungal Mycelia and Grazing Soil Invertebrates. *Fungal Ecol.* **2014,** *10,* 34–43.

Allen, C. D.; Macalady, A. K.; Chenchouni, H.; Bachelet, D.; McDowell, N.; Vennetier, M.; Kitzberger, T.; Rigling, A.; Breshears, D. D.; Hogg, E. T. A Global Overview of Drought and Heat-Induced Tree Mortality Reveals Emerging Climate Change Risks for Forests. *Forest Ecol. Manage.* **2010,** *259* (4), 660–684.

Allison, S. D.; Martiny, J. B. Resistance, Resilience, and Redundancy in Microbial Communities. *Proc. Natl. Acad. Sci.* **2008,** *105* (Supplement 1), 11512–11519.

Arndt, S.; Jørgensen, B. B.; LaRowe, D. E.; Middelburg, J.; Pancost, R.; Regnier, P. Quantifying the Degradation of Organic Matter in Marine Sediments: A Review and Synthesis. *Earth-Sci. Rev.* **2013,** *123,* 53–86.

Bennett, J. A.; Maherali, H.; Reinhart, K. O.; Lekberg, Y.; Hart, M. M.; Klironomos, J. Plant-Soil Feedbacks and Mycorrhizal Type Influence Temperate Forest Population Dynamics. *Sci.* **2017,** *355* (6321), 181–184.

Berg, M. P.; Kiers, E. T.; Driessen, G.; Van Der Heijden, M.; Kooi, B. W.; Kuenen, F.; Liefting, M.; Verhoef, H. A.; Ellers, J. Adapt or Disperse: Understanding Species Persistence in a Changing World. *Global Change Biol.* **2010,** *16* (2), 587–598.

Bressloff, P. C. *Stochastic Processes in Cell Biology,* vol 41; Springer, 2014.

Briones, M. J. I.; McNamara, N. P.; Poskitt, J.; Crow, S. E.; Ostle, N. J. Interactive Biotic and Abiotic Regulators of Soil Carbon Cycling: Evidence from Controlled Climate Experiments on Peatland and Boreal Soils. *Global Change Biol.* **2014,** *20* (9), 2971–2982.

Burns, R. G.; DeForest, J. L.; Marxsen, J.; Sinsabaugh, R. L.; Stromberger, M. E.; Wallenstein, M. D.; Weintraub, M. N.; Zoppini, A. Soil Enzymes in a Changing Environment: Current Knowledge and Future Directions. *Soil Biol. Biochem.* **2013,** *58,* 216–234.

Buyer, J. S.; Teasdale, J. R.; Roberts, D. P.; Zasada, I. A.; Maul, J. E. Factors Affecting Soil Microbial Community Structure in Tomato Cropping Systems. *Soil Biol. Biochem.* **2010,** *42* (5), 831–841.

Cardinale, B. J.; Duffy, J. E.; Gonzalez, A.; Hooper, D. U.; Perrings, C.; Venail, P.; Narwani, A.; Mace, G. M.; Tilman, D.; Wardle, D. A. Biodiversity Loss and Its Impact on Humanity. *Nature* **2012,** *486* (7401), 59–67.

Cayuela, M.; Van Zwieten, L.; Singh, B.; Jeffery, S.; Roig, A.; Sánchez-Monedero, M. Biochar's Role in Mitigating Soil Nitrous Oxide Emissions: A Review and Meta-Analysis. *Agric., Ecosyst, Environ.* **2014,** *191,* 5–16.

Chen, S.; Zou, J.; Hu, Z.; Chen, H.; Lu, Y. Global Annual Soil Respiration about Climate, Soil Properties and Vegetation Characteristics: Summary of Available Data. *Agric. Forest Meteorol.* **2014,** *198,* 335–346.

Classen, A. T.; Sundqvist, M. K.; Henning, J. A.; Newman, G. S.; Moore, J. A.; Cregger, M. A.; Moorhead, L. C.; Patterson, C. M. Direct and Indirect Effects of Climate Change on Soil Microbial and Soil Microbial-Plant Interactions: What Lies Ahead? *Ecosphere* **2015,** *6* (8), 1–21.

Cox, C. B.; Moore, P. D.; Ladle, R. J. *Biogeography: An Ecological and Evolutionary Approach;* John Wiley & Sons, 2016.

Cregger, M. A.; Sanders, N. J.; Dunn, R. R.; Classen, A. T. Microbial Communities Respond to Experimental Warming, But Site Matters. *Peer J* **2014,** *2,* e358.

Davidson, E. A.; Janssens, I. A. Temperature Sensitivity of Soil Carbon Decomposition and Feedbacks to Climate Change. *Nature* **2006,** *440* (7081), 165–173.

Devictor, V.; Mouillot, D.; Meynard, C.; Jiguet, F.; Thuiller, W.; Mouquet, N. Spatial Mismatch and Congruence between Taxonomic, Phylogenetic and Functional Diversity: The Need for Integrative Conservation Strategies in a Changing World. *Ecol. Lett.* **2010,** *13* (8), 1030–1040.

Dungait, J. A.; Hopkins, D. W.; Gregory, A. S.; Whitmore, A. P. Soil Organic Matter Turnover Is Governed by Accessibility Not Recalcitrance. *Global Change Biol.* **2012,** *18* (6), 1781–1796.

Evans, S. E.; Wallenstein, M. D. Climate Change Alters Ecological Strategies of Soil Bacteria. *Ecol. Lett.* **2014,** *17* (2), 155–164.

Fierer N Embracing the unknown, disentangling the complexities of the soil microbiome. Nature Reviews Microbiology 15 (10), 579.

Fischer, D.; Chapman, S.; Classen, A. T.; Gehring, C. A.; Grady, K.; Schweitzer, J.; Whitham, T. G. Plant Genetic Effects on Soils under Climate Change. *Plant Soil* **2014,** *379* (1–2), 1–19.

French, H. M. *The Periglacial Environment*; John Wiley & Sons, 2017.

Frey, S. D.; Lee, J.; Melillo, J. M.; Six J The Temperature Response of Soil Microbial Efficiency and Its Feedback to Climate. *Nat. Clim. Change* **2013,** *3* (4), 395–398.

Graham, E. B.; Knelman, J. E.; Schindlbacher, A.; Siciliano, S.; Breulmann, M.; Yannarell, A.; Beman, J.; Abell, G.; Philippot, L.; Prosser, J. Microbes as Engines of Ecosystem Function: When Does Community Structure Enhance Predictions of Ecosystem Processes? *Front. Microbiol.* **2016,** *7,* 214.

Guemiza, K.; Coudert, L.; Metahni, S.; Mercier, G.; Besner, S.; Blais, J-F. Treatment Technologies Used for the Removal of As, Cr, Cu, PCP and PCDD/F from Contaminated Soil: A Review. *J. Hazard. Mater.* **2017,** *333,* 194–214.

Guo, Q. Central-Marginal Population Dynamics in Species Invasions. *Front. Ecol. Evol.* **2014,** *2,* 23.

Hassani, M. A.; Durán, P.; Hacquard, S. Microbial Interactions within the Plant Holobiont. *Microbiome* **2018,** *6* (1), 58.

Hawkes, C. V.; Keitt, T. H. Resilience vs. Historical Contingency in Microbial Responses to Environmental Change. *Ecol. Lett.* **2015,** *18* (7), 612–625.

Hermy, M.; Verheyen, K. Legacies of the Past in the Present-Day Forest Biodiversity: A Review of Past Land-Use Effects on Forest Plant Species Composition and Diversity. In *Sustainability and Diversity of Forest Ecosystems*; Springer, 2007; pp 361–371.

IPCC. *The Physical Science Basis*; Cambridge University Press, 2007.

Jacott, C. N.; Murray, J. D.; Ridout, C. J. Trade-Offs in Arbuscular Mycorrhizal Symbiosis: Disease Resistance, Growth Responses and Perspectives for Crop Breeding. *Agronomy* **2017,** *7* (4), 75.

Jenny, H. A System of Quantitative Pedology. In *Factors of Soil Formation*; McGraw Hill: New York, 1941.

Kaisermann, A.; Maron, P.; Beaumelle, L.; Lata, J. Fungal Communities Are More Sensitive Indicators to Non-Extreme Soil Moisture Variations Than Bacterial Communities. *Appl. Soil Ecol.* **2015,** *86,* 158–164.

Keessstra, S. D.; Bouma, J.; Wallinga, J.; Tittonell, P.; Smith, P.; Cerdà, A.; Montanarella, L.; Quinton, J. N.; Pachepsky, Y.; Van Der Putten, W. H. The Significance of Soils and Soil Science towards Realization of the United Nations Sustainable Development Goals. Soil, 2016.

Kuzyakov, Y.; Blagodatskaya, E. Microbial Hotspots and Hot Moments in Soil: Concept & Review. *Soil Biol. Biochem.* **2015,** *83,* 184–199.

Kuzyakov Y, Gavrichkova O Time Lag between Photosynthesis and Carbon Dioxide Efflux from Soil: A Review of Mechanisms and Controls. *Global Change Biol.* **2010,** *16* (12), 3386–3406.

Langley, J. A.; Hungate, B. A. Plant Community Feedbacks and Long-Term Ecosystem Responses to Multi-Factored Global Change. *AoB Plants* **2014,** *6.*

Leff, J. W.; Jones, S. E.; Prober, S. M.; Barberán, A.; Borer, E. T.; Firn, J. L.; Harpole, W. S.; Hobbie, S. E.; Hofmockel, K. S.; Knops, J. M. Consistent Responses of Soil Microbial Communities to Elevated Nutrient Inputs in Grasslands across the Globe. *Proc. Natl. Acad. Sciences* **2015,** *112* (35), 10967–10972.

Lozupone, C. A.; Stombaugh, J. I.; Gordon, J. I.; Jansson, J. K.; Knight R Diversity, Stability and Resilience of the Human Gut Microbiota. *Nature* **2012,** *489* (7415), 220–230.

Manzoni, S.; Schimel, J. P.; Porporato, A. Responses of Soil Microbial Communities to Water Stress: Results from a Meta-Analysis. *Ecology* **2012,** *93* (4), 930–938.

McCormack, M. L.; Dickie, I. A.; Eissenstat, D. M.; Fahey, T. J.; Fernandez, C. W.; Guo, D.; Helmisaari, H. S.; Hobbie, E. A.; Iversen, C. M.; Jackson, R. B. Redefining Fine

Roots Improves Understanding of Below-Ground Contributions to Terrestrial Biosphere Processes. *New Phytol.* **2015,** *207* (3), 505–518.

Monceau, K.; Bonnard, O.; Thiéry, D. Vespa Velutina: A New Invasive Predator of Honeybees in Europe. *J. Pest Sci.* **2014,** *87* (1), 1–16.

Monks, P. S.; Archibald, A.; Colette, A.; Cooper, O.; Coyle, M.; Derwent, R.; Fowler, D.; Granier, C.; Law, K. S.; Mills, G. Tropospheric Ozone and Its Precursors from the Urban to the Global Scale from Air Quality to Short-Lived Climate Gorcer, 2015.

Moyano, F. E.; Manzoni, S.; Chenu, C. Responses of Soil Heterotrophic Respiration to Moisture Availability: An Exploration of Processes and Models. *Soil Biol. Biochem.* **2013,** *59*, 72–85.

Nadeem, S. M.;, Ahmad, M.; Zahir, Z. A.; Javaid, A.; Ashraf, M. The Role of Mycorrhizae and Plant Growth Promoting Rhizobacteria (PGPR) in Improving Crop Productivity under Stressful Environments. *Biotechnol. Adv.* **2014,** *32* (2), 429–448.

Nemergut, D. R.; Schmidt, S. K.; Fukami, T.; O'Neill, S. P.; Bilinski, T. M.; Stanish, L. F.; Knelman, J. E.; Darcy, J. L.; Lynch, R. C.; Wickey, P. Patterns and Processes of Microbial Community Assembly. *Microbiol. Mol. Biol. Rev.* **2013,** *77* (3), 342–356.

Nguyen, N. H.; Song, Z.; Bates, S. T.; Branco, S.; Tedersoo, L.; Menke, J.; Schilling, J. S.; Kennedy, P. G. FUNGuild: An Open Annotation Tool for Parsing Fungal Community Datasets by Ecological Guild. *Fungal Ecol.* **2016,** *20*, 241–248.

Peiffer, J. A.; Spor, A.; Koren, O.; Jin, Z.; Tringe, S. G.; Dangl, J. L.; Buckler, E. S.; Ley, R. E. Diversity and Heritability of the Maize Rhizosphere Microbiome under Field Conditions. *Proc. Natl. Acad. Sci.* **2013,** *110* (16), 6548–6553.

Peršoh, D. Plant-Associated Fungal Communities in the Light of Meta'omics. *Fungal Diversity* **2015,** *75* (1), 1–25.

Philippot, L.; Raaijmakers, J. M.; Lemanceau, P.; Van Der Putten, W. H. Going Back to the Roots: The Microbial Ecology of the Rhizosphere. *Nat. Rev. Microbiol.* **2013,** *11* (11), 789–799.

Rashid, M. I.; Mujawar, L. H.; Shahzad, T.; Almeelbi, T.; Ismail, I. M.; Oves, M. Bacteria and Fungi Can Contribute to Nutrients Bioavailability and Aggregate Formation in Degraded Soils. *Microbiol. Res.* 183, 26-41.

Rath, K. M.; Rousk, J. Salt Effects on the Soil Microbial Decomposer Community and Their Role in Organic Carbon Cycling: A Review. *Soil Biol. Biochem.* **2015,** *81*, 108–123.

Remigi, P.; Zhu, J.; Young, J. P. W.; Masson-Boivin, C. Symbiosis within Symbiosis: Evolving Nitrogen-Fixing Legume Symbionts. *Trends Microbiol.* **2016,** *24* (1), 63–75.

Schimel, J.; Schaeffer, S. M. Microbial Control Over Carbon Cycling in Soil. *Front. Microbiol.* **2012,** *3*, 348.

Sistla, S. A.; Schimel, J. P. Seasonal Patterns of Microbial Extracellular Enzyme Activities in an Arctic Tundra Soil: Identifying Direct and Indirect Effects of Long-Term Summer Warming. *Soil Biol. Biochem.* **2013,** *66*, 119–129.

Stockmann, U.; Adams, M. A.; Crawford, J. W.; Field, D. J.; Henakaarchchi, N.; Jenkins, M.; Minasny, B.; McBratney, A. B.; De Courcelles, V. d. R.; Singh, K. The Knowns, Known Unknowns and Unknowns of Sequestration of Soil Organic Carbon. *Agric., Ecosyst. Environ.* **2013,** *164*, 80–99.

Tedersoo, L.; Bahram, M.; Põlme, S.; Kõljalg, U.; Yorou, N. S.; Wijesundera, R.; Ruiz, L. V.; Vasco-Palacios, A. M.; Thu, P. Q.; Suija, A. Global Diversity and Geography of Soil Fungi. *Science* **2014,** *346* (6213), 1256688.

Tielbörger, K.; Bilton, M. C.; Metz, J.; Kigel, J.; Holzapfel, C.; Lebrija-Trejos, E.; Konsens, I.; Parag, H. A.; Sternberg, M. Middle-Eastern Plant Communities Tolerate 9 Years of Drought in a Multi-Site Climate Manipulation Experiment. *Nat. Commun.* **2014,** *5,* 5102.

Tscharntke, T.; Tylianakis, J. M.; Rand, T. A.; Didham, R. K.; Fahrig, L.; Batáry, P.; Bengtsson, J.; Clough, Y.; Crist, T. O.; Dormann, C. F. Landscape Moderation of Biodiversity Patterns and Processes-Eight Hypotheses. *Biol. Rev.* **2012,** *87* (3), 661-685.

Urakawa, R.; Shibata, H.; Kuroiwa, M.; Inagaki, Y.; Tateno, R.; Hishi, T.; Fukuzawa, K.; Hirai, K.; Toda, H.; Oyanagi, N. Effects of Freeze–Thaw Cycles Resulting from Winter Climate Change on Soil Nitrogen Cycling in Ten Temperate Forest Ecosystems throughout the Japanese Archipelago. *Soil Biol. Biochem.* **2014,** *74,* 82–94.

Urban, M. C. Accelerating Extinction Risk from Climate Change. *Science* **2015,** *348* (6234), 571–573.

Van Der Heijden, M. G.; Martin, F. M.; Selosse, M. A.; Sanders, I. R. Mycorrhizal Ecology and Evolution: The Past, the Present, and the Future. *New Phytol.* **2015,** *205* (4), 1406–1423.

van der Putten, W. H.; Bradford, M. A.; Pernilla Brinkman, E.; van de Voorde, T. F.; Veen, G. Where, When and How Plant–Soil Feedback Matters in a Changing World. *Funct. Ecol.* **2016,** *30* (7), 1109–1121.

Vandenkoornhuyse, P.; Quaiser, A.; Duhamel, M.; Le Van, A.; Dufresne, A. The Importance of the Microbiome of the Plant Holobiont. *New Phytol.* **2015,** *206* (4), 1196–1206.

von Wintersdorff, C. J.; Penders, J.; van Niekerk, J. M.; Mills, N. D.; Majumder, S.; van Alphen, L. B.; Savelkoul, P. H.; Wolffs, P. F. Dissemination of Antimicrobial Resistance in Microbial Ecosystems through Horizontal Gene Transfer. *Front. Microbiol.* **2016,** *7,* 173.

Wang, Q.; He, N.; Yu, G.; Gao, Y.; Wen, X.; Wang, R.; Koerner, S. E., Yu, Q. Soil Microbial Respiration Rate and Temperature Sensitivity along a North-South Forest Transect in Eastern China: Patterns and Influencing Factors. *J. Geophys. Res.: Biogeosci.* **2016,** *121* (2), 399–410.

Wu, Z.; Dijkstra, P.; Koch, G. W.; Peñuelas, J.; Hungate, B. A. Responses of Terrestrial Ecosystems to Temperature and Precipitation Change: A Meta-Analysis of Experimental Manipulation. *Global Change Biol.* **2011,** *17* (2), 927–942.

CHAPTER 3

The Vulnerability of Microbial Ecosystems in a Challenging Climate

BASHARAT A BHAT[1], LUBNA TARIQ[2], RAKEEB A MIR[2*],
ISHFAQ MAJEED[3], and MAAJID M. BANDH[3]

[1]Department of Bioresources, University of Kashmir, Srinagar, India

[2]Department of Biotechnology, BGSB University, Rajouri, India

[3]Department of Zoology, University of Kashmir, Srinagar, India

*Corresponding author. E-mail: rakeebahmad@gmail.com

ABSTRACT

Microbial communities are the crucial basis of the life forms on Earth. They catalyze various biogeochemical processes to drive the global nutrient cycles and play a pivotal role in ecosystem functioning. However, the critical role played by the microbes is not much appreciated in biodiversity conservation agenda, unlike that of the plant and animal diversity. Besides, microorganisms are essential for almost half of the primary biomass production on Earth. The effects of climatic change have been studied mainly in eukaryotes, and significantly less attention has been paid to these miniscule organisms, even though they are the significant stakeholders in balancing the ecosystems. Keeping in view their essential role in bringing the stability of biosphere, it is evident that they directly or indirectly regulate the existence of eukaryotes in different ecosystems. Drastic climatic changes have severely affected the existing microbial diversity. So the need of the hour is to devise strategies to prevent the adverse effects of climatic change on microbial diversity. This chapter delves deep into understanding the damaging effects of climate changes on the most vulnerable microbial ecosystems. The update may serve as a platform to initiate strategies to conserve the microbial gold mines to save the mother earth.

3.1 INTRODUCTION

A microbial community can be characterized as an assemblage of microbes, co-occurring, and potentially interacting in a given space and time. Despite their small size, microorganisms are the essential components of biosphere's ecological dynamics. They are not only the most diffuse forms of life but are also distinguished by an unparalleled functional and genetic diversity, which makes a profound contribution to the biogeochemical processes on Earth (Konopka, 2009; Deines et al., 2020). Microbial populations play a vital role in Earth's entire critical nutrient cycling and more so in deserts, where plants are scarce or even completely absent. Microbial species in the arid deserts find shelter inside transparent rocks (endoliths) as a tool for survival.

All ecosystems are centralized in microbial communities that comprise the microbes which interact strongly with each other in a microenvironment. However, in most ecosystems, the distribution of microbial species and the physicochemical properties are patchy. Even in combined planktonic oligotrophic ecosystems, marine snow may have a rich nutrient emphasis (Azam and Malfatti, 2007; Heinze, 2020), the microbial population's interaction is not insignificant but is essential for stringent advancement in the area. Given that microbial communities are of paramount importance in biogeochemical transformations, a thorough understanding of their dynamics would be crucial to know how the biosphere is modulating and reacting to future environmental conditions (O'Malley and Dupré, 2007).

Microbial diversity provides a vast reservoir of multiple processes and facilitates vital steps to maintain ecosystem balance through the processes of decomposition of organic matter and recycling of nutrients (Van Der Heijden et al., 2008; Joshi et al., 2019). Activities of soil microbes are affected directly or indirectly by climate change which feedbacks greenhouse gases to the atmosphere and thus contribute to global warming. Microbial diversity is essential for ecosystem stability through the role of microorganisms in structuring the food webs, recycling nutrients, and preserving functional repetitions as a "seed bank." Microorganisms modulate themselves in the natural environmental conditions and habitats to support the functioning of global ecosystems through interactive associations with other species. For example, prokaryotic (Archaea and Bacteria) and eukaryotic (fungi and other eukaryotes) species are present together, and decomposition of organic matter may be impaired without their diversity (Kaviya et al., 2019; Jobard et al., 2010).

One of the greatest scientific and political problems of the 21st century is the continuing global climate change triggered by the human-induced rise in greenhouse gases. Therefore, it is the need of the hour to understand the biological processes that control exchanges of carbon between land, oceans, and atmosphere and how these exchanges will react to climate change through input from the environment that could intensify or reduce regional and global climate change (Richter, 2014; Heimann and Reichstein 2008). Biofuels are thus gaining traction around the world for reducing both greenhouse gas emissions and petroleum-based fuel dependency (Dragone et al., 2020; Balat and Balat, 2009). As human activity is a significant reason for the increase in the organic waste level, this profoundly concerns the scientific community, some countries are clamping their legislation on recycling organic materials for animal feed (Mirabella et al., 2014; Pessiot et al., 2012).

Climate change is affecting almost all habitats on Earth with its ultimate effect on the diversity of plants and animals (Pecl et al., 2017; Parmesan, 2006). Ongoing climatic shifts are predicted to result in a drastic reduction in the variety of organisms. Extinction debts of many long-lived, gradually reproducing species populations are expected to decline in the coming years due to already occurring environmental changes (Dullinger et al., 2012). Crashes in the colonization of species are a few examples and its ranges are shifting in response to climate change (Menéndez et al., 2006). Because of their short generation time and dispersal capacities, soil microbes are expected to respond rapidly to climate change. However, the legacy effects described here as community properties that persist after the environmental alteration were witnessed in soil microbial communities which take up to 3 years to respond to drought and other ecological changes (Cuddington, 2011; Bradford et al., 2008; Evans and Wallenstein, 2012; Evans et al., 2014; Rousk et al., 2013; Giauque and Hawkes, 2016). The distribution of soil bacterial species is strongly influenced by soil properties like soil pH and nutrient availability, and these properties of soil change gradually over time (Wardle, 2006; Dequiedt et al., 2009; Lauber et al., 2009). The factors that drive shifts in soil bacterial communities may reflect historical climate (An et al., 2019; Svenning and Sandel, 2013; Chadwick and Chorover, 2001; Rounsevell et al., 1999; Kelly et al., 1998). Soil bacterial populations may also adapt to current climate change, and it may take years or decades to understand the full impact of the contemporary climate change. During field-based studies on climate change, mostly the aboveground responses like plant productivity, biomass, and composition were given great importance. At the same time, belowground answers typically related to microbial communities have received far less attention (Wilcox et al., 2015, 2017).

Given the close relationship between microbial communities and the functioning of soil, any disturbance in the composition of microbial communities as a result of climate change may interfere with soil functioning and thus the availability of ecosystem services (Bellard et al., 2012; McLaughlin, 2011). Therefore, it is typically essential to enhance our understanding of the role of altered precipitation regimes in controlling soil microbial communities to accurately forecast changes in the terrestrial ecosystem aspects related to future climate change scenarios (Maestre et al., 2015; Trivedi et al., 2019). Terrestrial ecosystems play a significant role in climate feedbacks because of the release and absorption of greenhouse gases like carbon dioxide, methane, and nitrous oxides while storing large amounts of carbon in living plants and soils, therefore acting as a significant global carbon sink (Jain et al., 2020).

Fundamental roles of terrestrial bacteria are in soil maintenance and its fertility. Soil bacterial communities and their associated processes usually are sensitive to climate (Jain et al., 2020; Singh et al., 2010). Different drivers of diversity are associated with the terrestrial and aquatic microorganisms that inhabit the same geographical area, for example, for birds (Acharya et al., 2011b), trees (Acharya et al., 2011a) or fishes (Bhatt et al., 2012). Several studies have unraveled that climatic factors describe the global diversity in latitudinal patterns (Tedersoo et al., 2014; Zhou et al., 2016) rather than regional diversity in altitudinal patterns (Bahram et al., 2012). The activity of soil microbes of feeding GHG's into the air result in global warming. It is influenced either directly or indirectly by climate change. Direct effects of climate change include the impact on soil microbes, temperature generation due to greenhouse gases, precipitation change, and extreme climatic events. In contrast, the indirect factors include changes in plant productivity due to climate-driven changes. Thus, by using direct and indirect effects of climate change, we can illustrate the importance of soil microbes and microbial metabolism in feedbacks of the carbon cycle and ill consequences of climate change (Jansson and Hofmockel, 2019; Pugnaire et al., 2019).

The occurrence of extreme weathering events like droughts and freezing has higher effects on microbes and their associated activities rather than overall changes in precipitation and temperature, which in turn are the essential consequences for nutrient flows and levels of carbon in diverse ecosystems (Schimel et al., 2007). These effects of stressors on the carbon cycle and microbial communities vary considerably across ecosystems. In alpine and arctic regions, the temperature changes have a substantial impact on the process of decomposition and the communities of microbes.

It has been predicted that by the year 2100, 25% of Earth's permafrost could thaw because of warming in climate. Due to this, a large amount of covered organic matter for microbial decomposition is freed (Anthony et al., 2018), thus providing positive feedback on climate change (Davidson and Janssens, 2006). It is shown that various belowground fungal communities exhibit higher degree drought tolerance at the driest end of the gradient. It in turn raises new questions about the possible impacts of disruption or weakening of plant–microbial interactions under climate change scenarios due to decoupling of both groups in response (Lambers and Oliveira, 2019). Understanding how the structure and functioning of microbial communities respond to droughts is essential for predicting climate change impacts on the global carbon cycle. The assembly of microbial communities is also highly vulnerable to short-term climate change (i.e., 2–3 years of experimentally induced drought), which can affect the provisions of essential microbial-mediated ecosystem services, such as decomposition and nutrient cycling (Pugnaire et al., 2019). The local environmental contexts are also modulated by global climatic change, which in turn will have significant impacts on soil microbial communities, including an increase or decrease in the relative abundance of some pathogenic taxa along a gradient of increasing precipitation (Ochoa-Hueso et al. 2018).

Viruses infect almost all forms of cellular life, including the bacteria, archaea, and microeukaryotes which form the basis of the ocean food web. Studies of marine viral infection (Weinbauer et al., 2003; Munn, 2019; Wilhelm and Matteson, 2008) and sediments have shown that cellular components including carbon and nutrients are released back into the microbial system by viral lysis (Middelboe and Glud, 2006; Siem-Jørgensen et al., 2008; Umani et al., 2010). Therefore, lysis decreases the potential for larger organisms to consume microbes and subsequent trophic transfer up the food webs (Burkholder et al., 2018; Weinbauer, 2004; Suttle, 2007). This book chapter attempts to highlight the importance of microbial diversity in balancing the ecosystems. Besides, the effect of global climatic change on microbial communities in varying habitats is also deeply discussed.

3.2   RHIZOBACTERIA AND THEIR IMPORTANCE

Rhizospheric microbes are an essential source of soil fertility and nutrient cycling. Their functions include being a good indicator of soil quality, plant pathogen suppression, residue decomposition, metal detoxification in plants, and pollutant degradation in soil (Bonanomi et al., 2018). They performthese functions by the oxidation of carbon-rich sources forming new biomass rich in carbon and nitrogen. They allow storage and cycling of nutrients like nitrogen fixation in leguminous plants (Niemeyer et al., 2012; Ju et al., 2020a). Rhizobacteria not only benefits plants but also improves soil quality as *Pseudomonas* spp. GHD-4 reduce the lead (Pb) concentration in the soil. They improve soil enzyme activities and grow more bacterial community diversity (Yu et al., 2019), for example, addition of *Bacillus subtilis* has been found to increase the activities of soil enzymes like dehydrogenase, $\beta-$ β-glucosidase, alkaline phosphatase and urease, which in turn increases soil quality and help in plant growth (Hidri et al., 2016). Rhizobacteria, thus not only help in phytoremediation but also in the fixation of nitrogen, solubilization of phosphorus, synthesis of phytohormone and release of siderophores. The Rhizobacteria and soil quality are interdependent on each other (Ashby et al., 2019; Hao et al., 2014).

Nitrogen fixation in leguminous plants is controlled by the interchange of C and N between the host plant and nitrogen-fixing bacteria (Saha et al., 2017; Khan et al., 2019). They also promote the biosynthesis of phytohormones (indole acetic acid, siderophore, and 1-amino cyclopropane carboxylate deaminase) in plants as reported in *Paenibacillus mucilaginous* and *Sinorhizobium meliloti* (Singh et al., 2019). Rhizobia and plant growth promoting rhizobacteria (PGPR) inoculation in plants can stimulate efficiency in retrieving soil nutrients, helping to cope with abiotic stresses which prevent plant diseases and reduce phytotoxicity by

Cu accumulation (Caddell et al., 2019; Ju et al., 2020b; Shukla et al., 2019; Ite and Ibok 2019; Sanchez-Hernandez, 2019),.

Rhizospheric microbial communities including Proteobacteria, Firmicutes, Actinobacteria, and Acidobacteria are the dominant species which act as primary agents of nutrient cycling and as indicators in heavy metal remediation (Ju et al., 2020; Sullivan and Gadd, 2019). The Rhizobia and PGPR increase the relative abundance of Firmicutes and Acidobacteria known for their tolerance to low substrate availability and extreme conditions (Kalam et al., 2017). Acidobacteria is known to work under the nutrient-deficient environments such as copper tail mining where it degrades the soft organic matter that is present. Thus, there is a mechanism involved in the studies of co-inoculation of Rhizobia and PGPR which promotes the growth of both plants by providing nutrients, and helps in soil phytoremediation (Hidri et al., 2016).

3.3 FUNGI AND THEIR IMPORTANCE

Fungi act as essential agents of deadwood decomposition using different wood biopolymers (hemicellulose, lignin, and cellulose) (Štercová, 2017). This makes fungi colonize wood and use nutrients present in the wood for their growth (Pastorelli et al., 2020). Bacteria also play an essential but lesser role in the degradation of deadwood than fungi (Purahong et al., 2016a) and it has been reported that bacteria decompose side products from incomplete degradation of deadwood by fungi (Bani et al., 2018). However, the two microbes work better in synergistic effect, as many studies involving the interaction of both bacteria and fungi have shown that in the interaction, each one makes the environment suitable for the other, thus help their coexistence (Novotna and Suárez, 2018). Decomposition by fungi lower the pH and generate reactive oxygen radicals creating a selective and harsh environment for bacterial colonization (Johnston et al., 2016; Kielak et al., 2016). Some bacteria may harm the fungal organisms by competing for energy and nutrients but may in turn help fungi with certain limiting nutrients like nitrogen, growth factors like vitamins and detoxification of certain compounds harmful to fungal growth (Johnston et al., 2016).

The enzymes like oxidoreductases and hydrolases produced by fungi, mineralize, or decompose plant cell wall polymers into simple compounds (Singh et al., 2020; Mäkelä et al., 2017; Chandra, 2019). Fungi are involved in the process of succession that makes the environment suitable for another community like Ascomycota – the first colonizers of dead wood for their better cellulose decomposition than lignin-like compounds (Pastorelli et al., 2020). The lignin present in deadwood is decomposed into simpler compounds by Basidiomycota as they appear later in the process (Walker and White, 2017; Asina et al., 2016). This way fungi breakdown complex molecules into simple substrates upon which bacterial growth establishes (Pastorelli et al., 2020; Barnhart-Dailey et al., 2019; Baker et al., 2019; Liu et al. 2017; Paliwal et al. 2019). The main factor influencing fungal inhabitation and quantity is the pH of the substrate involved, as it affects mycelial growth and enzyme production across different habitats including, soil, leaf, litter, and deadwood (Bani et al., 2018; Purahong et al., 2016b; Tláskal et al., 2016). Macro-fungal fruit bodies release ammonia from decomposition process which increases pH suitable for bacteria through ammonification, N-fixation, and denitrification (Purahong et al., 2016b; Stein and Klotz, 2016).

There occurs a difference in abiotic factors of soil environment between the hypersphere and the bulk soil, which leads to different soil bacterial and fungal associations (Hao et al., 2020). Ectomycorrhizal roots influence soil pH, soil carbon, nitrogen, and phosphorus content which have a substantial effect on bacterial communities. They also serve as hotspots for fungal–bacterial interactions and nutrient cycling. The carbon flow in hypersphere from fungal hyphae to bacteria occurs through consumption of hyphal exudates, grazing on living hyphae, or degradation of senescent hyphae (Hao et al., 2020). The studies on mutualistic interactions between fungi and bacteria reported that fungi select bacteria on the basis of their efficiency to utilize decomposed products to the maximum possible extant and also in many cases, bacteria provide carbon and nitrogen to the fungi (Dighton, 2006). Evidences also support that N-fixing and P-solubilizing bacteria in association with fungal hyphae enhance nutrient cycling in hyphosphere (Hao et al., 2020). *Penicillium* fungi are a ubiquitous group of soil microbes and are considered to be a key component in phosphorus cycling (Sharma et al., 2013), which mainly occurs through the release of organic anions with considerable differences between species (Zheng et al., 2017; Zhu et al., 2018; Tarafdar, 2019). Many strains have been developed as biofertilizers especially as plant phosphorus nutrition (Storer et al., 2018). Phosphonates have also been reported as the best source of phosphorus following bacterial C-P lyase activity in marine ecosystems (Karl and Björkman, 2015; Tapia-Torres et al., 2016).

3.4   STRATEGIES FOR MICROBIAL CONSERVATION

Microorganisms are an essential part of biodiversity that are employed in improving the productivity of crops and nurturing life. Microbes play an indispensable role in maintaining biogeochemical cycles and are involved in various ecosystem services. It is assumed that the existence of almost all life forms in the biosphere depends upon them. They are also employed as probiotics, biofertilizers, biopesticides, and are thus necessary for plant growth and development (Zaidi et al., 2009; Prakash et al., 2016). Despite their numerous roles in the biosphere, various anthropogenic activities (e.g., tilling, use of fertilizers, and agrochemicals) as well as climatic changes result in some alteration of microbial communities and pose a serious threat to their existence (Sharma et al., 2016). It has been seen that nitrogenous fertilizers result in a decrease of soil bacterial diversity and microbial carbon biomass (Dai et al., 2018). Keeping in view the importance and threats of microbes in the biosphere, there is an urgent need for their widespread conservation. In this regards, bioresource centers play an essential role by maintaining the microbial gene pools and making them available for supporting the R&D programs in microbiomes (Sharma et al., 2018; Ivshina and Kuyukina, 2018). Microbial resource centers play a vital role in the collection and long-term storage of microbes under controlled conditions by employing different methods (broadly classified into two categories viz., in situ conservation and ex situ conservation) for their conservation (Table 3.1.). Another technique for the preservation of microbial diversity is the in-factory conservation, which is mainly utilized for agro-industrial sectors (Sharma et al., 2016).

In situ conservation, also known as on-site conservation is the preservation of organisms in their natural habitats. These methods have the capability of long-term preservation of populations, species, and ecosystems. Conventionally, protected areas have been seen as the base or foundation for in situ conservation. A small fraction of bacterial diversity can be grown in the laboratory (Stewart, 2012), and there are plenty of microorganisms known to be non-cultivable. Hence, they can be preserved through the in situ mode of conservation.

Conservation of different life forms interacting with each other in a given environment can lead to preservation of microbes (Ivshina and Kuyukina, 2018). Preventing deforestation and assuring afforestation can prevent the soil containing diverse microflora from being washed away by the torrential rains. Moreover, avoiding pollution of water bodies can conserve diverse forms of phytoplankton, zooplankton, and other floating microbes (Mishra, 2015). Ecosystem and habitat preservation which are in the initial stages, need a sober attention to begin the conservation of microbes at the ecosystem or habitat level to ensure their conservation and availability for future applications.

TABLE 3.1 Conservation Methods for Microbial Communities.

Conservation methods	References
Subculturing; low temperature (−45 to −70°C)	Winters and Winn (2010)
Preservation using agar-bits; mineral oil overlay of slants-grown cultures	Nakasone et al. (2004)
Use of different sterile material like silica gel	Streeter (2003)
Cryopreservation	Fuller (2004), Cody et al. (2008), Smith et al. (2008), and Chian and Quinn (2010)
Lyophilization	Morgan et al. (2006) and Berner and Viernstein (2006)

Ex situ conservation also referred to as off-site conservation is considered as an efficient method for conserving microbial diversity. This method includes gene banks, culture collections, and microbial resource centers (Dulloo et al., 2017). There is a huge number of microbial centers around the globe where microorganisms are collected, identified, characterized and then stored/conserved as per the guidelines of the best practice of the Organization for Economic Development and Cooperation (OECD). In ex situ mode of conservation, the activities of microorganism are reduced/ stopped by imposing conditions including subculturing, preservation on agar beads, use of mineral oils, silica gel storage, spray-drying, cryopreservation, lyophilization, desiccation, and verification etc. (Flickinger, 2010). Among the above ex situ strategies, the lyophilization and cryopreservation are of paramount importance and are widely utilized to achieve long-term and stable storage of microorganism (Smith and Ryan, 2012; Singh and Baghela, 2017).

In-factory conservation is another approach for the conservation of microbes by keeping them in common conditions for practical purposes. Generally, two different conservation methods known as Dynamic conservation and Static conservation are applied in this mode of conservation. Dynamic protection does not impose many restrictions on the use of strains in comparison, while the stable conservation strategies are much restrictive and try to preserve strains under normal conditions to avoid any kind of change (Sharma et al., 2018).

3.5 CLIMATIC CHANGE AND ITS IMPACT ON MICROBIAL COMMUNITIES

Climate change is a complex worldwide issue that involves scientific, social, economic, political, moral, and ethical aspects (Mangodo et al., 2020). The alarming concentration of primary ozone-depleting substances, namely, methane, carbon dioxide, nitrous oxide, and chlorofluorocarbons in the environment mainly lead to the global climate changes (Mangodo et al., 2020). With the abrupt onset of climate changes, adaptability, which is very hard to accomplish under such conditions is needed (Mangodo et al., 2020; Ruess et al., 1999). The reaction of microbial diversity is clearly shown toward biotic and abiotic factors (Kardol et al., 2010), and subsequently, environmental change impacts are exact in the event of these microbes. Soil microorganisms are profoundly unique and respond quickly to such ecological conditions (Joergensen, 2010). However, the transient and spatial scales figure out which environmental factor is generally significant (Savage et al., 2009). The impacts of warming on microbial procedures are seen to be most noticeable at higher latitudes (Davidson et al., 2006).

3.5.1 IMPACTS ON TERRESTRIAL MICROBES

Climate change affects the physiology of decomposers and accordingly affects soil CO_2 efflux (Schindlbacher et al., 2011). Increase in temperature elevates fungal decomposition, and brings about lingering carbon dioxide outflow from the soil. In any case, higher temperatures additionally raise soil nitrogen levels that suppress the rates of fungal decay. Higher concentration of nitrogen availability negatively influences microbial diversity (Dutta and Dutta 2016). On the other hand, the stress exerted by the warmer climate has a direct negative impact on the biochemical reactions or processes taking place in the microorganisms (Zimmer, 2010). Among the various components of environmental change and their results that adjust the general abundance of microbes, precipitation patterns have the most severe impacts on their community composition (Castro et al., 2010). Contingent on the variables, which limit ecosystems productivity, adjustments in soil moisture levels and precipitation, could either upsurge or decrease the proportion of microbes and trigger moves in their community organization (Schimel et al., 1999; Williams, 2007; Chen et al., 2007). The impacts of snowfall on microbial networks and their metabolic exercises can significantly affect winter soil respiration (Aanderud et al., 2013). Snow-interceded changes in microbial ecosystem structure are significantly influenced through wintertime respiration dynamics (Aanderud et al., 2013).

Raised atmospheric levels of carbon dioxide cause soil organisms to transmit progressively intense harmful substances, for example, methane and nitrous oxide (Pathak and Pathak, 2012). The subsequent rise in CH_4 emissions, decreases the absorption of methane by soil microorganisms (up to 30%) (Pathak and Pathak, 2012; Phillips et al., 2001; Ineson et al., 1998). Besides, more elevated levels of carbon dioxide additionally lead to significant changes in the microbial diversity. The decomposing leaves in streams could have boarder consequences on the food chain as many microorganisms are a good spring of supplement for the small phytophagous organisms (Dutta and Dutta, 2016).

Due to extreme precipitation, the availability of water is affected, which increases the risk of droughts and floods and changes the snow-melt timings (Dutta and Dutta, 2016). Rainfall is an essential factor in determining the variability of moisture and respiratory activity in soils as well as sources or sinks of CO_2 (Aanderud et al., 2011; Shim et al., 2009). The shifts in precipitation regimes are quite significant (affected by a 20% increase or decrease in precipitation) because the level of moisture can determine the structure of the terrestrial microbial community. The change in precipitation patterns has an intense impact on the microbial community composition than any other factor of climate change, ultimately resulting in the alteration of overall abundance of microorganisms, such as fungi and bacteria (Castro et al., 2010). Oxygen availability increases in peatlands and wetlands, consequently elevate CO_2 flux due to drying up of soils (Singh et al., 2010; Fierer and Schimel, 2003). Several other environmental factors which could upturn or decline the ratio of fungi and bacteria and elicit alterations in their activity and composition include soil moisture levels (Schimel et al., 1999; Williams, 2007; Chen et al., 2007). The dominance of soil fungi can change due to minute changes in soil moisture (30% reduction in water-holding capacity) while bacterial communities remain intact (Classen et al., 2015). The composition and activity of soil microbial communities can also be affected by winter conditions. Winter soil respiration can be seriously affected by snow-mediated changes, which in turn will have effects on the microbial communities and their metabolic activities (Aanderud et al., 2013). Therefore, the flux of carbon dioxide (CO_2) from soils depends upon the variability of these conditions.

The critical consequences of climate change dynamics and varied snowfall has been anticipated for several terrestrial ecosystems also (Change, 2007; Henry, 2008). This is because thick snowpacks insulate soils from cold air temperatures and controls winter soil respiration, which in turn increases heterotrophic respiration (Mariko et al., 1994; Brooks et al., 1997; Rey et al., 2002). In coniferous forests, the microbial population under snow is susceptible to rising temperatures. Microbial activity is relatively higher in late winters because of shallow temperatures due to snowpacks which are essential for the development of snow molds. In these areas, snow molds contribute about 10–30% of total annual carbon dioxide production. Increasing temperatures characterized by subfreezing temperatures shorten the late winter period. As a result, snow molds emit a lesser amount of carbon dioxide (Dutta and Dutta, 2016).

Soils are shielded from the climatic variations compared with the aboveground vegetation, but indirect impacts of climatic change can occur to the associated soil communities through the medium of plants (Kardol et al., 2010; Durán et al., 2014). However, the arrangement of subsoil communities depends upon the effects of ecological situations acting on the related vegetation (Fierer and Jackson, 2006). The indirect consequences of climatic change arising due to plant community shifts sometimes may have counteractive effects and differ from direct impact on microbial community (Kardol et al., 2010). Plant–microbial relationships in soils are significant in determining the dynamics of soil respiration, but alterations in rainfall can have severe effects on such linkages (Aanderud et al., 2011; Yepez et al., 2007). Under high rainfall variations, plants decrease the sensitivity of respiration, but under contradictory conditions in mesic habitats, the function is reverse (Aanderud et al., 2011). Microbes consume carbon-rich liquids which are exuded by plants but under stressful situations, such as fluctuating temperatures, the nature of such secretions changes, leading to an alteration in the microbial secretions (Engelkes et al., 2008). Under the influence of climate change, indirect impacts can also arise from the modification of pathogenic soil activities of plants (Engelkes et al., 2008; Morriën et al., 2011).

Production of secondary metabolites such as polyphenols for defense also affects the litter input quality, which in turn alters the activities of microbial decomposer communities in such plants (Engelkes et al., 2008). Climate change can also change the phenology of roots and shoots, and consequently, the interaction in rhizospheric microbes is also markedly altered (Classen et al., 2005). Due to modifications in microbial diversity caused by climatic changes, plant functional characters are also changed (Lau and Lennon, 2011). In fact, under the influence of rising temperature, plants that remain in close association with soil communities migrate to higher altitudes (Classen et al., 2005). The net plant productivity increases, providing more substrates to heterotrophs due to increased soil temperatures (Trumbore, 1997). The discharge of organic acids, amino acids, and labile sugars from plant roots also increases quantitatively as well as qualitatively due to higher levels of CO_2 which in turn alters the dominance, diversity, growth, and activities of microbes that depend on nutrient availability (e.g., nitrogen) and change in CO_2 flows (Diaz et al., 1993; De Graaff et al., 2006; Bardgett et al., 2009).

Plant feedback to CO_2 can also affect the soil microbial respiration physiology (Singh et al., 2010). Nutrient availability in soil due to more significant mineralization of soil organic matter by microorganisms is also likely to increase by global warming (Ruess et al., 1999). Likewise, soil warming initiates nitrogen mineralization, which builds plant profitability. Besides, plant network synthesis regularly changes with warming (Walker et al., 2006; Harte et al., 2006; Hoeppner and Dukes, 2012). The increased temperature, along with supplement accessibility, can prompt changes in vegetation (Hicks et al., 2019; Hobbie, 1999), which influences the microbes.

3.5.2   IMPACTS ON AQUATIC MICROBES

Ocean warming has been bringing about significant changes in the diversity of aquatic microflora. The negative impacts of the oceanic variety are primarily influenced by carbon dioxide accumulation caused due to the ocean acidification. Expansion of oxygen-depleting substances also causes stratification, which have impacts on the microbial food networks and in the long run, biogeochemical cycles (Walsh, 2015). As the polar sea heats up, conditions in this biological ecosystem become more appropriate for microbes. In fact, because of the warming of the polar oceans, more organic matter is consumed, and subsequently, microbes become more dynamic (Zimmer, 2010). The temperature of ocean surface increases and consequently its thickness decreases, thus, leading to floating over the nutrient-rich cooler and deeper waters. Therefore, the supply of supplements is deficient from beneath, and the phytoplankton present in the upper layer suffers from starvation. Accordingly, the pumping of carbon is reduced, and primary production is grossly suppressed (Walsh, 2015).

A significant determinant of the impact of rise in temperature is cell size. In the Arctic regions, smaller phytoplankton species keep on flourishing. However, the bigger size phytoplanktons experience elimination under the regular rates of environmental change. This is because, in smaller cells, the surface-to-area proportion is higher, which makes them increasingly proficient in the attainment of supplements. Therefore, larger cells are sunk more rapidly, and the cell size of phytoplankton organisms would diminish due to abnormal increase in global warming. Consequently, due to the presence of smaller and more positive cells, the carbon pumped into the interior would decrease (Walsh, 2015).

Climatic change-induced global warming alters pH as well as wind patterns of seawater, which in turn can either stimulate or inhibit microbial growth (Zimmer, 2010). The oceans consume more gas and become more acidic due to the growing atmospheric carbon content, which has a significant effect on marine and coastal ecosystems (Dutta and Dutta, 2016). The higher carbon dioxide levels in the sea could also stimulate primary producers, triggering phytoplankton blooms. In the marine environment, depletion of oxygen and genesis of toxic sulfide through the anaerobic processes are caused due to decomposition of the resulting organic matter (Glöckner et al., 2012). The amount of available carbon dioxide (Hutchins et al., 2019) can potentially influence the biodiversity of critical ocean microbes. For example, in a nutrient-deficient oceanic environment, a bacterium *Trichodesmium* is responsible for the transformation of nitrogen gas into usable forms, and several other organisms utilize synthesized material. Although increased carbon dioxide concentrations improve the reproduction in microbes, it also leads to the depletion of nutrients and minerals, such as iron, and phosphorus. The resulting deficiency of resources in the oceans result in the extinction of bacteria and other organisms of the food chain, including fish (Authman et al., 2015). Differences in particular ocean microbes due to climate change influence the quantities of less water-soluble molecules, such as lipids in seawater, and thus the overall climate (Dutta and Dutta, 2016; Strååt et al., 2018; Häder and Sinha, 2018). Climate change also leads to a range of extensive damage to microbial diversity, which in turn affects the broad floral and faunal diversity.

Moreover, the surface temperature rise of 1°C–7°C in large freshwater bodies by doubling of CO_2 concentrations in the atmosphere (Magnuson et al., 1997) is greater in lakes than in oceans. This rise in temperature can have stark effects on the microbial world (Dutta and Dutta, 2016). As an outcome, algal blooms, which cause depletion of dissolved oxygen levels in water, are anticipated to increase by 20% in lakes. Algal bloom concentration is predicted to increase by 5% because of global warming, which can be toxic to animal and fish life (Dutta and Dutta, 2016). The proliferation rate of cyanobacterial blooms increases directly through high temperatures that leads to the growth of harmful cyanobacterium *Microcystis* during hot summers in eutrophic lakes (Joehnk et al., 2008). Further, during global warming, the CO_2 concentration changes and consequently alters the ratio of phosphorus, nitrogen, and carbon in the freshwater ecosystems (Hasler et al., 2016). Phosphorous decreases the nutrient content of phytoplankton in nutrient-poor water bodies at a point when the proportion of carbon exceeds that of nitrogen. These effects have a detrimental impact on the production of phytoplankton, which in turn affects the zooplankton and the subsequent productivity and community composition. These effects can spread to higher food chain levels. Higher CO_2 levels in water bodies rich in nutrients, speed up primary production and increase the abundance of nitrogen cyanobacteria (Hasler et al., 2016).

Global warming reduces the extent and magnitude of sea ice in the Arctic. As an outcome, the richness of ice algae, which exist in the supplement creamy ice, is probably going to be contrarily influenced. These ice algae are used up by zooplankton, which in turn are consumed by Arctic cod – the primary food source for various marine mammals like seals. So, any alteration in the population of ice algae due to global warming ultimately poses a threat to the polar bear populations (Srinivasan, 2018). Temperature increase causes ice melting, which likely causes salinity changes and affect water currents and stratification along with dramatic shifts in the microbial community.

Moreover, trapped organic matter and contaminants are releasing due to the melting of ice, which could affect microorganisms (Glöckner et al., 2012). Climate change can also affect the occurrence of storms and due to such prevalence in the Arctic flush, nutrients from soils to streams and oligotrophic lakes can elevate nutrient supply to aquatic bacteria. Greater availability of nutrients takes place under the influence of climate change. When environmental variability co-occurs with their growth rate, the ability of bacterial communities is hindered from swinging to an ideal activity for a given resource supply (Stieglitz et al., 2003; Vadstein, 2000). The growth and composition of bacterial communities can directly be affected by such changes (Judd et al., 2006). For example, development and design of bacterial communities of some aquatic ecosystems like some northern European lakes and coastal areas are also expected to increase due to higher temperature and increased precipitation (Jones, 1992; Eriksson Hägg et al., 2010). Higher nutrient concentrations and warmer temperatures can also accelerate bacterial productivity that reproduces at higher rates under such conditions (Ram and Sime-Ngando, 2008; White et al., 1991). This is because under such conditions of nutrient availability, bacteria delays cell division and increase in size due to enough resources (Shiomi and Margolin, 2007).

3.6 CONCLUSION AND FUTURE DIRECTIONS

Although the role of microorganisms in controlling the concentration of ozone harming substances is admirable. Yet, it cannot seem to be appropriately comprehended, due to faint attention paid by the researchers globally to this issue. The in-depth understanding regarding the adverse effects mediated by climate changes on most vulnerable microbial ecosystems will help researchers to shift focus on the importance of microflora. Whenever harnessed appropriately, microbes are the significant common assets for controlling environmental changes. However, if due attention is not paid, it may act as the most emerging accelerator to the issue. It is high time to research this aspect, comprehend the acting mechanisms more precisely, and thus suitably utilize it for developing solutions. The current chapter may pave the way to chalk out strategies for conservation of microbial diversity. Besides, a recent update may be useful to initiate procedures to investigate the importance of microbial communities globally. The microbial world needs to be deeply investigated to understand their importance in balancing almost all ecosystems and highlight their positive interference in all forms of life.

ACKNOWLEDGMENTS

This chapter was designed and initiated by RAM. It has been written by BAB, LT and edited and compiled by Dr RAM. The authors are very much thankful to Dr RAM for his assistance in the preparation of this chapter.

KEYWORDS

- **microbial diversity**
- **microbial ecosystem**
- **climate change**
- **nutrient cycling**

REFERENCES

Aanderud, Z. T.; Jones, S. E.; Schoolmaster Jr, D. R.; Fierer, N.; Lennon, J. T. Sensitivity of Soil Respiration and Microbial Communities to Altered Snowfall. *Soil Biol. Biochem.* **2013,** *57,* 217–227.

Aanderud, Z. T.; Schoolmaster, D. R.; Lennon, J. T. Plants Mediate the Sensitivity of Soil Respiration to Rainfall Variability. *Ecosystems* **2011,** *14* (1), 156–167.

Acharya, B. K.; Chettri, B.; Vijayan, L. Distribution Pattern of Trees along an Elevation Gradient of Eastern Himalaya, India. *Acta Oecologica* **2011a,** *37* (4), 329–336.

Acharya, B. K.; Sanders, N. J.; Vijayan, L.; Chettri, B. Elevational Gradients in Bird Diversity in the Eastern Himalaya: An Evaluation of Distribution Patterns and Their Underlying Mechanisms. *PLoS One* **2011b,** *6* (12).

An, J.; Liu, C.; Wang, Q.; Yao, M.; Rui, J.; Zhang, S.; Li, X. Soil Bacterial Community Structure in Chinese Wetlands. *Geoderma* **2019,** *337,* 290–299.

Andréasson, J.; Bergström, S.; Carlsson, B.; Graham, L. P.; Lindström, G. Hydrological Change–Climate Change Impact Simulations for Sweden. *AMBIO: J Human Environ.* **2004,** *33* (4), 228–234.

Andrino de la Fuente, A. *Carbon and Phosphorus Trading in the Arbuscular Mycorrhizal Symbiosis*; Hannover: Institutionelles Repositorium der Leibniz Universität Hannover, 2019.

Anthony, K. W.; von Deimling, T. S.; Nitze, I.; Frolking, S.; Emond, A.; Daanen, R.; Anthony, P.; Lindgren, P.; Jones, B.; Grosse, G. 21st-Century Modeled Permafrost Carbon Emissions Accelerated by Abrupt Thaw Beneath Lakes. *Nat. Commun.* **2018,** *9* (1), 1–11.

Ashby, M. N.; Belnap, C. P.; Kuchta, M. C.; Kunin, V.; Kostecki, C.; Lidstrom, U.; Shestakova, N.; Wood, L. Plant Growth-Promoting Microbes, Compositions, and Uses. Google Patents, 2019.

Asina, F.; Brzonova, I.; Voeller, K.; Kozliak, E.; Kubátová, A.; Yao, B.; Ji, Y. Biodegradation of Lignin by Fungi, Bacteria and Laccases. *Bioresour. Technol.* **2016,** *220,* 414–424.

Authman, M. M.; Zaki, M. S.; Khallaf, E. A.; Abbas, H. H. Use of Fish as Bio-Indicator of the Effects of Heavy Metals Pollution. *J. Aquacult. Res. Dev.* **2015,** *6* (4), 1–13.

Azam, F.; Malfatti, F. Microbial Structuring of Marine Ecosystems. *Nat. Rev. Microbiol.* **2007,** *5* (10), 782–791.

Bahram, M.; Põlme, S.; Kõljalg, U.; Zarre, S.; Tedersoo, L. Regional and Local Patterns of Ectomycorrhizal Fungal Diversity and Community Structure along an Altitudinal Gradient in the Hyrcanian Forests of Northern Iran. *New Phytol.* **2012,** *193* (2), 465–473.

Baker, P.; Tiroumalechetty, A.; Mohan, R. Fungal Enzymes for Bioremediation of Xenobiotic Compounds. In *Recent Advancement in White Biotechnology through Fungi*; Springer, 2019; pp 463–489.

Balat, M.; Balat, M. Political, Economic and Environmental Impacts of Biomass-Based Hydrogen. *Int. J. Hydr. Energy* **2009,** *34* (9), 3589–3603.

Bani, A.; Pioli, S.; Ventura, M.; Panzacchi, P.; Borruso, L.; Tognetti, R.; Tonon, G.; Brusetti, L. The Role of Microbial Community in the Decomposition of Leaf Litter and Deadwood. *Appl. Soil Ecol.* **2018,** *126,* 75–84.

Bardgett, R. D.; De Deyn, G. B.; Ostle, N. J. Plant-Soil Interactions and the Carbon Cycle. *J. Ecol. (Oxf.)* **2009,** *97* (5), 838–912.

Barnhart-Dailey, M. C.; Ye, D.; Hayes, D. C.; Maes, D.; Simoes, C. T.; Appelhans, L.; Carroll-Portillo, A.; Kent, M. S.; Timlin, J. A. Internalization and Accumulation of Model Lignin Breakdown Products in Bacteria and Fungi. *Biotechnol. Biofuels* **2019,** *12* (1), 175.

Bellard, C.; Bertelsmeier, C.; Leadley, P.; Thuiller, W.; Courchamp, F. Impacts of Climate Change on the Future of Biodiversity. *Ecol. Lett.* **2012,** *15* (4), 365–377.

Berner, D.; Viernstein, H. Effect of Protective Agents on the Viability of Lactococcus Lactis Subjected to Freeze-Thawing and Freeze-Drying. *Sci. Pharm.* **2006,** *74* (3), 137–149.

Bhatt, J. P.; Manish, K.; Pandit, M. K. Elevational Gradients in Fish Diversity in the Himalaya: Water Discharge Is the Key Driver of Distribution Patterns. *PloS one* **2012,** *7* (9).

Bonanomi, G.; Lorito, M.; Vinale, F.; Woo, S. L. Organic Amendments, Beneficial Microbes, and Soil Microbiota: Toward a Unified Framework for Disease Suppression. *Annu. Rev. Phytopathol.* **2018,** *56,* 1–20.

Bonfante, P.; Lanfranco, L. The Mycobiota: Fungi Take Their Place between Plants and Bacteria. *Curr. Opin. Microbiol.* **2019,** *49,* 18–25.

Bradford, M. A.; Davies, C. A.; Frey, S. D.; Maddox, T. R.; Melillo, J. M.; Mohan, J. E.; Reynolds, J. F.; Treseder, K. K.; Wallenstein, M. D. Thermal Adaptation of Soil Microbial Respiration to Elevated Temperature. *Ecol. Lett.* **2008,** *11* (12), 1316–1327.

Brooks, P. D.; Schmidt, S. K.; Williams, M. W. Winter Production of CO_2 and N_2O from Alpine Tundra: Environmental Controls and Relationship to Inter-System C and N Fluxes. *Oecologia* **1997,** *110* (3), 403–413.

Burkholder, J. M.; Shumway, S. E.; Glibert, P. M. Food Web and Ecosystem Impacts of Harmful Algae. *Harmful Algal Blooms* **2018,** 243–336.

Caddell, D. F.; Deng, S.; Coleman-Derr, D. Role of the Plant Root Microbiome in Abiotic Stress Tolerance. In *Seed Endophytes*; Springer, 2019; pp 273–311.

Castro, H. F.; Classen, A. T.; Austin, E. E.; Norby, R. J.; Schadt, C. W. Soil Microbial Community Responses to Multiple Experimental Climate Change Drivers. *Appl. Environ. Microbiol.* **2010,** *76* (4), 999–1007.

Chadwick, O. A.; Chorover, J. The Chemistry of Pedogenic Thresholds. *Geoderma* **2001,** *100* (3–4), 321–353.

Chandra, P. Fungal Enzymes for Bioremediation of Contaminated Soil. In *Recent Advancement in White Biotechnology through Fungi*; Springer, 2019; pp 189–215.

Change, I. C. *The Physical Science Basis*; Cambridge University Press, 2007.

Chen, M.-M.; Zhu, Y.-G.; Su, Y.-H.; Chen, B.-D.; Fu, B.-J.; Marschner, P. Effects of Soil Moisture and Plant Interactions on the Soil Microbial Community Structure. *Eur. J. Soil Biol.* **2007,** *43* (1), 31–38.

Chian, R.-C.; Quinn, P. *Fertility Cryopreservation*; Cambridge University Press, 2010.

Classen, A. T.; Sundqvist, M. K.; Henning, J. A.; Newman, G. S.; Moore, J. A.; Cregger, M. A.; Moorhead, L. C.; Patterson, C. M. Direct and Indirect Effects of Climate Change on Soil Microbial and Soil Microbial-Plant Interactions: What Lies Ahead? *Ecosphere* **2015,** *6* (8), 1–21.

Cody, W. L.; Wilson, J. W.; Hendrixson, D. R.; McIver, K. S.; Hagman, K. E.; Ott, C.; Nickerson, C. A.; Schurr, M. J. Skim Milk Enhances the Preservation of Thawed− 80 C Bacterial Stocks. *J. Microbiol. Methods* **2008,** *75* (1), 135–138.

Cuddington, K. Legacy Effects: The Persistent Impact of Ecological Interactions. *Biol. Theor.* **2011,** *6* (3), 203–210.

Dai, Z.; Su, W.; Chen, H.; Barberán, A.; Zhao, H.; Yu, M.; Yu, L.; Brookes, P. C.; Schadt, C. W.; Chang, S. X. Long-Term Nitrogen Fertilization Decreases Bacterial Diversity and Favors the Growth of Actinobacteria and Proteobacteria in Agro-Ecosystems across the Globe. *Global Change Biol.* **2018,** *24* (8), 3452–3461.

Davidson, E. A.; Janssens, I. A. Temperature Sensitivity of Soil Carbon Decomposition and Feedbacks to Climate Change. *Nature* **2006,** *440* (7081), 165–173.

Davidson, E. A.; Janssens, I. A.; Luo, Y. On the Variability of Respiration in Terrestrial Ecosystems: Moving Beyond Q10. *Global Change Biol.* **2006,** *12* (2), 154–164.

De Graaff, M. A.; Van Groenigen, K. J.; Six, J.; Hungate, B.; van Kessel, C. Interactions between Plant Growth and Soil Nutrient Cycling under Elevated CO_2: A Meta-Analysis. *Global Change Biol.* **2006,** *12* (11), 2077–2091.

Deines, P.; Hammerschmidt, K.; Bosch, T. C. Microbial Species Coexistence Depends on the Host Environment. *bioRxiv* **2020,** 609271.

Dequiedt, S.; Thioulouse, J.; Jolivet, C.; Saby, N. P.; Lelievre, M.; Maron, P. A.; Martin, M. P.; Prévost-Bouré, N. C.; Toutain, B.; Arrouays, D. Biogeographical Patterns of Soil Bacterial Communities. *Environ. Microbiol. Rep.* **2009,** *1* (4), 251–255.

Diaz, S.; Grime, J.; Harris, J.; McPherson, E. Evidence of a Feedback Mechanism Limiting Plant Response to Elevated Carbon Dioxide. *Nature* **1993,** *364* (6438), 616–617.

Dighton, J. *Fungi in Ecosystem Processes*, Vol. 31; CRC Press, 2016.

Dragone, G.; Kerssemakers, A. A.; Driessen, J. L.; Yamakawa, C. K.; Brumano, L. P.; Mussatto, S. I. Innovation and Strategic Orientations for the Development of Advanced Biorefineries. *Bioresour. Technol.* **2020,** *302*, 122847.

Dullinger, S.; Gattringer, A.; Thuiller, W.; Moser, D.; Zimmermann, N. E.; Guisan, A.; Willner, W.; Plutzar, C.; Leitner, M.; Mang, T. Extinction Debt of High-Mountain Plants under Twenty-First-Century Climate Change. *Nat. Clim. Change* **2012,** *2* (8), 619–622.

Dulloo, M. E.; Rege, J.; Ramirez, M.; Drucker, A. G.; Padulosi, S.; Maxted, N.; Sthapit, B.; Gauchan, D.; Thormann, I.; Gaisberger, H. Conserving Agricultural Biodiversity for Use in Sustainable Food Systems **2017**.

Durán, J.; Morse, J. L.; Groffman, P. M.; Campbell, J. L.; Christenson, L. M.; Driscoll, C. T.; Fahey, T. J.; Fisk, M. C.; Mitchell, M. J.; Templer, P. H. Winter Climate Change Affects Growing-Season Soil Microbial Biomass and Activity in Northern Hardwood Forests. *Global Change Biol.* **2014**, *20* (11), 3568–3577.

Dutta, H.; Dutta, A. The Microbial Aspect of Climate Change. *Energy Ecol. Environ.* **2016**, *1* (4), 209–232.

Engelkes, T.; Morriën, E.; Verhoeven, K. J.; Bezemer, T. M.; Biere, A.; Harvey, J. A.; McIntyre, L. M.; Tamis, W. L.; van der Putten, W. H. Successful Range-Expanding Plants Experience Less Aboveground and Below-Ground Enemy Impact. *Nature* **2008**, *456* (7224), 946–948.

Eriksson Hägg, H.; Humborg, C.; Mörth, C.-M.; Medina, M. R.; Wulff, F. Scenario Analysis on Protein Consumption and Climate Change Effects on Riverine N Export to the Baltic Sea. *Environ. Sci. Technol.* **2010**, *44* (7), 2379–2385.

Evans, S. E.; Wallenstein, M. D. Soil Microbial Community Response to Drying and Rewetting Stress: Does Historical Precipitation Regime Matter? *Biogeochemistry* **2012**, *109* (1–3), 101–116.

Evans, S. E.; Wallenstein, M. D.; Burke, I. C. Is Bacterial Moisture Niche a Good Predictor of Shifts in Community Composition under Long-Term Drought? *Ecology* **2014**, *95* (1), 110–122.

Fierer, N.; Jackson, R. B. The Diversity and Biogeography of Soil Bacterial Communities. *Proc. Natl. Acad. Sci.* **2006**, *103* (3), 626–631.

Fierer, N.; Schimel, J. P. A Proposed Mechanism for the Pulse in Carbon Dioxide Production Commonly Observed Following the Rapid Rewetting of a Dry Soil. *Soil Sci. Soc. Am. J.* **2003**, *67* (3), 798–805.

Flickinger, M. C. *Encyclopedia of Industrial Biotechnology: Bioprocess, Bioseparation, and Cell Technology, 7 Volume Set*; John Wiley & Sons, ISBN, 2010.

Fuller, B. J. Cryoprotectants: The Essential Antifreezes to Protect Life in the Frozen State. *CryoLetters* **2004**, *25* (6), 375–388.

Giauque, H.; Hawkes, C. V. Historical and Current Climate Drive Spatial and Temporal Patterns in Fungal Endophyte Diversity. *Fungal Ecol.* **2016**, *20*, 108–114.

Glöckner, F.; Stal, L.; Sandaa, R.; Gasol, J.; O'Gara, F.; Hernandez, F.; Labrenz, M.; Stoica, E.; Varela, M.; Bordalo, A. In JB Calewaert & N. McDonough. *Marine Microbial Diversity and Its Role in Ecosystem Functioning and Environmental Change, Marine Board Position Paper* **2012**, *17*.

Gómez-Sagasti, M. T.; Marino, D. PGPRs and Nitrogen-Fixing Legumes: A Perfect Team for Efficient Cd Phytoremediation? *Front. Plant Sci.* **2015**, *6*, 81.

Gupta, A.; Gupta, R.; Singh, R. L. Microbes and Environment. In *Principles and Applications of Environmental Biotechnology for a Sustainable Future*; Springer, 2017; pp 43–84.

Häder, D.-P.; Sinha, R. P. Effects of Global Climate Change on Cyanobacteria. *Aqua. Ecosyst. Chang. Clim.* **2018**, 45.

Hao, X.; Taghavi, S.; Xie, P.; Orbach, M. J.; Alwathnani, H.; Rensing, C.; Wei, G. Phytoremediation of Heavy and Transition Metals Aided by Legume-Rhizobia Symbiosis. *Int. J. Phytoremed.* **2014**, *16* (2), 179–202.

Hao, X.; Zhu, Y.-G.; Nybroe, O.; Nicolaisen, M. H. The Composition and Phosphorus Cycling Potential of Bacterial Communities Associated with Hyphae of Penicillium in Soil Are Strongly Affected by Soil Origin. *Front. Microbiol.* **2020,** *10*, 2951.

Harte, J.; Saleska, S.; Shih, T. Shifts in Plant Dominance Control Carbon-Cycle Responses to Experimental Warming and Widespread Drought. *Environ. Res. Lett.* **2006,** *1* (1), 014001.

Hasler, C. T.; Butman, D.; Jeffrey, J. D.; Suski, C. D. Freshwater Biota and Rising pCO_2? *Ecol. Lett.* **2016,** *19* (1), 98–108.

Heimann, M.; Reichstein, M. Terrestrial Ecosystem Carbon Dynamics and Climate Feedbacks. *Nature* **2008,** *451* (7176), 289–292.

Heinze, J. *Nature Underfoot: Living with Beetles, Crabgrass, Fruit Flies, and Other Tiny Life Around Us*; Yale University Press, 2020.

Henry, H. A. Climate Change and Soil Freezing Dynamics: Historical Trends and Projected Changes. *Clim. Change* **2008,** *87* (3–4), 421–434.

Hicks, L. C.; Rousk, K.; Rinnan, R.; Rousk, J. Soil Microbial Responses to 28 Years of Nutrient Fertilization in a Subarctic Heath. *Ecosystems* **2019,** 1–13.

Hidri, R.; Barea, J.; Mahmoud, O. M.-B.; Abdelly, C.; Azcón, R. Impact of Microbial Inoculation on Biomass Accumulation by Sulla Carnosa Provenances, and in Regulating Nutrition, Physiological and Antioxidant Activities of This Species under Non-Saline and Saline Conditions. *J. Plant Physiol.* **2016,** *201*, 28–41.

Hobbie, S. E. Temperature and Plant Species Control Over Litter Decomposition in Alaskan Tundra. *Ecol. Monogr.* **1996,** *66* (4), 503–522.

Hoeppner, S. S.; Dukes, J. S. Interactive Responses of Old-Field Plant Growth and Composition to Warming and Precipitation. *Global Change Biol.* **2012,** *18* (5), 1754–1768.

Hutchins, D. A.; Jansson, J. K.; Remais, J. V.; Rich, V. I.; Singh, B. K.; Trivedi, P. Climate Change Microbiology—Problems and Perspectives. *Nat. Rev. Microbiol.* **2019,** *17* (6), 391–396.

Ineson, P.; Coward, P.; Hartwig, U. Soil Gas Fluxes of N_2O, CH_4 and CO_2 Beneath Lolium Perenne under Elevated CO_2: The Swiss Free Air Carbon Dioxide Enrichment Experiment. *Plant Soil* **1998,** *198* (1), 89–95.

Insam, H.; Gómez-Brandón, M.; Ascher-Jenull, J. Recycling of Organic Wastes to Soil and Its Effect on Soil Organic Carbon Status. In *The Future of Soil Carbon*; Elsevier, 2018; pp 195–214.

Ite, A. E.; Ibok, U. J. Role of Plants and Microbes in Bioremediation of Petroleum Hydrocarbons Contaminated Soils. *Int. J.* **2019,** *7* (1), 1–19.

Ivshina, I. B.; Kuyukina, M. S. Specialized Microbial Resource Centers: A Driving Force of the Growing Bioeconomy. In *Microbial Resource Conservation*; Springer, 2018; pp 111–139.

Jacoby, R.; Peukert, M.; Succurro, A.; Koprivova, A.; Kopriva, S. The Role of Soil Microorganisms in Plant Mineral Nutrition—Current Knowledge and Future Directions. *Front. Plant sci.* **2017,** *8*, 1617.

Jain, P. K.; Purkayastha, S. D.; De Mandal, S.; Passari, A. K.; Govindarajan, R. K. Effect of Climate Change on Microbial Diversity and Its Functional Attributes. In *Recent Advancements in Microbial Diversity*; Elsevier, 2020; pp 315–331.

Jansson, J. K.; Hofmockel, K. S. Soil Microbiomes and Climate Change. *Nat. Rev. Microbiol.* **2019,** 1–12.

Jobard, M.; Rasconi, S.; Sime-Ngando, T. Diversity and Functions of Microscopic Fungi: A Missing Component in Aquatic Food Webs. *Aqua. Sci.* **2010,** *72* (3), 255–268.

Joehnk, K. D.; Huisman, J.; Sharples, J.; Sommeijer, B.; Visser, P. M.; Stroom, J. M. Summer Heatwaves Promote Blooms of Harmful Cyanobacteria. *Global Change Biol.* **2008,** *14* (3), 495–512.

Joergensen, R. G. Organic Matter and Microorganisms in Tropical Soils. In *Soil Biology and Agriculture in the Tropics*; Springer, 2010; pp 17–44.

Johnston, S. R.; Boddy, L.; Weightman, A. J. Bacteria in Decomposing Wood and Their Interactions with Wood-Decay Fungi. *FEMS Microbiol. Ecol.* **2016,** *92* (11).

Jones, R. I. The Influence of Humic Substances on Lacustrine Planktonic Food Chains. *Hydrobiologia* **1992,** *229* (1), 73–91.

Joshi, S.; Sahgal, M.; Tewari, S. K.; Johri, B. N. Tree Ecosystem: Microbial Dynamics and Functionality. In *Microbial Interventions in Agriculture and Environment*; Springer, 2019; pp 411–450.

Ju, W.; Jin, X.; Liu, L.; Shen, G.; Zhao, W.; Duan, C.; Fang, L. Rhizobacteria Inoculation Benefits Nutrient Availability for Phytostabilization in Copper Contaminated Soil: Drivers from Bacterial Community Structures in Rhizosphere. *Appl. Soil Ecol.* **2020a,** *150*, 103450.

Ju, W.; Liu, L.; Jin, X.; Duan, C.; Cui, Y.; Wang, J.; Ma, D.; Zhao, W.; Wang, Y.; Fang, L. Co-Inoculation Effect of Plant-Growth-Promoting Rhizobacteria and Rhizobium on EDDS Assisted Phytoremediation of Cu Contaminated Soils. *Chemosphere* **2020b,** 126724.

Judd, K. E.; Crump, B. C.; Kling, G. W. Variation in Dissolved Organic Matter Controls Bacterial Production and Community Composition. *Ecol.* **2006,** *87* (8), 2068–2079.

Kalam, S.; Das, S. N.; Basu, A.; Podile, A. R. Population Densities of Indigenous Acidobacteria Change in the Presence of Plant Growth Promoting Rhizobacteria (PGPR) in Rhizosphere. *J. Basic Microbiol.* **2017,** *57* (5), 376–385.

Kardol, P.; Cregger, M. A.; Campany, C. E.; Classen, A. T. Soil Ecosystem Functioning under Climate Change: Plant Species and Community Effects. *Ecol.* **2010,** *91* (3), 767–781.

Karl, D. M.; Björkman, K. M. Dynamics of Dissolved Organic Phosphorus. In *Biogeochemistry of Marine Dissolved Organic Matter*; Elsevier, 2015; pp 233–334.

Kaviya, N.; Upadhayay, V. K.; Singh, J.; Khan, A.; Panwar, M.; Singh, A. V. Role of Microorganisms in Soil Genesis and Functions. In *Mycorrhizosphere and Pedogenesis*; Springer, 2019; pp 25–52.

Kelly, E. F.; Chadwick, O. A.; Hilinski, T. E. The Effect of Plants on Mineral Weathering. *Biogeochemistry* **1998,** *42* (1–2), 21–53.

Khan, M. N.; Ijaz, M.; Ali, Q.; Ul-Allah, S.; Sattar, A.; Ahmad, S. Biological Nitrogen Fixation in Nutrient Management. In *Agronomic Crops*; Springer, 2019; pp 127–147.

Kielak, A. M.; Scheublin, T. R.; Mendes, L. W.; Van Veen, J. A.; Kuramae, E. E. Bacterial Community Succession in Pine-Wood Decomposition. *Front. Microbiol.* **2016,** *7*, 231.

Klug, J. L. Positive and Negative Effects of Allochthonous Dissolved Organic Matter and Inorganic Nutrients on Phytoplankton Growth. *Can. J. Fisheries Aqua. Sci.* **2002,** *59* (1), 85–95.

Konopka, A. What Is Microbial Community Ecology? *ISME J.* **2009,** *3* (11), 1223–1230.

Lambers, H.; Oliveira, R. S. Plant Water Relations. In *Plant Physiological Ecology*; Springer, 2019; pp 187–263.

Lau, J. A.; Lennon, J. T. Evolutionary Ecology of Plant–Microbe Interactions: Soil Microbial Structure Alters Selection on Plant Traits. *New Phytol.* **2011,** *192* (1), 215–224.

Lauber, C. L.; Hamady, M.; Knight, R.; Fierer, N. Pyrosequencing-Based Assessment of Soil pH as a Predictor of Soil Bacterial Community Structure at the Continental Scale. *Appl. Environ. Microbiol.* **2009,** *75* (15), 5111–5120.

Liu, S.-H.; Zeng, G.-M.; Niu, Q.-Y.; Liu, Y.; Zhou, L.; Jiang, L.-H.; Tan, X.-f.; Xu, P.; Zhang, C.; Cheng, M. Bioremediation Mechanisms of Combined Pollution of PAHs and Heavy Metals by Bacteria and Fungi: A Mini Review. *Bioresour. Technol.* **2017,** *224,* 25–33.

Maestre, F. T.; Delgado-Baquerizo, M.; Jeffries, T. C.; Eldridge, D. J.; Ochoa, V.; Gozalo, B.; Quero, J. L.; Garcia-Gomez, M.; Gallardo, A.; Ulrich, W. Increasing Aridity Reduces Soil Microbial Diversity and Abundance in Global Drylands. *Proc. Natl. Acad. Sci.* **2015,** *112* (51), 15684–15689.

Magnuson, J.; Webster, K.; Assel, R.; Bowser, C.; Dillon, P.; Eaton, J.; Evans, H.; Fee, E.; Hall, R.; Mortsch, L. Potential Effects of Climate Changes on Aquatic Systems: Laurentian Great Lakes and Precambrian Shield Region. *Hydrol. Proc.* **1997,** *11* (8), 825–871.

Mäkelä, M. R.; Bredeweg, E. L.; Magnuson, J. K.; Baker, S. E.; De Vries, R. P.; Hilden, K. Fungal Ligninolytic Enzymes and Their Applications. *Fungal Kingdom* **2017,** 1049–1061.

Mangodo, C.; Adeyemi, T.; Bakpolor, V.; Adegboyega, D. Impact of Microorganisms on Climate Change: A Review. *World News Natur. Sci.* **2020,** *31,* 36–47.

Mariko, S.; Bekku, Y.; Koizumi, H. Efflux of Carbon Dioxide from Snow-Covered Forest Floors. *Ecol. Res.* **1994,** *9* (3), 343–350.

McLaughlin, P. Climate Change, Adaptation, and Vulnerability: Reconceptualizing Societal–Environment Interaction within a Socially Constructed Adaptive Landscape. *Org. Environ.* **2011,** *24* (3), 269–291.

Mendes, R.; Garbeva, P.; Raaijmakers, J. M. The Rhizosphere Microbiome: Significance of Plant Beneficial, Plant Pathogenic, and Human Pathogenic Microorganisms. *FEMS Microbiol. Rev.* **2013,** *37* (5), 634–663.

Menéndez, R.; Megías, A. G.; Hill, J. K.; Braschler, B.; Willis, S. G.; Collingham, Y.; Fox, R.; Roy, D. B.; Thomas, C. D. Species Richness Changes Lag Behind Climate Change. *Proc. R. Soc. B: Biol. Sci.* **2006,** *273* (1593), 1465–1470.

Middelboe, M.; Glud, R. N. Viral Activity along a Trophic Gradient in Continental Margin Sediments Off Central Chile. *Marine Biol. Res.* **2006,** *2* (01), 41–51.

Mirabella, N.; Castellani, V.; Sala, S. Current Options for the Valorization of Food Manufacturing Waste: A Review. *J. Clean. Prod.* **2014,** *65,* 28–41.

Mishra, M. Microbial Diversity: Its Exploration and Need of Conservation. In *Applied Environmental Biotechnology: Present Scenario and Future Trends;* Springer, 2015; pp 43–58.

Morgan, C. A.; Herman, N.; White, P.; Vesey, G. Preservation of Microorganisms by Drying; A Review. *J. Microbiol. Methods* **2006,** *66* (2), 183–193.

Morriën, E.; Engelkes, T.; Van Der Putten, W. H. Additive Effects of Aboveground Polyphagous Herbivores and Soil Feedback in Native and Range-Expanding Exotic Plants. *Ecology* **2011,** *92* (6), 1344–1352.

Munn, C. B. *Marine Microbiology: Ecology & Applications;* CRC Press, 2019.

Nakasone, K. K.; Peterson, S. W.; Jong, S.-C. Preservation and Distribution of Fungal Cultures. In *Biodiversity of Fungi: Inventory and Monitoring Methods,* Vol. 3; Elsevier Academic Press: Amsterdam, 2004; pp 37–47.

Niemeyer, J. C.; Lolata, G. B.; de Carvalho, G. M.; Da Silva, E. M.; Sousa, J. P.; Nogueira, M. A. Microbial Indicators of Soil Health as Tools for Ecological Risk Assessment of a Metal Contaminated Site in Brazil. *Appl. Soil Ecol.* **2012**, *59*, 96–105.

Novotna, A.; Suárez, J. Molecular Detection of Bacteria Associated with Serendipita sp. a Mycorrhizal Fungus from the Orchid Stanhopea Connata Klotzsch in Southern Ecuador. *Bot. Lett.* **2018**, *165* (2), 307–313.

O'Malley, M. A.; Dupré, J. Size Doesn't Matter: Towards a More Inclusive Philosophy of Biology. *Biol. Phil.* **2007**, *22* (2), 155–191.

Ochoa-Hueso, R.; Collins, S. L.; Delgado-Baquerizo, M.; Hamonts, K.; Pockman, W. T.; Sinsabaugh, R. L.; Smith, M. D.; Knapp, A. K.; Power, S. A. Drought Consistently Alters the Composition of Soil Fungal and Bacterial Communities in Grasslands from Two Continents. *Global Change Biol.* **2018**, *24* (7), 2818–2827.

Paliwal, R.; Giri, K.; Rai, J. Microbial Ligninolysis: Avenue for Natural Ecosystem Management. In *Biotechnology: Concepts, Methodologies, Tools, and Applications*; IGI Global, 2019; pp 1399–1423.

Parmesan, C. Ecological and Evolutionary Responses to Recent Climate Change. *Annu. Rev. Ecol. Evol. Syst.* **2006**, *37*, 637–669.

Pastorelli, R.; Paletto, A.; Agnelli, A. E.; Lagomarsino, A.; De Meo, I. Microbial Communities Associated with Decomposing Deadwood of Downy Birch in a Natural Forest in Khibiny Mountains (Kola Peninsula, Russian Federation). *Forest Ecol. Manage.* **2020**, *455*, 117643.

Pathak, A.; Pathak, R. Microorganisms and global warming. *Int. J. Appl. Microbiol. Sci.* **2012**, *1*, 21–23.

Pecl, G. T.; Araújo, M. B.; Bell, J. D.; Blanchard, J.; Bonebrake, T. C.; Chen, I.-C.; Clark, T. D.; Colwell, R. K.; Danielsen, F.; Evengård, B. Biodiversity Redistribution under Climate Change: Impacts on Ecosystems and Human Well-Being. *Science* **2017**, *355* (6332), caai9214.

Pessiot, J.; Nouaille, R.; Jobard, M.; Singhania, R.; Bournilhas, A.; Christophe, G.; Fontanille, P.; Peyret, P.; Fonty, G.; Larroche, C. Fed-Batch Anaerobic Valorization of Slaughterhouse By-Products with Mesophilic Microbial Consortia without Methane Production. *Appl. Biochem. Biotechnol.* **2012**, *167* (6), 1728–1743.

Phillips, R. L.; Whalen, S. C.; Schlesinger, W. H. Influence of Atmospheric CO_2 Enrichment on Methane Consumption in a Temperate Forest Soil. *Global Change Biol.* **2001**, *7* (5), 557–563.

Prakash, O.; Sharma, R.; Singh, P.; Yadav, A. Strategies for Taxonomical Characterization of Agriculturally Important Microorganisms. In *Microbial Inoculants in Sustainable Agricultural Productivity*; Springer, 2016; pp 85–101.

Prasad, M.; Srinivasan, R.; Chaudhary, M.; Choudhary, M.; Jat, L. K. Plant Growth Promoting Rhizobacteria (PGPR) for Sustainable Agriculture: Perspectives and Challenges. In *PGPR Amelioration in Sustainable Agriculture*; Elsevier, 2019; pp 129–157.

Pugnaire, F. I.; Morillo, J. A.; Peñuelas, J.; Reich, P. B.; Bardgett, R. D.; Gaxiola, A.; Wardle, D. A.; Van Der Putten, W. H. Climate Change Effects on Plant-Soil Feedbacks and Consequences for Biodiversity and Functioning of Terrestrial Ecosystems. *Sci. Adv.* **2019**, *5* (11), eaaz1834.

Purahong, W.; Arnstadt, T.; Kahl, T.; Bauhus, J.; Kellner, H.; Hofrichter, M.; Krüger, D.; Buscot, F.; Hoppe, B. Are Correlations between Deadwood Fungal Community Structure,

Wood Physico-Chemical Properties and Lignin-Modifying Enzymes Stable across Different Geographical Regions? *Fungal Ecol.* **2016a,** *22,* 98–105.

Purahong, W.; Wubet, T.; Lentendu, G.; Schloter, M.; Pecyna, M. J.; Kapturska, D.; Hofrichter, M.; Krüger, D.; Buscot, F. Life in Leaf Litter: Novel Insights into Community Dynamics of Bacteria and Fungi during Litter Decomposition. *Mol. Ecol.* **2016b,** *25* (16), 4059–4074.

Ram, A. S. P.; Sime-Ngando, T. Functional Responses of Prokaryotes and Viruses to Grazer Effects and Nutrient Additions in Freshwater Microcosms. *ISME J.* **2008,** *2* (5), 498–509.

Rey, A.; Pegoraro, E.; Tedeschi, V.; De Parri, I.; Jarvis, P. G.; Valentini, R. Annual Variation in Soil Respiration and Its Components in a Coppice Oak Forest in Central Italy. *Global Change Biol.* **2002,** *8* (9), 851–866.

Richter, B. *Beyond Smoke and Nirrors: Climate Change and Energy in the 21st Century*; Cambridge University Press, 2014.

Rounsevell, M.; Evans, S.; Bullock, P. Climate Change and Agricultural Soils: Impacts and Adaptation. *Clim. Change* **1999,** *43* (4), 683–709.

Rousk, J.; Smith, A. R.; Jones, D. L. Investigating the Long-Term Legacy of Drought and Warming on the Soil Microbial Community across Five European Shrubland Ecosystems. *Global Change Biol.* **2013,** *19* (12), 3872–3884.

Ruess, L.; Michelsen, A.; Schmidt, I. K.; Jonasson, S. Simulated Climate Change Affecting Microorganisms, Nematode Density and Biodiversity in Subarctic Soils. *Plant Soil* **1999,** *212* (1), 63–73.

Saha, B.; Saha, S.; Das, A.; Bhattacharyya, P. K.; Basak, N.; Sinha, A. K.; Poddar, P. Biological Nitrogen Fixation for Sustainable Agriculture. In *Agriculturally Important Microbes for Sustainable Agriculture*; Springer, 2017; pp 81–128.

Sanchez-Hernandez, J. C. *Bioremediation of Agricultural Soils*; CRC Press, 2019.

Savage, K.; Davidson, E. A.; Richardson, A. D.; Hollinger, D. Y. Three Scales of Temporal Resolution from Automated Soil Respiration Measurements. *Agric. Forest Meteorol.* **2009,** *149* (11), 2012–2021.

Schimel, J. P.; Gulledge, J. M.; Clein-Curley, J. S.; Lindstrom, J. E.; Braddock, J. F. Moisture Effects on Microbial Activity and Community Structure in Decomposing Birch Litter in the Alaskan Taiga. *Soil Biol. Biochem.* **1999,** *31* (6), 831–838.

Schimel, J.; Balser, T. C.; Wallenstein, M. Microbial Stress-Response Physiology and Its Implications for Ecosystem Function. *Ecology* **2007,** *88* (6), 1386–1394.

Schindlbacher, A.; Rodler, A.; Kuffner, M.; Kitzler, B.; Sessitsch, A.; Zechmeister-Boltenstern, S. Experimental Warming Effects on the Microbial Community of a Temperate Mountain Forest Soil. *Soil Biol. Biochem.* **2011,** *43* (7), 1417–1425.

Sharma, S. B.; Sayyed, R. Z.; Trivedi, M. H.; Gobi, T. A. Phosphate Solubilizing Microbes: Sustainable Approach for Managing Phosphorus Deficiency in Agricultural Soils. *SpringerPlus* **2013,** *2* (1), 587.

Sharma, S. K.; Gupta, A. K.; Shukla, A. K.; Ahmad, E.; Sharma, M. P.; Ramesh, A. Microbial Conservation Strategies and Methodologies: Status and Challenges. *Indian J. Plant Genet. Resour.* **2016,** *29* (3), 340–342.

Sharma, S. K.; Singh, S. K.; Ramesh, A.; Sharma, P. K.; Varma, A.; Ahmad, E.; Khande, R.; Singh, U. B.; Saxena, A. K. Microbial Genetic Resources: Status, Conservation, and Access and Benefit-Sharing Regulations. In *Microbial Resource Conservation*; Springer, 2018; pp 1–33.

Shim, J. H.; Pendall, E.; Morgan, J. A.; Ojima, D. S. Wetting and Drying Cycles Drive Variations in the Stable Carbon Isotope Ratio of Respired Carbon Dioxide in Semi-Arid Grassland. *Oecologia* **2009,** *160* (2), 321–333.

Shiomi, D.; Margolin, W. A Sweet Sensor for Size-Conscious Bacteria. *Cell* **2007,** *130* (2), 216–218.

Shukla, P. K.; Misra, P.; Maurice, N.; Ramteke, P. W. Heavy Metal Toxicity and Possible Functional Aspects of Microbial Diversity in Heavy Metal-Contaminated Sites. In *Microbial Genomics in Sustainable Agroecosystems*; Springer, 2019; pp 255–317.

Siem-Jørgensen, M.; Glud*, R. N.; Middelboe, M. Viral Dynamics in a Coastal Sediment: Seasonal Pattern, Controlling Factors and Relations to the Pelagic–Benthic Coupling. *Marine Biol. Res.* **2008,** *4* (3), 165–179.

Singh, B. K.; Bardgett, R. D.; Smith, P.; Reay, D. S. Microorganisms and Climate Change: Terrestrial Feedbacks and Mitigation Options. *Nat. Rev. Microbiol.* **2010,** *8* (11), 779–790.

Singh, D.; Ghosh, P.; Kumar, J.; Kumar, A. Plant Growth-Promoting Rhizobacteria (PGPRs): Functions and Benefits. In *Microbial Interventions in Agriculture and Environment*; Springer, 2019; pp 205–227.

Singh, R. K.; Tripathi, R.; Ranjan, A.; Srivastava, A. K. Fungi as Potential Candidates for Bioremediation. In *Abatement of Environmental Pollutants*; Elsevier, 2020; pp 177–191.

Singh, S.; Baghela, A. Cryopreservation of Microorganisms. In *Modern Tools and Techniques to Understand Microbes*; Springer, 2017; pp 321–333.

Smith, D.; Ryan, M. Implementing Best Practices and Validation of Cryopreservation Techniques for Microorganisms. *Sci. World J.* **2012,** *2012*.

Smith, D.; Ryan, M. J.; Stackebrandt, E. The Ex Situ Conservation of Microorganisms: Aiming at a Certified Quality Management. In *Biotechnology*; EOLSS Publisher: Oxford, 2008.

Souza, R. d.; Ambrosini, A.; Passaglia, L. M. Plant Growth-Promoting Bacteria as Inoculants in Agricultural Soils. *Genet. Mol. Biol.* **2015,** *38* (4), 401–419.

Srinivasan, P. *Carbon Budget and Cost Analysis of C2CNT: A Carbon Dioxide to Solid Carbon Conversion Process*; The George Washington University, 2018.

Stein, L. Y.; Klotz, M. G. The Nitrogen Cycle. *Curr. Biol.* **2016,** *26* (3), R94–R98.

Štercová, L. Importance of Fungal Decomposition of Wood in the Ecosystems of Natural Forests. **2017.**

Stewart, E. J. Growing Unculturable Bacteria. *J. Bacteriol.* **2012,** *194* (16), 4151–4160.

Stieglitz, M.; Shaman, J.; McNamara, J.; Engel, V.; Shanley, J.; Kling, G. W. An Approach to Understanding Hydrologic Connectivity on the Hillslope and the Implications for Nutrient Transport. *Global Biogeochem. Cycles* **2003,** *17* (4).

Storer, K.; Coggan, A.; Ineson, P.; Hodge, A. Arbuscular Mycorrhizal Fungi Reduce Nitrous Oxide Emissions from N_2O Hotspots. *New Phytol.* **2018,** *220* (4), 1285–1295.

Strååt, K. D.; Mörth, C.-M.; Undeman, E. Future Export of Particulate and Dissolved Organic Carbon from Land to Coastal Zones of the Baltic Sea. *J. Marine Syst.* **2018,** *177*, 8–20.

Streeter, J. Effect of Trehalose on Survival of Bradyrhizobium Japonicum during Desiccation. *J. Appl. Microbiol.* **2003,** *95* (3), 484–491.

Sullivan, T. S.; Gadd, G. M. Metal Bioavailability and the Soil Microbiome. In *Advances in Agronomy*; Academic Press Inc., 2019.

Suttle, C. A. Marine Viruses—Major Players in the Global Ecosystem. *Nat. Rev. Microbiol.* **2007,** *5* (10), 801–812.

Svenning, J. C.; Sandel, B. Disequilibrium Vegetation Dynamics under Future Climate Change. *Am. J. Bot.* **2013,** *100* (7), 1266–1286.

Tapia-Torres, Y.; Rodríguez-Torres, M. D.; Elser, J. J.; Islas, A.; Souza, V.; García-Oliva, F.; Olmedo-Álvarez, G. How to Live with Phosphorus Scarcity in Soil and Sediment: Lessons from Bacteria. *Appl. Environ. Microbiol.* **2016,** *82* (15), 4652–4662.

Tarafdar, J. Fungal Inoculants for Native Phosphorus Mobilization. In *Biofertilizers for Sustainable Agriculture and Environment*; Springer, 2019; pp 21–40.

Tedersoo, L.; Bahram, M.; Põlme, S.; Kõljalg, U.; Yorou, N. S.; Wijesundera, R.; Ruiz, L. V.; Vasco-Palacios, A. M.; Thu, P. Q.; Suija, A. Global Diversity and Geography of Soil Fungi. *Science* **2014,** *346* (6213), 1256688.

Tláskal, V.; Voříšková, J.; Baldrian, P. Bacterial Succession on Decomposing Leaf Litter Exhibits a Specific Occurrence Pattern of Cellulolytic Taxa and Potential Decomposers of Fungal Mycelia. *FEMS Microbiol. Ecol.* **2016,** *92* (11), fiw177.

Trivedi, C.; Delgado-Baquerizo, M.; Hamonts, K.; Lai, K.; Reich, P. B.; Singh, B. K. Losses in Microbial Functional Diversity Reduce the Rate of Key Soil Processes. *Soil Biol. Biochem.* **2019,** *135,* 267–274.

Trumbore, S. E. Potential Responses of Soil Organic Carbon to Global Environmental Change. *Proc. Natl. Acad. Sci.* **1997,** *94* (16), 8284–8291.

Umani, S. F.; Malisana, E.; Focaracci, F.; Magagnini, M.; Corinaldesi, C.; Danovaro, R. Disentangling the Effect of Viruses and Nanoflagellates on Prokaryotes in Bathypelagic Waters of the Mediterranean Sea. *Marine Ecol. Progr. Ser.* **2010,** *418,* 73–85.

Vadstein, O. Heterotrophic, Planktonic Bacteria and Cycling of Phosphorus. In *Advances in Microbial Ecology*; Springer, 2000; pp 115–167.

Van Der Heijden, M. G.; Bardgett, R. D.; Van Straalen, N. M. The Unseen Majority: Soil Microbes as Drivers of Plant Diversity and Productivity in Terrestrial Ecosystems. *Ecol. Lett* **2008,** *11* (3), 296–310.

Walker, G. M.; White, N. A. Introduction to Fungal Physiology. *Fungi: Biol. App.* **2017,** 1–35.

Walker, M. D.; Wahren, C. H.; Hollister, R. D.; Henry, G. H.; Ahlquist, L. E.; Alatalo, J. M.; Bret-Harte, M. S.; Calef, M. P.; Callaghan, T. V.; Carroll, A. B. Plant Community Responses to Experimental Warming across the Tundra Biome. *Proc. Natl. Acad. Sci.* **2006,** *103* (5), 1342–1346.

Walsh, D. Consequences of Climate Changes on Microbial Life in the Ocean. *Microbiol Today (Nov 2015 issue). Microbiology Society, London,* **2015.**

Ward, D. *The Biology of Deserts*; Oxford University Press, 2016.

Wardle, D. A. The Influence of Biotic Interactions on Soil Biodiversity. *Ecol. Lett.* **2006,** *9* (7), 870–886.

Weinbauer, M. G. Ecology of Prokaryotic Viruses. *FEMS Microbiol. Rev.* **2004,** *28* (2), 127–181.

Weinbauer, M. G.; Brettar, I.; Höfle, M. G. Lysogeny and Virus-Induced Mortality of Bacterioplankton in Surface, Deep, and Anoxic Marine Waters. *Limnol. Oceanogr.* **2003,** *48* (4), 1457–1465.

White, P. A.; Kalff, J.; Rasmussen, J. B.; Gasol, J. M. The Effect of Temperature and Algal Biomass on Bacterial Production and Specific Growth Rate in Freshwater and Marine Habitats. *Microbial Ecol.* **1991,** *21* (1), 99–118.

Wilcox, K. R.; Shi, Z.; Gherardi, L. A.; Lemoine, N. P.; Koerner, S. E.; Hoover, D. L.; Bork, E.; Byrne, K. M.; Cahill Jr, J.; Collins, S. L. Asymmetric Responses of Primary Productivity to Precipitation Extremes: A Synthesis of Grassland Precipitation Manipulation Experiments. *Global Change Biol.* **2017**, *23* (10), 4376–4385.

Wilcox, K. R.; von Fischer, J. C.; Muscha, J. M.; Petersen, M. K.; Knapp, A. K. Contrasting Above-and Below-Ground Sensitivity of Three Great Plains Grasslands to Altered Rainfall Regimes. *Global Change Biol.* **2015**, *21* (1), 335–344.

Wilhelm, S. W.; Matteson, A. R. Freshwater and Marine Virioplankton: A Brief Overview of Commonalities and Differences. *Freshwater Biol.* **2008**, *53* (6), 1076–1089.

Williams, M. A. Response of Microbial Communities to Water Stress in Irrigated and Drought-Prone Tallgrass Prairie Soils. *Soil Biol. Biochem.* **2007**, *39* (11), 2750–2757.

Winters, R. D.; Winn, W. C. A Simple, Effective Method for Bacterial Culture Storage: A Brief Technical Report. *J. Bacteriol. Virol.* **2010**, *40* (2), 99–101.

Yepez, E. A.; Scott, R. L.; Cable, W. L.; Williams, D. G. Intraseasonal Variation in Water and Carbon Dioxide Flux Components in a Semiarid Riparian Woodland. *Ecosystems* **2007**, *10* (7), 1100–1115.

Yu, S.; Bai, X.; Liang, J.; Wei, Y.; Huang, S.; Li, Y.; Dong, L.; Liu, X.; Qu, J.; Yan, L. Inoculation of Pseudomonas sp. GHD-4 and Mushroom Residue Carrier Increased the Soil Enzyme Activities and Microbial Community Diversity in Pb-Contaminated Soils. *J. Soils Sediments* **2019**, *19* (3), 1064–1076.

Zaidi, A.; Khan, M.; Ahemad, M.; Oves, M. Plant Growth Promotion by Phosphate Solubilizing Bacteria. *Acta Microbiol. et Immunol. Hungarica* **2009**, *56* (3), 263–284.

Zhang, T.; Wang, N.-F.; Liu, H.-Y.; Zhang, Y.-Q.; Yu, L.-Y. Soil pH Is a Key Determinant of Soil Fungal Community Composition in the Ny-Ålesund Region, Svalbard (High Arctic). *Front. Microbiol.* **2016**, *7*, 227.

Zheng, B.-X.; Hao, X.-L.; Ding, K.; Zhou, G.-W.; Chen, Q.-L.; Zhang, J.-B.; Zhu, Y.-G. Long-Term Nitrogen Fertilization Decreased the Abundance of Inorganic Phosphate Solubilizing Bacteria in an Alkaline Soil. *Sci. Rep.* **2017**, *7* (1), 1–10.

Zhou, J.; Deng, Y.; Shen, L.; Wen, C.; Yan, Q.; Ning, D.; Qin, Y.; Xue, K.; Wu, L.; He, Z. Temperature Mediates Continental-Scale Diversity of Microbes in Forest Soils. *Nat. Commun.* **2016**, *7*, 12083.

Zhu, J.; Li, M.; Whelan, M. Phosphorus Activators Contribute to Legacy Phosphorus Availability in Agricultural Soils: A Review. *Sci. Total Environ.* **2018**, *612*, 522–537.

Zimmer, C. The Microbe Factor and Its Role in Our Climate Future. Retrieved December **2010**, 15, 2015.

CHAPTER 4

Climate Change and Its Influence on Microbial Diversity, Communities, and Processes

IRTEZA QAYOOM[1], HAIKA MOHI-UD-DIN[1], AQSA KHURSHEED[1], AASHIA ALTAF[1], and SUHAIB A. BANDH[2*]

[1]*Sri Pratap College, Cluster University Srinagar, India*

[2]*Government Degree College, DH Pora Kulgam, Jammu and Kashmir, India*

Corresponding author. E-mail: qirteza@gmail.com

ABSTRACT

Climate change has been occurring throughout the Earth's geological past. However, the cause of alarming concern in the present age is the rapid and unprecedented rate of climate changes taking place primarily because of the anthropogenic activities. Although the effect of climate change on higher life forms and ecosystems have been well documented, it remains undervalued for the lower life forms i.e., microbes, despite their unequivocal role in the biosphere through the support they provide to the higher life forms. Microbes play a key part in determining the atmospheric concentrations of greenhouse gases and have likely responded to the climate change. Climate change exerts abiotic stress on environmental microbiomes that can lead to the shifts in the geographical range of microbial species; affect their distribution, biodiversity and abundance. This chapter highlights the impact of climate change on microbial diversity, communities and processes across different habitats.

4.1   INTRODUCTION

The increase in anthropogenic activities on Earth has led to suggestions that we have entered into a human-dominated geological epoch, the Anthropocene, where human activities have a significant impact on many key earth processes from biogeochemical to the evolution of life (Lewis and Maslin, 2015). Since the Industrial Revolution, the changes have shifted from a local to a global scale (Ruuskanen, 2019). One of the most noticeable global anthropogenic impacts is climate change (Hegerl et al., 2007), which happens primarily due to the effects of higher greenhouse gas (GHG) concentration in the atmosphere (US EPA, 2016). Climate change is a complex 21st century global problem, as can be seen from the fact that, for the first time in human history, in 2013, atmospheric carbon dioxide was above 400 ppm (Dutta and Dutta, 2016). The increase in GHGs has resulted in the rise in global average temperature called global warming. Global warming is expected to have some severe global implications on both natural and artificial ecosystems. Global climate change has already resulted in the acidification of oceans, amplification of extreme weather events, rise in sea level, and melting of permafrost and glaciers (IPCC, 2018).

The anthropogenic climate change has led to the extinction of animal and plant species, biodiversity loss (Barnosky et al., 2011; Crist and Engelman, 2017; Johnson et al., 2017; Pecl et al., 2017; Cavicchioli et al., 2019) and land degradation. Although invisible to the naked eye, climate change also has a significant effect on microbes (Piffaretti et al., 2018). The impact of climate change on higher life forms has properly been studied and documented. However, on the contrary, discussions on climate change usually ignore the tiniest microscopic organisms (Cavicchioli et al., 2019), even though they play a significant role in nutrient cycling, energy flow, biodiversity, and total biomass of the ecosystems. It is due to the insufficient understanding that microbial activities are not considered adequately in most climate change models (Dupre, 2008; Dutta and Dutta, 2016). This is a significant oversight because microbial communities are generally the first responders to environmental perturbations and can either augment or buffer environmental shifts via, often complex, lively, and negative feedback loops (Labbate et al., 2016). Microorganisms hold a greater significance in the context of climate change as they play an essential role in biogeochemical processes like carbon and nitrogen cycle and act both as producers and consumers of GHGs, which in turn is responsible for the problem under debate (Microbiology Online, 2015; Dutta and Dutta, 2016). Besides this, microbes are central to the ecosystem functioning and also support higher life forms. Therefore, any changes in their diversity, abundance, or functional capacity will also affect the resilience of all other organisms and thus their ability to respond to any climatic eventuality (Maloy et al., 2016; Cavicchioli et al., 2019).

Climate change has influenced microorganisms in both terrestrial and aquatic ecosystems. However, the impact of climate change on microbial communities and their functions may vary between different ecosystems and regions. In terrestrial biomes, climate shifts alter the geographical range of microbial species; affect their distribution, diversity, and abundance; and influence microbe–microbe and plant–microbe interactions. It involves the microbial community structure and composition either directly or indirectly through the alteration of plant composition, plant litter, and root exudates (Classen et al., 2015; Jain et al., 2020). Thus, altering the soil processes, especially Nitrogen and Carbon cycling are mediated by these microbes (Margesin and Niklinska, 2019).

Furthermore, moisture and temperature fluctuations due to climate variability affect the microbial processes, including the rate of respiration, fermentation, methanogenesis, and decomposition (Weiman, 2015). In polar soil environments, such as in Arctic permafrost, changes, for example, warming, melting, thawing, drying, and changes in plant communities, are expected to drive changes in the composition and diversity of microbial communities (Jansson and Tas, 2014). The elevated temperature and levels CO_2 also affect the marine planktonic communities, thus, affecting the productivity, nitrification, denitrification, and photosynthetic C fixation mediated by them (Glockner et al., 2012). Climate change also influences processes like oceanic stratification, mixing, the availability of nutrients, thermohaline circulation, and extreme weather events. These changes upset marine microbiota in significant ways, including some significant changes in productivity, oceanic food webs, and carbon exports and burial in the sea bed (Hurd, 2018; Gao, 2012; Boyd, 2013; Portner, 2014; Brennan and Collins, 2015; Hutchins et al., 2016; Hutchins and Fu., 2017; Rintoul et al., 2018; Cavicchioli et al., 2019). Furthermore, there are cyanobacterial algal blooms that exemplify a significant deleterious effect of climate change in aquatic ecosystems (Dutta and Dutta, 2016) like eutrophic lakes, reservoirs, and estuaries.

4.2   EFFECT OF CHANGING CLIMATE ON SOIL MICROBES

According to UN Framework Convention on Climate Change (UNFCCC), climate change is directly or indirectly caused due to human activity that changes the global atmospheric composition as well as due to the natural climate variability observed over comparable periods. As one of the most important sciences of recent years, climate change has influenced various environments globally, regionally, and locally, with soil microbial environment as no exception. Soil microbes play a significant role in the decomposition of organic soil materials, and the idea is that global warming will increase the rates of heterotrophic microbial activities and thereby emit more carbon dioxide into the atmosphere (Jenkison et al., 1991; Davidson and Janssens, 2006). Global warming increases net carbon exchange from the ground to air (Jenkinson et al., 1991; Schimel et al., 1994). Further interaction between the members of the community can be beneficial, detrimental, or can have no outcome, and these associations may change with ecological pressures (Vandenkoornhuse et al., 2015). Soil life interacts in many different ways with each other and thus forms and retain environmental properties (Berg et al., 2010; van der Putten et al., 2013). Considering their significance in defining ecosystem characteristics, soil microbe–microbe and soil plant–microbe interactions respond to environmental changes and provide insights into global ecological capabilities, such as net primary productivity and the stockpiling of soil carbon (Ostle et al., 2009; Berg et al., 2010; Fisher et al., 2014).

Relative abundance and ramification of soil communities get changed due to climate change since members of soil microbial community contrast in their physiology, temperature affectability, and developmental rates (Gray et al., 2011; Briones et al., 2014; Delgado-Baquerizo et al., 2014; Whitaker et al., 2014). Microbes live working together with a vast number of different other species, for example, some lucrative and pathogenic species which have next to zero impact on complex communities. Because natural communities are composed of organisms with altogether different life-history attributes and dispersal capacity, it is improbable that the entirety of the microbial community will react to climatic factors in a similar way (Sundqvist et al., 2015).

Researchers believe that changes in environmental variables such as barometric and warming are likely to interact with changes in the ecosystem properties and behavior. Nonetheless, it is less evident that microbial networks are regulating ecosystem processes. Soil microbial networks are responsible for supporting habitats through carbon (C) cycling. Their exercises are governed by biotic and abiotic factors, for example, the amount and nature of litter inputs, temperature, and wetness. Atmospheric and climatic changes affect both biotic and abiotic drivers in ecosystems and the reaction of the environment to these changes (Castro et al., 2009). Environmental change factors, such as atmospheric CO_2 and modified precipitation regimes can conceivably have both immediate and roundabout effects on soil microbial networks. Expanded vegetative development and soil carbon stockpiling under raised carbon dioxide focus have appeared in various tests. Everything depends on the reaction of soil forms, including those which track mineralization of nitrogen (N) against climate change, to ecosystems' capacity to maintain expanded carbon stock. These soil procedures are mediated by microbial networks whose actions and assembly may likewise react to developing atmospheric carbon dioxide (Austin et al., 2009). In soil systems, microbes intervene in the rotting of plant matter, utilization, and production of trace gases, the transformation of metals, and plant development (Panikov, 1999). As we experience uncommon ecological effects due to environmental change, microorganisms react, adjust, and develop in the changing environmental factors. Since they have generation times as short as a couple of hours, they do as such at higher rates than most different life forms, and this makes microorganisms sentinels for understanding the impacts of environmental change on natural frameworks and worldwide biogeochemical cycles. Researchers can consider the effects of environmental change on microbes to both comprehend and ideally anticipate the future impacts of environmental change on all types of life.

4.3 EFFECT OF CHANGING CLIMATE ON FRESHWATER MICROBES

Freshwater environments make up <1% of Earth's surface, yet it contains 10% of all the defined species (Strayer and Dungeon, 2010; Dudgeon et al., 2019; Benateau et al., 2019). The freshwater ecosystems play a vital role in holding the overall diversity and performing various ecosystem services (Hassan et al., Jan 2020). Still, at the same time, some studies show that freshwater biodiversity is exceptionally susceptible to climate change (Heino et al., 2009). The present rate of climate change is unknown and biological responses to these changes are quick at the species, community, and ecosystem levels. The concern that how will be its effects on freshwater species in the next few decades hase been projected to be grimmer (Xenopoulous et al., 2005). Moreover, the water temperature of rivers, streams, and lakes also changes in the growing season, which directly influences freshwater organisms (Magnuson et al., 2000). Further, invasion of non-native species due to climate change in the new environments results in loss of native biodiversity, altered habitats, and altered biogeochemical cycles. Also, the tropical phytoplankton–zooplankton linkage gets affected by the increase in water temperature (Al Sayed et al., 2007; Eissa et al., 2011). Temperature rise affects the growth rate of freshwater microalgal growth rates, thus resulting in their bloom accumulation, drought intensification, storms, vertical stratification, and salinization that results in the dislocation of nutrients in a water body, and thereby promotes the formation of algal blooms (Hans W Paerl and Jurgen Marxsen, March 2016). Impacts on algae due to climate change are considered to be contributory to getting the past, as well as present anomalies are regarded as valuable pointers of climate change. With the increasing temperature, the metabolic activity of protists increases resulting in the negative implications on the other microbes in the water body. Since temperature-sensitive protist populations respond differently to weather, the whole community gets disturbed by the changing temperature scenarios. However, it is known that microbes/phytoplanktons have a higher tolerance to a higher temperature (Huertas et al., 2011). Many experiments indicate that microbes in aquatic biofilms are impacted by climate change in terms of their structure and function (Marxsen et al., 2010). The temperature variations in the density of water influence water stratification, and thus, the transportation of microbial nutrients (Glokner et al., 2012) occurs. For each 10°C increase in temperature (Morita, 1974), microbial metabolic rates double. Bacterioplankton in aquatic systems play an essential role in nutrient cycles. Still, variation in water temperature produces consequences on both productivities as well as the abundance of the bacterioplankton (Starosak and smith, 2004). With the increasing temperature, metabolic rates of bacteria increase, which subsequently increases the energy demand of the bacterial populations. Under the low availability of food, an increase in temperature exerts negative impacts on the growth of bacteria.

In lake hydrology, climate change gives rise to increased evaporation and precipitation rates, which results in water level variations in Inland waters, thereby affecting the discharge of ecosystem greenhouse gases in the atmosphere, and hence effecting microbial communities (Weise et al., 2016). Research shows that global climate change in alpine lake habitats is influenced, for example, by ice thaw, permafrost melting. Growing summer waters temperatures, which are impacting their biodiversity (Holzappfel and Vinebrook, 2005; Packer et al., 2008), as well as by changes in the environment, causing turbidity shifts in the bacterioplankton population of alpine lakes (Peku and Sonmamga, 2016). Thus, climate change has adverse effects on the world's freshwater microbes, and hence there is an urgent need to apply precautionary measures to minimize these devastating damages.

4.4 EFFECTS OF CHANGING CLIMATE ON MARINE MICROBES

In the marine environment, microbes are ubiquitously present (Sunagawa et al., 2015) and cover a diversity of microorganisms, including bacteria, archaea, viruses, fungi, and protists (Glöckner et al., 2012). They inhabit all types of ocean habitats from the deep-sea to the surface. With an estimated 10^4–10^6 cells/mm, their biomass, along with environmental complexity and high turnover rates, is the source of their tremendous genetic diversity (Whitman et al., 1998; Sunagawa et al., 2015). These microbes play an essential role in biogeochemical routes, such as carbon and nutrient cycling (Falkowski et al., 1998; Sunagawa et al., 2015), and form the basis of marine food webs (Cavicchioli et al., 2019). Therefore, they are vital for the functioning of such marine ecosystems (Jain et al., 2020). The microbes and their communities drive and react to changes in the environment, such as climate change associated with temperature shifts, carbon chemistry, oxygen contents, the content of nutrients, and changes in ocean stratification (Doney et al., 2012; Hutchins et al., 2019). These variations have a significant consequence for marine microbiome, potentially altering the microbial diversity, function, and community dynamics. High levels of atmospheric CO_2 as a result of climate change has caused the absorption of more carbon dioxide into the seas that disturb the marine carbonate buffering capacity (Hutchins and Fu, 2017), and consequently makes it acidic by lowering its pH (Dutta and Dutta, 2016). This decrease in pH in turn, accelerates the dissolution of calcium carbonate, which elevates dissolved CO_2 and decreases the pH further, causing drastic effects of acidity in the oceans (Glöckner et al., 2012; Dutta and Dutta, 2016) called as Ocean Acidification (OA). Microbes are vital components to oceanic conditions and drive biogeochemical processes, such as the marine carbon cycle and nitrogen cycle (Arnosti, 2011; Azam and Malfatti, 2007; Das and Mangwani, 2015). They are responsive to any change and are easily reactive to environmental stress because of their intimate contact and interaction with the environment. Thus, OA influences microbial diversity, primary production, trace gas emission in the oceans, and microbial processes, such as the extracellular enzymes, quorum sensing activity, and nitrogen cycling (Das and Mangwani, 2015). Moreover, OA prevents archaea and bacterial nitrification (Hutchins et Fu, 2017; Hutchins et al., 2019) and alters their genetic expression to promote cell maintenance instead of growth (Bunse et al., 2016). Higher CO_2 levels have also shown to influence cyanobacteria, which are the most abundant marine photoautotrophic organism (Burns et al., 2005; Das and Mangwani, 2015). Comprehensive work into the impact of higher pCO_2 on various cyanobacterial species, for example, *Trichodesmium* sp.,.*Crocosphaera watsonii* sp., *Synechococcus* sp. suggested that these species increase their growth (Fu et al., 2007; Barcelos e Ramos et al., 2007; Kranz et al., 2009, 2010; Das and Mangwani, 2015). However, in *Nodularia spumigena*, the growth rate was observed to be decreased with an increase in pCO_2 (Czerny et al., 2009; Das and Mangwani, 2015). OA can also affect the microbial cell process and its extracellular enzyme activity, for example, increased activity of bacterial protease, a-glucosidase, and b-glucosidase was identified at high levels of pCO_2 (Grossart et al., 2006). Piontek et al. (2010) reported similar findings that found increased microbial polysaccharide degradation under acidified conditions.

The oceans absorb more than 90% of the additional heat energy from the warming of the climate, which increases the ocean warm-up, especially in high-altitude areas (Petrou et al., 2016; Basu and Mackey, 2018). Ocean warming can affect marine biota either by directly affecting the rate of biological processes or by indirectly affecting the stratification, which, in effect, impacts the nutrient supply and light availability to organisms in mixed layers (Beardall et al., 2009; Basu and Mackey, 2018). Temperature affects the phytoplankton physiology by changing the metabolic rates, such as enzyme reaction rate and efficiency (Basu and Mackey, 2018). Other effects include alterations in phenology, shrinking in cell size, and changes in the phytoplankton-driven carbon cycle (Taucher et al., 2015; Basu and

Mackey, 2018). The sensitivity of phytoplankton toward climate-related environmental factors is species-specific, which results in the difference in competition among species, thus changes the phytoplankton composition with a significant consequence for the entire aquatic food web of a given habitat (Hader and Gao, 2017).

Warming in marine environments has also intensifed multiple harmful algal blooms (HABs) in many regions of middle and higher latitudes by changing the geographical and phenogenic conditions of some HABs (Trainer et al., 2019). Increased temperature increases the production of toxins, growth rates, and ecological dominance of eukaryotic HABs. Besides, it can also increase the growth rate of toxic Cyanobacteria, and the same is well-documented (Rapala et al., 1997; Paerl and Huisman, 2008; Davis et al., 2009; Paerl and Paul, 2012; Brutemark et al., 2015; Lürling et al., 2017; Walls et al., 2018; Burford et al., 2019; Griffith and Gobbler, 2019).

The rise in sea surface temperature also influences water density, which impacts the ocean stratification and currents, thereby affecting the mobility of microbial nutrients (Glockner et al., 2012; Dutta and Dutta, 2016). The predicted shift in stratification and vertical water mass mixtures caused by oceanic warming influence the morphology, phenology, and productivity of single-cell marine organisms, including autotrophic protists, bacteria and archaea, or hetcrotrophic microbes (Sarmento et al., 2010; Danovaroet al., 2011; Lewandowska et al., 2014; Mojica et al., 2015). Moreover, the climate change significantly affects the functional processes and associated biogeochemical cycles performed by deep-sea benthic archaea (Hutchins and Fu, 2017; Cavicchioli et al., 2019). Warming and acidification also change the structure of planktonic communities by influencing the trophic interaction chains of plankton based food webs (Gaedke et al., 2010; Murphy et al., 2019).

Environmental variables associated with climate also alter the coral microbiome and contribute to the increased coral richness, decreased evenness, and lower stability (McDevitt-Irwin et al., 2017; van Oppen and Blackall, 2019; Vanwonterghem and webster, 2020). Several studies have shown a dominant change from the relatively beneficial bacterial taxa with elevated temperatures and acidification to opportunistic and potentially pathogenic classes, such as *Alteromonadaceae* and *Vibrionaceae* (Thurber et al., 2009; Bourne et al., 2016; Litmann et al., 2011; McDevitt-Irwin et al., 2017; O'Brien et al., 2016; Vanwontergh. and Webs, 2016). Additionally,

OA is also known to decrease calcification of many calcifying organisms (Gao et al., 2019). Recently, climate change variables, such as acidification, nutrient availability, sea surface warming, and deoxygenation has also been observed to have potential implications on marine viruses (Yang and Zhang, 2015), which subsequently affect carbon sequestration, food webs, biogeochemical cycle, and metabolic balance of oceans (Danovaro et al., 2011).

4.5   EFFECT OF CHANGING CLIMATE ON BELOWGROUND MICROBES

The rhizosphere is a functional interface grounded by the establishment of a highly complex and varied molecular dialogue between plants and microorganisms involving the transfer of nutrients and the interactions of signaling molecules from the plant roots (Prosser et al., 2006; van Elsas et al., 2012; Gksarmiri et al., 2017). It is of paramount significance for ecosystem services like C and water cycling, nutrient trapping, cropping, carbon harvesting, and carbon stockpiling (Adl, 2016; Ahkami et al., 2017). The functional and taxonomic diversity of the rhizospheric microbial communities (RMCs) is excessively affected by biotic and abiotic factors, like root exudates, nutrient rivalries, edaphic properties, and climate changes (Fierer et al., 2012; Pennanen et al., 1999; Rousk et al., 2010; Uroz et al., 2012; Wei et al., 2017). The influence on rhizosphere ecology by global climate change, which includes increased temperatures and unusual weather patterns due to higher carbon dioxide levels of atmosphere, will consequently alter ecosystem functions by various direct and indirect ways (Ahkami et al., 2017). Several studies indicate that climate change fluctuates species interactions, thus contributes to change in biological diversity and functioning of land ecosystems (Walther et al., 2002; Gottfried et al., 2012; Langley and Hungate, 2014; Classen et al., 2015) but fewer studies focus on soil microbial communities (Schimel et al., 2007; de Vries et al., 2012; Classen et al., 2015). The effect of climate change on soil microbial communities differs highly; thus, only a few trends are evident. The direct adverse effects of high levels of CO_2 on below-ground organisms are improbable given that soil biota adapts to a higher concentration of CO_2 present in the soil pore spaces than found in the atmosphere (Holmstrup et al., 2017).

Niklaus et al. (2003) recounted no significant direct consequence of increased CO_2 levels on microbial community or microbial nitrogen and carbon over a prolonged period of 6 years in a temperate grassland given an increase in plant productivity and a variation in soil aggregation (Balser et al., 2010). The Meta-analysis by Blankinship et al. (2011) showed that soil biota has a positive reaction to the increase of carbon dioxide (Amthor et al., 2001) which would help grow litter, root exudates, and act as critical sources of food for decomposing food webs (Arnones et al., Couteaux et al., 2000; Holmstrup et al., 2017). Moreover, growing levels of CO_2 have beneficial effects on arbuscular and ectomycorrhizal fungi as well as varying effects on peptidoglycan recognition proteins (PGRPs) and endophytic fungi (Compant et al., 2010). Fewer studies address the impact of high CO_2 levels on specific bacterial taxa, which vary in their responses as *Betaproteobacteria* and *Bacterioidetes* (copiotrophic bacteria) consume labile C, require high nutrients with high growth rates (Fierer et al., 2007), and show growth with increased levels of CO_2. In contrast, Acidobacteria offers slow growth rates, and outcompete copiotrophs at lower nutrient availability (Fierer et al., 2007) and higher CO_2 levels (De Vries et al., 2015).

Warming has an almost similar effect as elevated CO_2 level as it causes an increase in microbial abundance. Archaea or specific taxonomic or functional groups of soil microbial communities show variable responses to warming, for example. Warming causes a decrease in bacterial diversity (Sheik et al., 2011; Deslippe et al., 2012) while the variety of fungal communities remains unaffected (Anderson et al., 2013; de Vries et al., 2015). Climate warming may cause the prevalence and intensification of wildfires, particularly in forest biomes (Dooley and Treseder, 2012) and consequent reduction in soil microbial and fungal biomass. Even so, not all bacteria are resistant to fire or all fungi are prone to fire since Izzo et al. (2006), through the greenhouse experiments, illustrated that certain species of ectomycorrhizal fungi could compete effectively for roots in response to fire (Tobin and Janzen et al., 2008). Declining mycorrhizal fungal biomass is uniquely related to ecosystem changes to nitrogen, including alterations in the microbial population, composition, and decomposition (Balser et al., 2001; Treseder, 2004; Klironomos et al., 1997; Henriksen and Breland, 1999; Balser et al., 2010). Some studies have shown changes in soil microbes in response to extreme climatic conditions like droughts. For example, the droughts increase fungal abundance, which in turn reduces the bacterial diversity (Schimel et al., 2007; De Vries and Shade, 2013;

de Vries et al., 2015). Also, elevated temperature affects the microbial community composition since several microbial groups prefer a specific range of temperatures for growth and activity (Fierer and Schimel, 2003; Singh et al., 2010; Dutta and Dutta, 2016). Increased temperatures have both positive and negative impacts on plant-associated useful microbes, which differ with the system and temperature range (Compant et al., 2010).

4.6   BACTERIAL COMMUNITY RESPONSES TO INCREASED TEMPERATURE

As a general trend, the planet is getting warmer each day due to climate change, and this climate chaos is characterized by multiple predictors like the glacial meltdown and the severe weather events (Toussaint, 2019). Climate change led to increased temperatures and increased precipitations, thereby increasing the inflow of nutrients, which leads to an increase in overall heterotrophy (Dinasquet et al., 2012). Bacterial communities have been found to adapt to the hotter temperatures, thereby accelerating their respiration rates and releasing more carbon to the atmosphere, hence speeding up the phenomenon of climate warming. Although bacteria occur ubiquitously, yet the prime natural territory of bacteria include soils, subsurface sediments, and oceans. Bacterial population accounts for almost 20–50% of the total mass of all organisms on the surface of the Earth (Toussaint, 2019). In marine environments, heterotrophic bacteria occupy a principal place in the food chain, where interactions and different metabolic activities between niches are managed by temperature (Sarmento et al., 2016). The attainable effects of the high temperature have resulted in a heterotrophic rather than autographic output which has contributed to a rise in heterotrophic output (Hoppe et al., 2002; Muren et al., 2005; Summer and Lengfell, 2008). However, different marine processes like bacterial respiration, bacterial production, growth efficiencies get affected in warmer oceans. In poikilothermic species like bacteria, on the other hand, the rate of metabolism increases with the temperatures, which can have an unfavourable effect upon the growth of bacteria if the food supply is insufficient (Clarke and Fraser, 2004). The fate of the metabolism of heterotrophic bacteria is decided by the amount of unbalanced organic carbon obtained from primary producers. However, in autotrophs, the production of metabolic rate remains unsettled (Beadle, 2018).

Most of the prokaryotes use energy to perform respiration and release carbon dioxide to the environment, increasing temperatures, thereby causing a double whammy effect, which permits them to task more effectively and generates an immense amount of carbon to the globe (Thomas et al., 2019). Some studies conclude that at medium temperature range (below 45°C), prokaryotes show robust responses to climate change, whereas, at a higher temperature range (above 45°C), no such response observed. Erosion of organic permafrost soil due to climate change in oceans is digested by bacteria, releasing a large amount of carbon dioxide in the atmosphere hence exceeding the enrichment of blooms in the arctic region. According to some reports, in the Southern Ocean around Antarctica, 40% of the carbon sequestered by the oceans is taken up by marine micro planktons after which they descent on the seafloor, thereby heating the planet. In 2012, due to increasing global temperature, significantly less stretch of arctic sea ice undulated into the marine ecosystems, as a result of which a good number of diatoms and bacteria in the sea got heaped at the bottom due to which entire food chain got disrupted (Benuyn, 2019).

Moreover, terrestrial systems are no less pertinent to the topic of global climate change. The soil is a massive pool of carbon and is considered significant for determining the trends in climate change (Raich and Potter, 1995; Dutta et al., 2016). Temperature is considered as one of the essential factors in managing and forming of soil microbial communities. However, minimal knowledge is available on how weather affects soil bacteria (Pretikanen et al., 2004). The comparative richness, as well as the purpose of soil bacterial communities, have been found to change due to the increasing temperatures (Schindlbaches), thereby influencing various processes like nitrogen fixation, nitrification, methanogenesis, and denitrification. Climate change causes changes in the enzymatic activities of soil bacteria (Wang and Post, 2012; Wang et al., 2013) as bacterial biochemical reactions take place with fewer efficiencies, hence adding more and more carbon dioxide to the atmosphere (Zimmer, 2010). It has been observed that soil bacteria can either intensify or lessen the global warming rate (Singh et al., 2010). Climate change employs direct as well as indirect influences on terrestrial bacterial communities (Bardgett et al., 2008). Direct impact takes place because of increasing temperatures, which increases respiration rates of bacteria, thus resulting in carbon transfer from soil to air (Rustad et al., 2001; Davidson and Janssens, 2006) while the indirect impact includes a change in function, structure, and heterogeneity of plant growth in the environment

(Bardgett et al., 2008). Due to enhanced photosynthesis, carbon travels from plants to roots, then to mycorrhizal fungi, and eventually into the soil, by removing organic compounds, leading to a rise in the soil carbon level that raises bacterial communities' breathing rates and thus releases more carbon into the atmosphere.

Researchers performed various experiments to check the response of bacterial activity at different temperatures in the soil, for example, at a temperature of 25–30°C, activity of bacteria was found to be 14 times above 0°C. However, above this temperature, the bacterial activity was found to decrease. Some studies reported that the rate of microbial growth at constant temperature is relatively higher than that of the fluctuating temperature (Howell et al., 1971; Biederbeck et al., 1973). In bamboo soils, an increase in temperature resulted in an increase in soil respiration, which in one way or the other affected the ground bacterial diversity (Lin et al., 2017). Similarly, some experiments were performed to check soil microbial reaction at increased temperatures, in vegetated and bare plots. It was observed that bacterial community structure changed significantly in vegetated fields. In contrast, no such changes were observed for bacterial communities in infertile soils, thereby indicating a strong linkage between community structure and ecosystem function that consequently stresses the significance of plant–soil microbial interaction in mediating feedback to future global warming (Koyama et al., 2018). Thus, considering the observed uncertainties due to global heating, it is high time to prioritize the ways to reduce the release of more and more greenhouse gases into the atmosphere.

4.7   POTENTIAL IMPACTS OF CLIMATE CHANGE ON MICROALGAE

Ecologically, microalgae are the critical drivers of photosynthetic carbon fixation and primary production (Yong et al., 2016). Increased carbon dioxide in the atmosphere and climate warming has affected marine biosphere, with pH shifts, available biomass, stable water columns, light-emitting systems and nutrient regimes (Guinder and Molinero, 2013) and have led to various variations on microalgal growth, photosynthesis, homeostasis, respiratory problems, oxidative damage, and trophic transfer (Yong et al., 2016).

4.7.1   EFFECT OF TEMPERATURE ON MICROALGAE

Increased environmental temperature reduces photosynthetic rate, affects PSII viability, thylakoid membrane fluidity, and production of biomass in microalgae (Zidarova and Pouneva, 2006; Yong et al., 2016). Moreover, it also changes the biochemical characteristics of microalgae (Teoh et al., 2005). High temperature inhibits the photosynthesis mechanism as it is sensitive to heat and maybe disrupted before other stress symptoms are observed (Camejo et al., 2005; Lee et al., 2017). Also, it decreases algae's potential for light consumption, thus reducing carbon fixation and electron transport. This limitation causes light inhibition by damage to the PSII (Levasseur et al., 1990; Anning et al., 2001; Salleh and McMinn, 2011) as it is the most thermosensitive component of the photosynthetic apparatus (Foyer et al., 1994; Allakhverdiev et al., 2008; Barati et al., 2017). Davison (1991) proposed that the cellular concentration of RuBisCo and other Kelvin cycle enzymes are also modulated at high temperatures, which in effect lowers the PSII quantum efficiency.

Temperature is also an imperative environmental factor regulating the growth of photosynthetic organisms (Yong et al., 2016). Temperature fluctuation affects microalgae by inducing changes in the chemical, physiological, and molecular activities (Falkowski and Oliver, 2007; Li et al., 2011; Anuwar et al., 2020). Such changes are an essential response to adapt to the conditions for their long-term survival (Anuwar et al., 2020). Each microalgal species has an optimum growth temperature and beyond this range, the growth decreases due to heat stress, enzyme denaturation (salvucci et al., 2004; Barati et al., 2017) or alterations in proteins involved in photosynthesis (Ratkowsky et al., 1983; Ras et al., 2013). However, microalgae of the same taxonomic group with varying latitudes (Yong et al., 2016) respond differently to the temperature regimes in terms of their particular growth rate, for example, some microalgae may have a limited temperature range indicating their sensitivity to temperature. Other species have a more comprehensive temperature range, which suggests that they are likely to be immune to adjust to the climate (Ras et al., 2013; Anuwar et al., 2020). Thus, increased temperature could therefore have a different effect on different species, causing increased metabolic activity and growth of individual species and pushing others outside their optimum temperature, thereby changing the composition and competition of species (Beardall and Raven, 2004). Climate warming also influences the phytoplankton composition, size, and structure and favors those that are best suited to changing climate conditions (Winder and Sommer, 2012). Thus, a different response by phytoplankton to temperature change can shift their composition, which will eventually have a broad impact on the structure and function of an ecosystem (Winder and Sommer, 2012).

Increased surface water temperature as a result of climate warming enhances the stratification. It decreases the thickness of the upper mixing layer (UML), which results in nutrient depletion due to the partial upwelling of nutrients from deep water layers (Hader and Gao, 2017). Decreased availability of nutrients with elevated UV exposure to phytoplankton may have consequences on phytoplankton productivity (Gao et al., 2012b) and community assemblages (Beardall et al., 2009). Since phytoplankton is the basis of the food web, any shift in their community will result in the reduction of energy transfer efficiency in the food chain, which consequently weakens the functioning of an aquatic ecosystem (Acevedo-Trejos et al., 2015; Li et al., 2016). Climate change can also indirectly affect phytoplankton through the increase of grazing pressure by heterotrophs, which strongly disturb the community composition, diversity, and temporal dynamics of phytoplankton (Winder and Sommer, 2012).

4.7.2 EFFECT OF CARBON DIOXIDE ON MICROALGAE

Acidification of oceans due to climate change can alter the photosynthesis and photosynthetic growth of organisms (Wu et al., 2008; Hader and Gao, 2017). Several studies have reported such effects from increased CO_2 concentrations, but the findings are variable (Coad et al., 2016). Owing to differing requirements by different species for inorganic carbon due to varying activity of carbon concentrating mechanisms (CCMs) in their cells (Giordano et al., 2005) and which can be seen as a different response to photosynthesis or growth (Clark and Flynn, 2000; Beardall et al., 2009). In diatoms, for example, there have been documented positive, negative, and neutral OA effects (Gao and Campbell, 2014; Li et al., 2016) that may be caused by the difference between species or phenotypes (Gao et al., 2019). The different responses by photosynthetic organisms to CO_2 (Hu et al., 2018) will lead to changes in the species composition (Beardall and Raven, 2004; Raven et al., 2005b; Beardall and Stojkovic, 2006; Beardall et al., 2009).

4.7.3   EFFECT OF LIGHT INTENSITY ON MICROALGAE

In euphotic areas, microalgae are prone to significant light shifts, and is further aggravated by the sea surface stratification caused by climate change. The disturbance of light intensity may be harmful to photosynthesis, thereby affecting microalgal communities (Yong et al., 2016). Studies have found that microalgae change their chlorophyll content, photosystem ratios, photosystem antenna dimensions, biomolecular ratios, and nutrient uptake as the light intensity rises (Norici et al., 2011; Ma et al., 2015; Meneghesso et al., 2016). However, Increased UVB exposure is also found to induce the formation of cyclobutane pyrimidine dimer (CPDs), which inhibits the new synthesis of various cellular components, including those which are crucial for production, maintenance, and growth (Beardall and Raven, 2004). Moreover, the weakening of the thylakoid structure and loss of photosynthetic pigment induced (Hader et al., 2015; Hader and Gao, 2017) by UVB is another consequence. In cyanobacteria, for example, it causes filament fragmentation (Rastogi et al., 2014b) and degeneration of the thylakoid light-harvesting antenna structure, known as phycobilisomes (Hader and Gao, 2015; Hader and Gao, 2017).

4.7.4   EFFECT OF MULTIPLE ENVIRONMENTAL FACTORS ON MICROALGAE

A combination of numerous stressors, such as temperature, salinity stress, UV radiation, and nutrient limitation also affects the physiology and metabolism of microalgae. Various studies have reported the combined effects of different stressors to comprehensively envisage the impact of climate change on microalgae (Yong et al., 2016). They examined the effects of temperature, supply of nutrients, UV radiations, and CO_2 by Beardall et al. (2014), and indicated that UV-B is a significant stressor affecting the influence of other environmental factors on marine phytoplankton. Irradiance, temperature, and photoperiod effects on microalgae have been identified as species-dependent (Singh and Singh, 2014; Yong et al., 2016) factors. However, it remains unclear if the processes and variables involved are interdependent.

4.8 THERMAL ADAPTATION OF MICROBES TO ELEVATED TEMPERATURE

Environmental factors determine the physiology and evolutionary adaptations of life forms. Among environmental factors, the temperature is the only factor that passes through the physical barrier due to which organisms are unable to screen themselves from temperature variation. However, they do so in the case of extreme pH or salinity conditions where biological membranes maintain a steep concentration gradient. Instead, each biomolecule in microbes must be adapted to the temperature they grow (Enggvist, 2018). Different microbial adaptations studies concentrate on thermophilic genomes and biomolecules (Wang et al., 2015; Stetter, 1996; Enggvist, 2018). Various microbes perceive a shift in environmental temperatures initially through the cellular membrane as variation in environmental temperature has been shown to change both the fatty acid composition and the unsaturation level of membrane lipids in multiple species of bacteria and archaea (Matsuno et al., 2009; Mykytczuk et al., 2010; Paulucci et al., 2011). Some microbes thrive at high-temperature ranges on account of the thermoadaptation molecular mechanism, which helps to withstand the thermal inactivity of the cell components (Pandey et al., 2016). They include hyperthermophiles that grow up to 105°C and thermophiles, which grow upto 50°C and 60°C, indicating that they differ in their optimum temperature range but possess similar adaptations to withstand harsh conditions (Reed et al., 2013). Thermophiles are restricted mostly to the archaeal domain, and among bacteria, they are present in two orders, namely, Thermotogales and Aquificales (Egorova and Antranikian, 2005; Hartman et al., 2019). They occupy different ecological niches like deep-sea hydrothermal vents, terrestrial hot springs, and other extreme geographical/geological sites like volcanic sites, tectonically active faults along with decomposing matters, such as the compost and deep organic landfills (Panda et al., 2019). Environmental disturbance, like the variation in temperature, gives rise to the evolution of the genome, which develops the potential of tolerance toward higher temperatures as a result of which bacteria can survive under elevated temperatures. The bacterial thermal adaptation is achieved through horizontal gene transfer (HGT), gene loss, or gene mutations (Averhoff and Muller, 2010). For example, in the *Thermotoga maritima* and *Aquifex aeolicus*, hyperthermophilic properties were achieved through HGT from Archaea (Aravind et al., 1998; Nelson et al., 1999; Wang et al., 2015; Forterre et al., 2000; Brochier-Armanet and Forterre, 2006; Pollo et al., 2015). Also, the biomolecules, such as proteins and nucleic acids provide the remarkable capacity to hyperthermophiles to survive in such extreme environments (Pandey et al., 2016). Membrane lipids contain ether linkage; that is further branched, saturated, and are of high molecular weight, which results in the high melting temperature of membrane lipids (Koga et al., 1993; Solanki and Gupta, 2013). Diverse species of microbes, as well as bacteria, have developed molecular strategies to perceive shifts in temperature. Almost all molecules of the cell, such as proteins, lipids, RNA, and DNA, can act as thermosensors that can sense variation in environmental temperature and initiate suitable response pathways (Shapiro and Cowen, 2012). High temperatures intensify deviation in protein conformation that reveals hydrophobic residues and regions usually concealed within their native structure. Consequently, it leads to nonfunctional intermolecular interactions, which form a larger insoluble structure called protein aggregates (Dobson, 2004; Govers et al., 2018). Thermophilic proteins maintain their structure and function at high temperatures through many hydrophobic residues, disulphide bonds, and ionic interactions (Reed et al., 2013). Elevated temperature leads to enzyme denaturation; hence they cease to function, which impedes the metabolism. Besides, it remarkably enhances membrane fluidity, which causes cell disruption. Their membrane lipid for stopping denaturation is more saturated than mesophilic. It thus gives them the required amount of fluidity that is important to the membrane work, as a consequence of which heat-filling thermophiles thrive at high temperatures (Rampelotto, 2010) largely between 15°C and 40°C (Ulrih et al., 2009; Rampelotto, 2010). Another vital thermoadaptive mechanism is achieved through reverse gyrase enzymes that create supercoiled DNA, which results in the higher melting point of DNA (Imanaka, 2011; Hartman et al., 2019). Also, thermophilic methanotrophs adapt to increased temperatures through several ways such as the composition of their fatty acids is unique since it is dominated by saturated (C16: 0), methylated fatty acids (C9-ome-16:0), and their cyclic derivatives (C17sus), which aid to stabilize the structure of cell membrane and therefore enhances their fluidity. Another thermal adaptation in representatives of the genus *Methylocaldum* is the formation of melanin (Medvedkova et al., 2008). At higher temperatures, toxic aromatic intermediates are formed, which are eliminated by polymerization to form melanin (Tichonova and Kravchenko, 2019). Increased temperature initiates cellular heat shock reaction that leads to the formation of heat shock proteins, which maintain homeostasis of protein and assures cell survival (Lindquist, 1986; Wilkening et al., 2018). Thermostable enzymes that prevent denaturation are vital for thermophiles and hyperthermophiles. Furthermore, these microbes have chaperone proteins, which play a protective role in protein folding and preserve their native form (Keenleyside, 2019). Besides, chaperone proteins can act as temperature sensors in different microbes, for example, as part of an oligomer complex in *Saccharomyces cerevisiae*, the small heat shock protein Hsp26 changes from low to high-affinity chaperone state with an increasing temperature (Franzmann et al., 2008). Soil organisms can also tolerate differences in the temperature or adjust to changes in temperature range, in which case physiology and function of these species can alter, and aggregation can result in these evolutionary responses impacting the size of the ecosystem (de Angelis et al., 2019).

4.9 THE RESPONSE OF MICROBIAL COMMUNITIES TO CLIMATE CHANGE AND DRYING STRESS

Microbes have been changing the climate and vice-versa since Earth's beginning. Researchers can examine the impacts of environmental variation on microorganisms to mutually comprehend and ideally envision the future effects of environmental change on all life forms. Atmospheric models venture that precipitation forms will probably heighten, later on, bringing about the expanded span of dry seasons and developed recurrence of enormous soil, rewetting occasions, which are unpleasant to the microorganisms that drive soil biogeochemical cycling (Evans et al., 2012). Microbial groups synchronize numerous subterranean carbon cycling forms; in this way, the effect of environmental change on the structure and capacity of soil microbial networks could, one by one, impact the discharge or ability of carbon in soils (Cregger et al., 2012). Climate change has already become more and more commonly established, particularly in the Mediterranean region (Gibelin and Deque, 2003; IPCC, 2014a, 2014b; Sawain et al., 2014). It is expected that the frequency and intensity of these scandalous climate events will keep on increasing (Planton et al., 2008; IPCC, 2012, 2013). For more than 12,000 years, the global climate has remained stable, and such firmness is vital to human existence (NASA,

2015a). The soil system is a massive reservoir of complex carbon and a powerful driver of climate change trends (Raich and Potter, 1995). Roughly 60% of the world's methane production is consumed by microorganisms (Zimmerman and Labonte, 2015). Although frozen grounds have slowly transformed into wetlands, the former have become a favorite route (Arizona University, 2014). However, microorganisms can also improve optimistic climate change inputs. It is evident from the fact that the dim surface of marine phytoplankton assimilates more excellent sunlight rates when Arctic Ocean ice is liquefied, which can intensify water heating up to 20% faster than the expected climate models. Under more exceedingly awful situations, an extra 10% of ocean ice could vanish, and there could be around 50% without ice days during summers (Piotrowski, 2015).

Physical soil characteristics like porosity, weather tolerance, and capacity to retain water can change at extreme climate occasions, thereby changing microbe physical habitats. Such physical procedures may be critical in driving a drought microbial response (Navarro-Garcia et al., 2012). Shrinking and slacking and rising are the major complete annihilation processes that occur when dry soil is rewetted. At the same time, organic matter settles and aggregates against those processes by improving coherence and decreasing the rate of weathering (Sullivan, 1990; Chenu et al., 2000).

Microbial processes have a significant role in the global flow of the most critical biogenic greenhouse gasses (carbon dioxide, methane, or nitrous). They would likely respond rapidly to changes in the climate. It is necessary to understand the instruments used by microbes to control terrestrial greenhouse gas flows to enhance the atmospheric models (Singh et al., 2010). These soil microbial networks are unpredictably associated with the activity of the environment, as they play a significant role in carbon and nitrogen cycling (Vries et al., 2013). Worldwide change is testing plant and animal populations with novel ecological conditions and increased biometrical carbon dioxide fixations, warmer temperatures, and modified precipitation systems. Sometimes contemporary or quick development can enhance the impacts of worldwide change (Lau et al., 2012). Future climates will probably be portrayed by increasingly variable precipitation, both dry spell term and precipitation occasions and are relied on to increment (Huntington et al., 2006). Like nutrient accessibility, moisture stress has likewise been appeared to drive trade-offs among gatherings of characteristics in bacterial segregates (Lennon et al., 2012). The planning and magnitude of precipitation occasions are relied upon to change in future decades, bringing about to linger dry seasons and more significant precipitation occasions. Even though precipitation shifts are affected by microbial community composition and efficiency, it is uncertain if Texa receives techniques that boost under new regimes (Evans et al., 2014).

4.10 CHANGE IN SOIL NEMATODE COMMUNITY STRUCTURE FOLLOWING CLIMATE CHANGE

Nematodes are multicellular, vermiform, invertebrate animal species, mostly microscopic with worldwide distribution that ranges from terrestrial to aquatic habitats and include both free-living and parasitic forms (Hailu and Hailu, 2019). Despite their global presence their tiny size obscures their significance. Nonparasitic and beneficial nematodes consume fungi, bacteria, and other microscopic animals (Swart et al., 2017). In soil, nematodes constitute various trophic and ecological groups, such as primary feeders (herbivores), secondary consumers (bacterivorous and fungivores), and tertiary scavengers (carnivores, omnivores, and predatory nematodes), which play a crucial part in soil food webs (Bonger and Ferris, 1999; Thakur et al., 2014; Yogesh et al., 2020) and are directly associated with essential ecosystem functions like primary production, primary and secondary consumption, and decomposition, etc. Soil nematodes play a vital role in soil nutrient cycling and promote N mineralization (Yadav et al., 2018). On account of such functions, these communities act as suitable model systems for the analysis of climate change impact on belowground productivity (Singh and Prasad, 2016). However, the study of climate change effect on soil nematodes is restricted, with much work directed toward the impact of a single atmospheric component in a controlled environment (Singh and Prasad, 2016). Moreover, plant-parasitic nematodes have gained much attention in most research work, but nematodes that are useful for the soil environment have remained less focused (Ugarte and Zaborski, 2020). Landesman et al. (2010) demonstrated the highly sensitive existence of soil nematodes in annual precipitation alterations by increased precipitation has induced a rise in the total number of soil nematodes during their 1-year plumage manipulation experiment on sandy forest soils in New Jersey pinelands (USA). Also, sensitivity to drought was reported in nematode families. However, the effect was most significant on Plectidae, while Cephalobidae and Qudsianematidae had no significant impact, which might be due to the relative anhydrobiotic potential of these families. Also, bacteria-feeding nematodes, as compared with all other trophic groups, are highly susceptible to droughts. It shows the potential for a change in the rates of N mineralization by increasing bacterivorous grazing due to altered precipitation patterns (Ingham et al., 1985: Hunt et al., 1987; Landesman et al., 2010). Increased rainfall causes the dispersal of nematodes in soil and also increases the rates of primary production, which consequently enhances the accessibility of nutrients for nematodes (Munteanu et al., 2017). A soil moisture film significantly affects nematode activity (Wallace, 1973). Hence soil moisture, relative humidity, and related environmental factors have a direct impact on nematode survival. Some nematodes exhibit a swift response to an unfavourable environment through some behavioral and physiological responses. For example, *Steinernema carpocapsae* switches between aerobic and anaerobic metabolism to adjust to shifts in the levels of O_2 in soil (Shih et al., 1996). The immediate response of soil nematodes induced by environmental extremes includes aggregation that likely protects them against desiccation (Cooper et al., 1971; Mc Sorley, 2003) and developmental dormancy and slide capacity (Dauer stage) to survive under unfavourable conditions as well as in prolonged food scarcity (McSorley, 2003; Schratzberger et al., 2019). Extreme precipitation regimes from year to year alter nematode community trophic structure that tips the predator–prey balance (Todd et al., 1999; Stevnbak et al., 2012; Ruan et al., 2012; Franco et al., 2019). Although the response of nematodes to temporal variation in precipitation may differ across different ecosystems as nematode communities are resistant to water shortage in arid grasslands (Freckman et al., 1987; Moorhead, 1987; Vandegehuchte et al., 2015), which is not the case in mesic greens (Sylvain et al., 2014; Franco et al., 2019). The study of the effect of temperature and soil moisture change on the nematode fauna in a semiarid grassland indicates that nematode population density is affected by both temperature and soil moisture content (Bakonyi and Nagy, 2000; Munteanu, 2017).

It is reported that under climate change scenarios, changing precipitation may occur synchronously with atmospheric N deposition (Song et al., 2016) and may have an interactive effect on terrestrial ecosystems (Liu et al., 2020). The study of interactive effects of water and nitrogen addition on soil nematode communities suggests that water incorporation causes a remarkable increase in nematode abundance and generic richness. The relatively lot of different trophic groups remain unaffected. However, nitrogen addition had no impact on the number of soil nematodes, although it caused a decrease in their generic richness and caused considerable changes in nematode community composition (Song et al., 2016). They reported that high precipitation adversely affected soil nematode community, whereas N deposition had a positive effect on soil nematode communities.

Furthermore, increased precipitation counteracts the positive effects of N deposition on the nematode community when both co-occur in temperate forest ecosystems (Liu et al., 2020). Soil nematode community has been proven to be affected by temperature (Simmons et al., 2009; Li et al., 2013; Dong et al., 2013; Ruess et al., 1999; Yan et al., 2017), yet limited studies account for the impact of warming on soil nematode communities with different results. For instance, experimental warming of soils (+1.7°C and +3.4°C) did not affect individual nematode trophic groups in a temperate-boreal forest ecotone. However, there was an increase in the ratio between microbial-feeding and plant-feeding nematodes (Thakur et al., 2014), whereas all nematodes together with root herbivores decreased with 1°C rise in temperature (Stevnbak et al., 2012; Johnson and Jones, 2017). The response of nematodes to high CO_2 levels is complex, depending on trophic groups: no impact on prairie soil nematodes has been recorded (Freckman et al., 1987), but cotton rhizosphere has dwindled (Runity et al., 1994), or forests (Hoeksema et al., 2000), grasslands (Hungate et al., 2000) and pasture soils (Yeates and Orchard, 1993; Yeates et al., 1997, 2003; Colagiero and Ciancio, 2011) have increased. Elevated CO_2 levels reduced the numbers of bacterial feeders, raised fungal feeders, and predators in the forest (Neher et al., 2004) or prairie soils (Yeates et al., 2003; Colagiero and Ciancio, 2011). Increasing temperatures and a rise in CO_2 levels might have a direct impact on plant-parasitic nematodes by interference with their developmental rate and subsistence strategies, and an indirect effect by modification of host plant physiology (Somasekhar and Prasad, 2012).

4.11 CLIMATE CHANGE IMPACTS ON THE FUNCTIONAL ASPECT OF MICROBES

4.11.1 CLIMATE CHANGE IMPACTS ON THE PRACTICAL ASPECTS OF TERRESTRIAL MICROBES

Climate change is a hot topic across the world, and it has become a significant problem in the world. The concentrations of greenhouse gases are continually increasing due to different human and natural factors, which lead to adverse effects on microbial structure, function, and metabolic activities (Abatenh et al., 2018). Various biotic and abiotic factors, such as root exudates, nutrient competition, changing climate, and edaphic factors have an intense impact on the functional diversity of many terrestrial communities (Fierer et al., 2003; Pennanen et al., 1999; Rousk et al., 2010; Uroz et al., 2012; Wei et al., 2011). Climate change results in changes in the diversity and processes of soil microbial communities (Shade et al., 2012). As different microbial groups perform different ecosystem functions like nitrogen fixation, nitrification, denitrification and methanogenesis. Shifts in microbial community composition lead to the change in the ecosystem functions (Isobel et al., 2011; Bakjen et al., 2012; Salles et al., 2012; Bodelier et al., 2001). With the increase in temperature, terrestrial microbial processing, turnover, and activity rate increase (Castro et al., 2010). Thus, it becomes clear that global heating plays a significant role in changing the relative abundance, together with functional aspects of soil microbial communities, like the rate of growth, temperature sensitivity, and physiology (Classen et al., 2015). Climate change improves the soil microbial activity, thus increases the soil respiration (Rustad et al., 2001; Wu et al., 2011), which corresponds to temperatures, but it is, in turn, repressed in high or low humidity (Luo and Zhou, 2006). Enzyme activity is another significant aspect observed in this regard, as the enzyme production rates get affected by variations in moisture and temperature, thereby effecting enzyme efficiency, substrate availability, and microbial efficiency (Allison and Vitousek, 2005). However, carbon-degrading enzymes in terrestrial microbes are more sensitive than nitrogen degrading enzymes (Wallenstein et al., 2009; Stone et al., 2012).

Also, due to increasing temperatures, carbon efflux in the soil gets influenced due to changes in the physiology of decomposers (Schindlbacher et al., 2011). Growing temperatures would likely lead to fungal breakdown leading to increased carbon dioxide release from the ground. Although high temperatures raise the level of nitrogen in soils, yet it affects the microbial activity and diversity and also subdues the rate of fungal decomposition (American Society for Microbiology, 2008). Whereas, in terrestrial ecosystems, bacterial biochemical reactions also take place less efficiently as a result of a warmer climate. Climate change precipitation shifts are also significant in that soil moisture content variations to determine the structure and extent of decomposition of the terrestrial microbial populations (Fierer and Schimel, 2003; Singh et al., 2010). In terrestrial fungal communities, even minimal amounts of change may change the dominance of organisms, but terrestrial bacterial communities remain uninfected (Classen et al., 2015).

Extreme drought reduces terrestrial microbial activity, resulting in a soil carbon loss that is linked to decreased enzyme activity leading to organic material decomposition (Nardo et al., 2011). Increased deposits of carbon or nitrogen also impacts the soil processes through the interaction of soil-food networks and protist populations (Lindberg and Bengtslon, 2005; Lec and Boeck, 2013; Perdue and Crossley, 1989; Kardol et al., 2010). Substantial attention is also paid to the effect of high temperatures on the metabolism of microbial species. Several studies showed that microbial interactions and modification intercede overtime for the direct effects of temperature on the physiology of microbial communities (Bradford, 2013; Frey et al., 2013; Hagerty et al., 2014; Karhu et al., 2014).

4.11.2   CLIMATE CHANGE IMPACTS ON THE FUNCTIONAL ASPECT OF AQUATIC MICROBES

Global changes to the ecosystems would affect not only animals and plants but also microscopic organisms. Kathleen Treseder of the University of California, Irvine, at the 108th General Meeting of the American Society for Microbiology, said microorganisms have different core capabilities for ecosystems around the world, and we are only now starting to realize the effects the worldwide change is having on them. Such microorganisms lose their habitat as global temperatures increase, and glaciers recede. They will probably go extinct until we research them and gain a deeper understanding of their achievements, Schmidt says. It must be expressed here that ozone-depleting substances like carbon dioxide, methane, and nitrogen dioxide dominatingly start from microorganisms (Singh et al., 2010). The microbial world is a vital carbon and other biogeochemical process resources that need consideration in the field of environmental change (Walsh, 2015).

No less important are aquatic ecosystems to the question of environmental change. Seas are called the biological pump and go about as a CO_2 sink (Glöckner et al., 2012). Over the decades, the upper marine layer has accumulated CO_2 and other ozone harming gases, while the deep-sea is the largest carbon store facility in the world (Sarmiento and Gruber, 2002). Approximately 6×10^{12} kg of CO_2 are treated by marine species every year in the oceanic biological system (Field et al., 1998). Since their emergence in the ocean, more than 3.5 billion years ago and after their growth, microorganisms have produced and expelled ozone-depleting substances (Zimmer, 2010). Approximately, 93% of carbon dioxide is absorbed in the planet by the oceans, which roughly ranges about 9×10^{13} kg (90 billion tons), while around 6×10^{12} kg (6 billion tons) of carbon dioxide is anthropogenic (Stewart, 2003). The dynamics of the oceanic carbon cycle is mostly dominated by smaller nanoplanktons, picoplankton, and archaea (Stewart, 2003). Also, these systems, through improvements in the processing and breathing of natural material, affect global geochemistry, a major environmental factor in fundamental ways (Suttle, 2007).

KEYWORDS

- **environmental microbiomes**
- **communities**
- **environmental factors**
- **anthropogenic activities**
- **geographical ranges**

REFERENCES

Abastenh, E.; Gizaw, B.; Tsegaye, Z.; Jefera, G.Microbial Function on Climate Change. *Environ. Pollut. Clim. Change* **2018,** *2* (1), 1000147.

Adam, P.; Piotrowski, A.; Maciej, J.; Napiorkowski.; B.; Jaroslaw, J. Napiorkowski Marzena Osuch.; Comparing Various Artificial Neural Network Types for Water Temperature Prediction in Rivers. *J. Hydrol.* **2015,** *529,* 302–315.

Adl, S. Rhizosphere, Food Security, and Climate Change: A Critical Role for Plant-Soil Research. *Rhizosphere* **2016,** *1,* 1-3.

Ahkami, A. H.; White III, R.; Handakumbura, P.; Jansson, C. Rhizosphere Engineering: Enhancing Sustainable Plant Ecosystem Productivity. *Rhizosphere* **2017,** *3* (2), 233–243.

Eissaa, A. E.; Zaki, M. M. The Impact of Global Climate Change on Aquatic Environment. *Procedia Environ. Sci.* **2007 (2011)** *24,* 251–259.

Allison, S. D.; Vitousek, P. M. Responses of Extracellular Enzymes to Simple and Complex Nutrient Inputs. *Soil Biol. Biochem.* **2005,** *37,* 937–944.

Amthor, J. S. Effects of Atmospheric CO_2 Concentration on Wheat Yield: Review of Results from Experiments Using Various Approaches to Control CO_2 Concentration. *Field Crops Res.* **2001,** *73,* 1–34.

Anderson, I. C.; Drigo, B.; Keniry, K.; Ghannoum, O.; Chambers, S. M.; Tissue, D. T.; Cairney, J. W. G. Interactive Effects of Preindustrial, Current and Future Atmospheric CO2 Concentrations and Temperature on Soil Fungi Associated with Two Eucalyptus Species. *FEMS Microbiol. Ecol.* **2013,** *83,* 425–437.

Aravind, L.; Tatusov, R. L.; Wolf, Y. I.; Walker, D. R.; Koonin, E. V. Evidence for Massive Gene Exchange between Archaeal and Bacterial Hyperthermophiles. *Trends Genet.* **1998,** *14,* 442–444.

Arnone, J. A., et al. Dynamics of Root Systems in Native Grasslands: Effects of Elevated Atmospheric CO2. *New Phytol.* **2000,** *147,* 73–86.

Averhoff, B.; Muller, V. Exploring Research Frontiers in Microbiology: Recent Advances in Halophilic and Thermophilic Extremophiles. *Res. Microbiol.* **2010,** *161,* 506–514.

Azam, F.; Malfatti, F. Microbial Structuring of Marine Ecosystems. *Nat. Rev. Microbiol.* **2007,** *5,* 782–791.

Bakonyi, G.; Nagy, P. Temperature- and Moisture-Induced Changes in the Structure of the Nematode Fauna of a Semiarid Grassland, Patterns and Mechanisms. *Global Change Biol.* **2000,** *6,* 697–707.

Balser, T. C.; Kinzig, A.; Firestone, M. K. Linking Soil Microbial Communities and Ecosystem Functioning. In *The Functional Consequences of Biodiversity*: *Empirical Progress and Theoretical Extensions*; Kinzig, A.; Pacala S.; Tilman D., Eds.; Princeton University Press; Princeton, NJ, 2001; pp 265–294.

Balser, T.C.; Gutknecht, J.L.M.; Liang, C. How Will Climate Change Impact Soil Microbial Communities? In *Soil Microbiology and Sustainable Crop Production*; Dixon, G.; Tilston, E., Eds.; Springer: Dordrecht, 2010.

Barcelos e Ramos, J.; Biswas, H.; Schulz, K.G.; LaRoche, J.; Riebesell, U. Effect of Rising Atmospheric Carbon Dioxide on the Marine Nitrogen Fixer Trichodesmium. *Global Biogeochem. Cy*. **2007,** *21* (2). http://dx.doi.org/10.1029/2006GB002898

Bardgett, R. D. Microbial Contributions to Climate Change through Climate Change through Carbon Cycle Feedbacks. *ISME* **2008,** *2,* 805–814.

Bardgett, R. D.; Manning, P.; Franciska, E. M.; De Vries, T. Hierarchical Responses of Plant–Soil Interactions to Climate Change: Consequences for the Global Carbon Cycle. *J. Ecol.* **2013,** *101* (2), 334–343.

Bardgett, R. D.; Streeter, T. C.; Bradford, S. et al. Soil Microbes Complete Effectively with Plants for Organic Nitrogen Inputs to Temperate Grasslands. *Ecology* **2013,** *47,* 171–175.

Barnosky, A. D. et al. Has the Earth's Sixth Mass Extinction Already Arrived? *Nature* **2011,** *471,* 51–57.

Basu, S.; Mackey, K. R. M. Phytoplankton as Key Mediators of the Biological Carbon Pump: Their Responses to a Changing Climate. *Sustainability* (*Switzerland*). **2018,** *10.* doi: 10.3390/su10030869

Beardall, J.; Raven, J. A. The Potential Effects of Global Climate Change on Microalgal Photosynthesis Growth and Ecology. *Phycologia* **2004,** *43* (1), 26–40.

Beardall, J.; Stojkovic, S.; Larsen, S. Living in a High CO2 World: Impacts of Global Climate Change on Marine Phytoplankton. *Plant Ecol. Divers.* **2009,** *2,* 191–205. [CrossRef].

Benateau, S.; Gaurdard, A.; Stamm, C. Climate Change and Freshwater Ecosystem. *Federal Office Environ. Hydrol.* **2019,** *54,* 1–110.

Sonja, B.; Bernd, W.; Duanne, A. W., Martin, M. Geochemistry on Sediment Profile C01014. Pangea. *Data Pub. Earth Environ. Sci.* **2010,** *74,* 23–25.

Blankinship, J.; Niklaus, P.; Hungate, B. A Meta-Analysis of Responses of Soil Biota to Global Change. *Oecologia* **2011,** *165,* 553–565.

Bonger, T.; Ferris, H. Nematode Community Structure as a Bioindicator in Environmental Monitoring. *Trends Ecol. Evol.* **1999,** *14* (6), 224–228.

Bourne, D. G.; Morrow, K. M.; Webster, N. S. Insights into the Coral Microbiome: Underpinning the Health and Resilience of Reef Ecosystems. *Annu. Rev. Microbiol.* **2016,** *70,* 317–340.

Boyd, P. W. Framing Biological Responses to a Changing Ocean. *Nat. Clim. Change* **2013,** *3,* 530–533.

Bradford, M. A.; Davis, S.; Frey, R.; Maddox, J. Thermal Adaptation of Soil Respiration to Elevated Temperatures. *Ecol. Lett.* **2008,** *11,* 1316–1327.

Bradford, M. A.; Frey, S. D.; Mellillo, J. M et al. Thermal Adaptations of Soil Microbial Respiration to Elevated Temperatures. *Ecol Lett.* **2013,** *11,* 1316–1327.

Brennan, G.; Collins, S. Growth Responses of a Green Alga to Multiple Environmental Drivers. *Nat. Clim. Change* **2015,** *5,* 892–897.

Briones, M. J. I.; McNamara, N. P.; Poskitt, J.; Crow, S. E.; Ostle, N. J. Interactive Biotic and Abiotic Regulators of Soil Carbon Cycling: Evidence from Controlled Climate Experiments on Peatland and Boreal Soils. *Global Change Biol.* **2014,** *20,* 2971–2982.

Brochier-Armanet, C.; Forterre, P. Widespread Distribution of Archaeal Reverse Gyrase in Thermophilic Bacteria Suggests a Complex History of Vertical Inheritance and Lateral Gene Transfers. *Archaea* **2006,** *2* (2), 83–93.

Brutemark, A.; Engström-Öst, J.; Vehmaa, A,; Gorokhova, E. Growth, Toxicity and Oxidative Stress of a Cultured Cyanobacterium (*Dolichospermum* sp.) under Different CO_2/pH and Temperature Conditions: Environmental Stress for Cyanobacteria. *Phycol. Res.* **2015,** *63,* 56–63. doi: 10.1111/pre.12075

Bunse, C., et al. Response of Marine Bacterioplankton pH Homeostasis Gene Expression to Elevated CO_2. *Nat. Clim. Change* **2016,** *5,* 483–491.

Burford, M. A.; Carey, C. C.; Hamilton, D. P.; Huisman, J.; Paerl, H. W.; Wood, S. A.; Wulff, A. Perspective: Advancing the Research Agenda for Improving Understanding of Cyanobacteria in a Future of Global Change. *Harmful Algae* **2019,** *91,* Article 101601.

Burns, R. A.; MacDonald, C. D.; McGinn, P. J.; Campbell, D. Inorganic Carbon Repletion Disrupts Photosynthetic Acclimation Tolow Temperature in the Cyanobacterium Synechococcus Elongates S1. *J. Phycol.* **2005,** *41* (2), 322–334.

Castro, H. F.; Classen, A. T.; Austin, E. E.; Norby, R. J.; Schadt, C. W. Soil Microbial Community Responses to Multiple Experimental Climate Change Drives. *Appl. Environ. Microbiol.* **2010,** *76,* 999–1007.

Cavicchioli, R.; Ripple, W. J.; Timmis, K. N. et al. Scientists' Warning to Humanity: Microorganisms and Climate Change. *Nat. Rev. Microbiol.* **2019,** *17,* 569–586. https://doi.org/10.1038/s41579-019-0222-5

Cavvicchioli, R.; Ripple, W.; Webster, S. Scientists Warning to Humanity Microorganisms and Climate Change. *Nat. Rev. Microbiol.* **2019,** *17,* 569–586.

Chenu, C., Bissonnais, Y. Le., Arrouays, D. Organic Matter Influence on Clay Wettability and Soil Aggregate Stability. *Soc. Sci. Soc. Am. J.* **2000,** *64* (4), 1479–1486.

Clarke, A.; Fraser, K. P. P. Why Does Metabolism Scale with Temperature. *Funct. Ecol* **2004,** *18*, 243–251.

Classen, A. T.; Sundquist, M. K.; Hemming, T. A., et al. Direct and Indirect Effects of Climate Change on Soil Microbial and Soil Microbial Plant Interaction. *Ecosphere* **2015a,** *6* (8), 1–21.

Classen, A. T.; Sundqvist, M. K.; Henning, J. A.; Newman, G. S.; Moore, J. A. M.; Cregger, M. A.; Moorhead, L. C.; Patterson, C. M. Direct and Indirect Effects of Climate Change on Soil Microbial and Soil Microbial-Plant Interactions: What Lies Ahead? *Ecosphere* **2015b,** *6* (8), 130. http://dx.doi.org/10.1890/ES15-00217.1

Classen, A. T.; Sundqvist, M. K.; Henning, J. A.; Newman, G. S.; Moore, J. A. M.; Cregger, M. A., et al. Direct and Indirect Effects of Climate Change on Soil Microbial and Soil Microbial-Plant Interactions: What Lies Ahead? *Ecosphere* **2015c,** *6* (8), art130. Retrieved from https://doi.org/10.1890/ES15-00217.1. doi:10.1890/ES15-00217.1.

Classen, A. T.; Sundqvist, M. K.; Henning, J. A.; Newman, G. S.; Moore, J. A. M.; Cregger, M.A.; Moorhead, L.C.; Patterson, C. M. Direct and Indirect Effects of Climate Change on Soil Microbial and Soil Microbial-Plant Interactions: What Lies Ahead? *Ecosphere* **2015d,** *6* (8), [130].

Classen, A.; Sundquist, K. M.; Hemming, J. A., et al. Direct and Indirect Effects of Climate Change in Soil Microbial and Soil Community Plant Interactions. *Ecosphere* **2015,** *6* (0), 1–21.

Coad, T.; McMinn, A.; Nomura, D.; Martin, A. Effect of Elevated CO2 Concentration on Microalgal Communities in Antarctic Pack Ice. *Deep Sea Res. II*: *Topical Stud. Oceanogr.* **2016,** *131*. doi: 10.1016/j.dsr2.2016.01.005

Colagiero, M.; Ciancio, dr A. Climate Changes and Nematodes: Expected Effects and Perspectives for Plant Protection. J. Zoo. Redia. **2011,** *94*, 113–118.

Colatriano, D.; Ramachandran, A.; Yergeau, E.; Maranger, R.; Gélinas, Y.; Walsh, D. A. Metaproteomics of Aquatic Microbial Communities in a Deep and Stratified Estuary. *Proteomics* **2015,** *15* (20), 3566–3579.

Compant, S.; Van Der Heijden, M. G. A.; Sessitsch, A. Climate Change Effects on Beneficial Plant–Microorganism Interactions. *FEMS Microbiol. Ecol.* **2010,** *73* (2), 197–214.

Cooper, A. F. J. R.; Van Gundy, S. D. Senescence, Quiescence, and Cryptobiosis. In *Plant Parasitic Nematodes*, Vol. II; Zuckerman, B. M.; Mai, W. F.; Rohde, R. A.; Eds.; Academic Press: New York, 1971; pp 297–318.

Couteaux, M. M.; Bolger, T. Interactions between Atmospheric CO2 Enrichment and Soil Fauna. *Plant Soil* **2000,** *224*, 123–134.

Cregger, M. A.; Schadt, C. W.; McDowell, N. G.; Pockman, W. T.; Classena, A.T.. Response of the Soil Microbial Community to Changes in Precipitation in a Semiarid Ecosystem. *Appl. Environ. Microbiol.* **2012,** *78*, 1–24.

Crist, E.; Mora, C.; Engelman, R. The Interaction of Human Population, Food Production, and Biodiversity Protection. *Science* **2017,** *356*, 260–264.

Czerny, J.; Barcelos e Ramos, J.; Riebesell, U. Influence of Elevated CO2 Concentrations on Cell Division and Nitrogen Fixation Rates in the Bloom-Forming Cyanobacterium Nodularia Spumigena. *Biogeosciences* **2009,** *6*, 1865–1875.

Danovaro, R.; Corinaldesi, C.; Dell'anno, A.; Fuhrman, J. A.; Middelburg, J. J.; Noble, R. T.; Suttle, C. A. Marine Viruses and Global Climate Change. *FEMS Microbiol. Rev.* **2011,** *35* (6), 993–1034. https://doi.org/10.1111/j.1574-6976.2010.00258.x

Das, S.; Mangwani, N. Ocean Acidification and Marine Microorganisms: Responses and Consequences. *Oceanologia* **2015,** *57,* 349–361. [CrossRef].

Davidson, E. A.; Janssens, I. A. Temperature Sensitivity of Soil Carbon Decomposition and Feedback to Climate Change. *Nature* **2006,** *440,* 165–173.

Davis, T. W.; Berry, D. L.; Boyer, G. L.; Gobler, C. J. The Effects of Temperature and Nutrients on the Growth and Dynamics of Toxic and Non-Toxic Strains of Microcystis during Cyanobacteria Blooms. *Harmful Algae* **2009,** *8,* 715–725. doi: 10.1016/j.hal.2009.02.004

De Angelis, K.; Chowdhury, P. R.; Pold, G.; Romero-Olivares, A.; Frey, S. Microbial Responses to Experimental Soil Warming: Five Testable Hypotheses. In *Ecosystem Consequences of Soil Warming*: *Microbes, Vegetation, Fauna and Soil Biogeochemistry*; Mohan, J. E., Ed.; Academic Press, 2019; pp 141–156.

de Vries, F. T.; Bardgett, R. D. Climate Change Effects on Soil Biota in the UK. Biodiversity Climate Change Impacts Report Card Technical Paper, 2015.

De Vries, F. T.; Shade, A. Controls on Soil Microbial Community Stability under Climate Change. *Front. Microbiol.* **2013,** *4,* 265.

Delgado-Baquerizo, M.; Maestre, F. T.; Escolar, C.; Gallardo, A.; Ochoa, V.; Gozalo, B.; Prado-Comesana, A. Direct and Indirect Impacts of Climate Change on Microbial and Biocrust Communities Alter the Resistance of the N Cycle in a Semiarid Grassland. *J. Ecol.* **2014,** *102,* 1592–1605.

Deslippe, J. R.; Hartmann, M.; Simard, S. W.; Mohn, W. W. Long-Term Warming Alters the Composition of Arctic Soil Microbial Communities. *FEMS Microbiol. Ecol.* **2012,** *82,* 303–315.

Dinasquet, J.; Riemam, L.; Lunna, S. S.; Anderson, A. Effect of Resource Availability on Bacterial Community Responses to Increased Temperature. *Aqua. Microbial Ecol.* **2012,** *68*: 131–142.

Dobson, C. M. Principles of Protein Folding, Misfolding and Aggregation. *Semin. Cell Dev. Biol.* **2004,** *15,* 3–16.

Doney, S. C.; Ruckelshaus, M.; Duffy, J. E.; Barry, J. P.; Chan, F.; English, C. A.; Galindo, H. M.; Grebmeier, J. M.; Hollowed, A. B.; Knowlton, N., et al. Climate Change Impacts on Marine Ecosystems. *Annu. Rev. Marine Sci.* **2012,** *4,* 11–37.

Dong, Z.; Hou, R.; Chen, Q.; Ouyang, Z.; Ge, F. Response of Soil Nematodes to Elevated Temperature in Conventional and No-Tillage Cropland Systems. *Plant Soil* **2013,** *373,* 907–918.

Dooley, S.; Treseder, K. The Effect of Fire on Microbial Biomass: A Meta-Analysis of Field Studies. *Biogeochemistry* **2012,** *109* (1–3), 49–61.

Dupre, J. Climate Change and Microbes: Influence in Numbers, 2008. http://environmentalresearchweb.org/cws/article/opinion/37020. Accessed 15 Dec 2015.

Dutta, et al. *The Microbe Factor and Its Role in Our Climate Future*; Springer Verlag: Berlin Heidelberg **2016,** *1* (4), 209–232.

Dutta, H.; Dutta, A. The Microbial Aspect of Climate Change. *Energ. Ecol. Environ.* **2016,** *1,* 209–232. https://doi.org/10.1007/s40974-016-0034-7.

Dutta, H.; Dutta, D. A. *The Microbial Aspect of Climate Change*; Springer Verlag: Berrlin Heidelberg **2016,** *1* (4), 209–232.

Dutta, H.; Hans, W. P.; Jurgen, M. *Temperature Effects on Marine Microorganisms*; University Park Press, 2010, pp 75–79.

Egorova K.; Antranikian, E. 2005. Industrial Relevance of Thermophilic Archaea. *Curr. Opin. Microbiol.* **2005**, *8*(6), 649–655.

Engqvist.; M.K.M. (2018). Correlating enzyme annotations with a large set of microbial growth temperatures reveals metabolic adaptations to growth at diverse temperatures. BMC Microbiol 18: 177: 2-14.

Evans, S. E.; Wallenstein, M. D. Climate Change Alters Ecological Strategies of Soil Bacteria. *Ecol. Lett.* **2013,** *17* (2), 155–164.

Evans, S. E.; Wallenstein, M. D. Soil Microbial Community Response to Drying and Rewetting Stress: Does Historical Precipitation Regime Matter? *Biogeochemistry* **2012,** *109*, 101–116.

Falkowski, P. G.; Barber, R. T.; Smetacek, V. Biogeochemical Controls and Feedbacks on Ocean Primary Production. *Science* **1998,** *281*, 200–206. doi: 10.1126/science.281. 5374.200; pmid: 9660741

Falkowski, P G.; Oliver, M. J. Mix and Match: How Climate Selects Phytoplankton. *Nat. Rev. Microbiol.* **2007,** *5*, 813–819.

Fierer, N.; Bradford, M. A.; Jackson, R. B. Toward an Ecological Classification of Soil Bacteria. *Ecology* **2007,** *88*, 1354–1364.

Fierer, N.; Lauber, C. L.; Ramirez, K. S.; Zaneveld, J.; Bradford, M. A.; Knight, R. Comparative Metagenomic, Phylogenetic and Physiological Analyses of Soil Microbial Communities across Nitrogen Gradients. *ISME J.* **2012,** *6* (5), 1007–1017.

Fierer, N.; Pennanen, T., et al. The Diversity and Biogeography of Soil Bacterial Communities. *Proc. Natl. Acad. Sci.* **2006,** *103*, 621–631.

Fierer, N.; Schimel, J. P. A. Proposed Mechanism for the Pulse in Carbon Dioxide Production Commonly Observed Following the Rapid Rewetting of a Dry Soil. *Soil Sci. Soc. Am. J.* **2003,** *67*, 798–805.

Fisher, C. R.; Demopoulos, A. W. J.; Cordes, E. E.; Baums, I. B.; White, H. K.; Bourque, J. R. Coral Communities as Indicators of Ecosystem-Level Impacts of the Deepwater Horizon Spill. *Bioscience* **2014,** *64* (9), 796–807.

Forterre, P.; Bouthier, De, La Tour, C.; Philippe, H.; Duguet, M. Reverse Gyrase from Hyperthermophiles: Probable Transfer of a Thermoadaptation Trait from Archaea to Bacteria. *Trends Genet.* **2000,** *16* (4), 152–154.

Franco, A. L. C.; Gherardi, L. A.; de Tomasel, C. M., et al. Drought Suppresses Soil Predators and Promotes Root Herbivores in Mesic, But Not in Xeric Grasslands. *Proc. Natl. Acad. Sci. USA.* **2019,** *116* (26), 12883–12888.

Franzmann, T. M.; Menhorn, P.; Walter, S.; Buchner, J. Activation of the Chaperone Hsp26 Is Controlled by the Rearrangement of Its Thermosensor Domain. *Mol. Cell* **2008,** *29*, 207–216.

Freckman, D. W.; Moore, J. C.; Hunt, H. W.; Elliot, E. T. The Effects of Elevated CO_2 and Climate Change on Soil Nematode Community Structure of Prairie Soil. *Bull. Ecol. Soc. Am.* **1991,** *72* (Suppl.), 119.

Freckman, D. W.; Whitford, W. G.; Steinberger, Y. Effect of Irrigation on Nematode Population Dynamics and Activity in Desert Soils. *Biol. Fertil. Soils* **1987,** *3–3*, 3–10.

Fu, F. X.; Warner, M. E.; Zhang, Y.; Feng, Y.; Hutchins, D. A. Effects of Increased Temperature and CO_2 on Photosynthesis, Growth, and Elemental Ratios in Marine

Synechococcus and Prochlorococcus (Cyanobacteria). *J. Phycol.* **2007,** *43* (3), 485–496.

Gaedke, U.; Ruhenstroth-bauer, M.; Wiegand, I.; Tirok, K.; Aberle, N.; Breithaupt, P., Sommer, U. Biotic Interactions May Overrule Direct Climate Effects on Spring Phytoplankton Dynamics. *Global Change Biol.* **2010,** *16,* 1122–1136. https://doi.org/10.1111/j.1365-2486.2009.02009.x

Gao, K., et al. Rising CO2 and Increased Light Exposure Synergistically Reduce Marine Primary Productivity. *Nat. Clim. Change* **2012,** *2,* 519–523.

Gao, K.; Beardall, J.; Hader, D-P.; Hall-Spencer, J. M.; Gao, G.; Hutchins, D. A. Effects of Ocean Acidification on Marine Photosynthetic Organisms under the Concurrent Influences of Warming, UV Radiation, and Deoxygenation. *Front. Marine Sci.* **2019,** *6* (JUN), [322]. https://doi.org/10.3389/fmars.2019.00322

Gao, K.; Campbell, D. Photophysiological Responses of Marine Diatoms to Elevated CO2 and Decreased pH: A Review. *Funct. Plant Biol.* **2014,** *41,* 449–459.

Gao, K.; Xu, J.; Gao, G.; Li, Y.; Hutchins, D. A.; Huang, B.; Wang, L.; Zheng, Y.; Jin, P.; Cai, X.; Häder, D-P.; Li, W.; Xu, K.; Liu, N.; Riebesell, U. Rising CO2 and Increased Light Exposure Synergistically Reduce Marine Primary Productivity. *Nat. Clim. Change* **2012,** *2,* 519–523. https://doi.org/10.1038/nclimate1507

Garciaab, F. N.; Casermeiroc, M. A.; Schimeld, J. P. When Structure Means Conservation: Effect of Aggregate Structure in Controlling Microbial Responses to Rewetting Events. *Soc. Biol. Biochem.* **2012,** *44* (1), 1–8.

Giordano, M.; Beardall, J.; Raven, J A. CO2 Concentrating Mechanisms in Algae: Mechanisms, Environmental Modulation and Evolution. *Annu. Rev. Plant Biol.* **2005,** *56,* 99–131.

Gkarmiri, K.; Mahmood, S.; Ekblad, A.; Alström, S.; Högberg, N.; Finlay, R. Identifying the Active Microbiome Associated with Roots and Rhizosphere Soil of Oilseed Rape. *Appl. Environ. Microbiol.* **2017,** *83* (22), e01938–17.

Glöckner, F. O.; Stal, L.; Sandaa, R. A.; Gasol, J. M.; O'Gara, F.; Hernandez, F.; Labrenz, M.; Stoica, E.; Varela, M.M.; Bordalo, A.; Pitta, P. Marine Microbial Diversity and Its Role in Ecosystem Functioning and Environmental Change. In *Marine Board Position Paper 17*; Calewaert, J. B.; McDonough, N., Eds.; Marine Board-ESF, Ostend: Belgium, 2012; pp 13–25.

Glokner, F. O.; Stal, L.; Sandaa, R. A., et al. *Marine Microbial Diversity and Its Role in Ecosystem Functioning and Environmental Change*; Board ESF: Ostend Belgium, 2010; pp 13–25.

Gobler, C. J.; Doherty, O. M.; Hattenrath-Lehmann, T. K.; Griffith, A. W.; Kang, Y.; Litaker, R. W. Ocean Warming since 1982 Has Expanded the Niche of Toxic Algal Blooms in the North Atlantic and North Pacific Oceans. *Proc. Natl. Acad. Sci.* 2017, *114,* 4975–4980. https://doi.org/10.1073/pnas.1619575114.

Gottfried, M.; Pauli, H.; Futschik, A.; Akhalkatsi, M.; Barančok, P.; Allonso, J. L. B.; Coldea, G.; Dick, J.; Erschbamer, B.; Kazakis, G. Continent-Wide Response of Mountain Vegetation to Climate Change. *Nat. Clim. Change* **2012,** *2,* 111–115.

Govers, S. K.; Mortier, J.; Adam, A.; Aertsen, A. Protein Aggregates Encode Epigenetic Memory of Stressful Encounters in Individual Escherichia Coli Cells. *PLoS Biol.* **2018,** *16* (8).

Griffith, A. W.; Gobler, C.J. Harmful Algal Blooms: A Climate Change Co-Stressor in Marine and Freshwater Ecosystems. *Harmful Algae* **2019,** *91,* 101590.

Grossart, H.P.; Allgaier, M.; Passow, U.; Riebesell, U. Testing the Effect of CO2 Concentration on the Dynamics of Marine Heterotrophic Bacterioplankton. *Limnol. Oceanogr.* **2006,** *51,* 1–11.

Grote, J.; Schott, T.; Bruckner, C. G.; Glöckner, F. O.; Jost, G.; Teeling, H.; Labrenz, M.; Jürgens, K. Genome and Physiology of a Model Epsilonproteobacterium Responsible for Sulfide Detoxification in Marine Oxygen Depletion Zones. *PNAS* **2015,** *109* (2), 506–510.

Guinder, V.; Molinero, J. C. Climate Change Effects on Marine Phytoplankton, 2013. doi: 10.1201/b16334-4

Häder, D.-P.; Gao, K. The Impacts of Climate Change on Marine Phytoplankton. In *Climate Change Impacts on Fisheries and Aquaculture*; Phillips, B. F.; Pérez-Ramírez, M., Eds.; 2017. doi: 10.1002/9781119154051.ch27

Hailu, F. A.; Hailu, Y. A. Agro-Ecological Importance of Nematodes (Round Worms). *Acta Sci. Agric.* **2019,** *4,* 156–162.

Hartman, K.; Bahun, M.; Šnajder, M.; Urih, N. P. Molecular Adaptation to High Temperatures: Pernisine from the Archaeon Aeropyrum Pernix K1. *Biologia Serbica* **2019,** *41* (2), 44–50.

Hassan, B.; Qadri, H.; Niamat, A.; Khan, N. *Impact of Climate Change on Freshwater Ecosystem and Its Sustainable Management*; Springer Nature: Singapore, 2020; pp 105–121.

Hegerl, G. C.; Zwiers, F. W.; Braconnot, P.; Gillett, N. P.; Luo, Y.; Marengo Orsini, J. A., et al. Understanding and Attributing Climate Change, 2007.

Heino, J.; Virkkala, R.; Toivonen, H. Climate Change and Freshwater Biodiversity: Detected Patterns, Future Trends and Adaptations in the Northern Regions. *Res. Prog. Biodiversity* **2008,** *84,* 39–54.

Henriksen, T.; Breland, T. Nitrogen Availability Effects on Carbon Mineralization, fungal and Bacterial Growth, and Enzyme Activities during Decomposition of Wheat Straw in Soil. *Soil Biol. Biochem.* **1999,** *31,* 1121–1134.

Hiltpold, I.; Johnson, S. N.; Le Bayon, R. C.; Nielsen, U. N. Climate Change in the Underworld: Impats for Soil-Dwelling Invertebrates. In *Global Climate Change and Terrestrial Invertebrates*; Johnson, S. N.; Jones, T. H., Ed.; John Wiley & Sons, Ltd.; 2017; pp 201–205.

Hoeksema, J. D.; Lussenshop, J.; Teeri, J. A. Soil Nematodes Indicate Food Web Responses to Elevated Atmospheric CO2. *Pedobiologia* **2000,** *44,* 725–735.

Holmstrup, M.; Damgaard, C.; Schmidt, I. K., et al. Long-Term and Realistic Global Change Manipulations Had Low Impact on Diversity of Soil Biota in Temperate Heathland. *Sci. Rep.* **2017,** *7,* 41388.

Holzappfel; Vinebrook. Scientists Warning to Humanity Microorganisms. *Nat. Rev. Microbiol.* **2005,** *17,* 569–586.

Hoppe, H. G.; Breithaupt, P.; Waltherk, K. R. Climate Warming in Winter Affects the Coupling Between Phytoplankton and Bacteria during Spring Bloom: A Mesocosm Study. *Aqua. Microb. Ecol.* **2008,** *51,* 105–115.

Hungate, B. A.; Jaeger, C. H.; Gamara, G.; Chapin, F. S.; Field, C. B. Soil Microbiota in Two Annual Grasslands: Responses to Elevated Atmospheric CO2. *Oecologia* **2000,** *123,* 589–598.

Hunt, H. W.; Coleman, D. C.; Ingham, E. R.; Ingham, R. E.; Elliott, E. T.; Moore, J. C.; Rose, S. L.; Reid, C. P. P.; Morley, C. R. The Detrital Food Web in a Shortgrass Prairie. *Biol. Fert. Soil* **1987**, *3*, 57–68.

Hurd, C. L.; Lenton, A.; Tilbrook, B.; Boyd, P. W. Current Understanding and Challenges for Oceans in a Higher-CO2 World. *Nat. Clim. Change* **2018**, *8*, 686–694. doi: 10.1038/s41558-018-0211-0

Hutchins, D. A.; Boyd, P. W. Marine Phytoplankton and the Changing Ocean Iron Cycle. *Nat. Clim. Change* **2016**, *6*, 1072–1079. doi: 10.1038/nclimate3147.

Hutchins, D. A.; Fu, F. X. Microorganisms and Ocean Global Change. *Nat. Microbiol.* **2017**, *2*, 17508. doi: 10.1038/nmicrobiol.2017.58.

Hutchins, D. A.; Jansson, J. K.; Remais, J. V.; Rich, V. I.; Singh, B. K.; Trivedi, P. Climate Change Microbiology—Problems and Perspectives. *Nat. Rev. Microbiol.* **2019**, *17* (6), 391–396. https://doi.org/10.1038/s41579-019-0178-5.

Imanaka, T. Molecular Bases of Thermophily in Hyperthermophiles. *Proc. Japan Acad. B Phys. Biol. Sci.* **2011**, *87* (9), 587–602.

Ingham, R.E.; Trofymow, J.A.; Ingham, E.R.; Coleman, D.C. Interactions of Bacteria, Fungi, and Their Nematode Grazers: Effects on Nutrient Cycling and Plant Growth. *Ecol. Monogr.* **1985**, *55*, 119–140.

IPCC. Global Warming of 1.5 C An IPCC Special Report on the Impacts of Global Warming of 1.5 C above Pre-Industrial Levels and Related Global Greenhouse Gas Emission Pathways In *The Context of Strengthening the Global Response to the Threat of Climate Change, Sustainable Development, and Efforts to Eradicate Poverty*; Masson Delmotte, V.; Zhai, P.; Pörtner, H. O.; Roberts, D.; Skea, J.; Shukla, P. R., et al. In Press.

Izzo, A.; Canright, M.; Bruns, T. D. The Effects of Heat Treatments on Ectomycorrhizal Resistant Propagules and Their Ability to Colonize Bioassay Seedlings. *Mycol. Res.* **2006**, *110*, 196–202.

Jain, P. K.; Purkayastha, S. D.; De Mandal, S.; Passari, A. K.; Govindarajan, R. K. Effect of Climate Change on Microbial Diversity and Its Functional Attributes. *Rec. Adv. Microb. Diversity* **2010**, 315–331. doi:10.1016/b978-0-12-821265-3.00013-x

Jansson, J. K.; Taş, N. The Microbial Ecology of Permafrost. *Nat. Rev. Microbiol.* **2014**, *12*, 414–425. doi: 10.1038/nrmicro3262

Jenkinson, D. S.; Adams. D. E.; Wild, A. Model Estimates of Co2 Emissions from Soil in Response to Global Warming. *Nature* **1991**, *351*, 304–306.

Johnson, C. N., et al. Biodiversity Losses and Conservation Responses in the Anthropocene. *Science* **2017**, *356*, 270–275.

Kaura, A.; Singha, J.; Vigb, A. P.; Dhaliwal, S. S.; Rupa, P. J. Cocomposting with and without Eisenia Fetida for Conversion of Toxic Paper Mill Sludge to a Soil Conditioner. *Bioresour. Technol.* **2010**, *101* (21), 8192–8198.

Kaye, J.; Twenty-Six Authors. Microbes and Climate Change, 2016.

Keenleyside, W. Microbiology: Canadian Edition. Open Stax Microbiology, 2019. http://cnx.org/ contents/5CTdmJL@7.1:ryt9cF1D@13/Preface.

Klironomos, J. N.; Rillig, M. C.; Allen, M. F.; Zak, D. R.; Kubiske, M.; Pregitzer, K. S. Soil Fungal-Arthropod Responses to Populus Tremuloides Grown under Enriched Atmospheric CO_2 under Field Conditions. *Global Change Biol.* **1997**, *3*, 473–478.

Koga, Y.; Nishihara, M.; Morii, H.; Akagawam. Ether Polar Lipids of Methanogenic Bacteria, Structures, Comparative Aspects and Biosynthesis. *Microbiol. Mol. Biol. Rev.* **1993,** *57* (1), 164–182.

Kranz, S. A.; Levitan, O.; Richter, K. U.; Prášil, O.; Berman-Frank, I.; Rost, B. Combined Effects of CO2 and Light on the N2-Fixing Cyanobacterium Trichodesmium IMS101: Physiological Responses. *Plant Physiol.* **2010,** *154* (1), 334–345.

Kranz, S.; Sültemeyer, D.; Richter, K. U.; Rost, B. Carbon Acquisition in Trichodesmium: The Effect of pCO2 and Diurnal Changes. *Limnol. Oceanogr.* **2009,** *54* (2), 548–559.

Labbate, M.; Seymour, J. R.; Lauro, F.; Brown, M. V. Editorial: Anthropogenic Impacts on the Microbial Ecology and Function of Aquatic Environments. *Front. Microbiol.* **2016,** *7*, 1044. https://doi.org/10.3389/fmicb.2016.01044

Landesman, W.; Treonis, A.; Dighton, J. Effects of a One-Year Rainfall Manipulation on Soil Nematode Abundances and Community Composition. *Pedobiologia* **2010,** *54*, 87–91.

Langley, J. A.; Hungate, B. A. Plant Community Feedbacks and Long-Term Ecosystem Responses to Multi-Factored Global Change. *AoB Plants* **2014,** *6*, 12.

Lassen, A. T.; Sundqvist, M. K.; Henning, J. A.; Newman, G. S.; Moore, J. A. M.; Cregger, M. A.; Moorhead, L. C.; Patterson, C. M. Direct and Indirect Effects of Climate Change on Soil Microbial and Soil Microbial-Plant Interactions: What Lies Ahead? *Ecosphere* **2015,** *6* (8), 1–21.

Lau, J. A.; Lennon, J. T. Rapid Responses of Soil Microorganisms Improve Plant Fitness in Novel Environments. *PNAS* **2012,** *109* (35), 14058–14062.

Lewandowska, A. M.; Boyce, D. G.; Hofmann, M.; Matthiessen, B.; Sommer, U.; Worm, B. Effects of Sea Surface Warming on Marine Plankton. *Ecol. Lett.* **2014,** *17* (5), 614–623.

Lewis, S. L.; Maslin, M. A. Defining the Anthropocene. *Nature* **2015,** *519*, 171–180. doi: 10.1038/nature14258.

Li, Q., et al. Nitrogen Addition and Warming Independently Influence the Belowground Micro-Food Web in a Temperate Steppe. *Plos One* **2013,** *8*, e60441.

Li, W.; Xu, X.; Fujibayashi, M.; Niu, Q.; Tanaka, N.; Nishimura, O. Response of Microalgae to Elevated CO_2 and Temperature: Impact of Climate Change on Freshwater Ecosystems. *Environ. Sci. Pollut. Res. Int.* **2016,** *23* (19), 19847–19860. https://doi.org/10.1007/s11356-016-7180-5.

Li, X.; Hu, H. Y.; Zhang, Y. P. Growth and Lipid Accumulation Properties of a Freshwater Microalga Scenedesmus sp. under Different Cultivation Temperature. *Bioresour. Technol.* **2011,** *102*, 3098–3102.

Lindquist, S. The Heat-Shock Response. *Annu. Rev. Biochem.* **1986,** *55*, 1151–1191.

Littman, R.; Willis, B. L.; Bourne, D. G. Metagenomic Analysis of the Coral Holobiont during a Natural Bleaching Event on the Great Barrier Reef. *Environ. Microbiol. Rep.*, **2011,** *3*, 651–660.

Liu, T.; Mao, P.; Shi, L.; Wang, Z., et al. Contrasting Effects of Nitrogen Deposition and Increased Precipitation on Soil Nematode Communities in a Temperate Forest. *Soil Biol. Biochem.* **2020,** *148*, 107869.

Luo, y.; Zhou, X. *Soil Respiration and the Environment*; Academic Press: London, 2006.

Lürling, M.; van Oosterhout, F.; Faassen, F. Eutrophication and Warming Boost Cyanobacterial Biomass and Microcystins. *Toxins* **2017,** *9*, 64. doi: 10.3390/toxins9020064

Ma, R.; Lu, F.; Bi, Y.; Hu, Z. Effects of Light Intensity and Quality on Phycobiliprotein Accumulation in the Cyanobacterium Nostoc Sphaeroides Kützing. *Biotechnol. Lett.* **2015,** *37,* 1663–1669.

Magnuson, J. J.; Robertson, D. M.; Benson, B. J.; Wynne, R. H. Historical Trends in Lake and River Ice in Northern Hemisphere. *Science* **2000,** *289,* 1743–1746.

Maloy, S.; Moran, M. A.; Mulholland, M. R.; Sosik, H. M.; Spear, J. R. Microbes and Climate Change: Report on an American Academy of Microbiology and American Geophysical Union Colloquium held in Washington, DC, in March 2016. *Am. Soc. Microbiol.* **2017.**

Margesin, R.,.; Niklinska, M. A. Editorial: Elevation Gradients: Microbial Indicators of Climate Change? *Front. Microbiol.* **2019,** *10,* 2405. https://doi.org/10.3389/fmicb.2019.02405

Margesin, R.; Niklinska, M. A. Editorial: Elevation Gradients: Microbial Indicators of Climate Change? *Front. Microbiol.* **2019,** *10,* 2405. doi: 10.3389/fmicb.2019.02405

Maria, L. M.; Castro, C.; Hall, C. M. Crustal Noble Gases in Deep Brines as Natural Tracers of Vertical Transport Processes in the Michigan Basin. *Adv. Earth Space Sci.* **2009,** *10* (6), 1–24.

Marxsen, J., et al. Climate Change and Microbial Ecology. *Curr. Res. Future Trends* **2010,** *34,* 1–204.

Matsuno, Y., et al. Effect of Growth Temperature and Growth Phase on the Lipid Composition of the Archaeal Membrane from Thermococcus Kodakaraensis. *Biosci. Biotechnol. Biochem.* **2009,** *73,* 104–108.

McDevitt-Irwin, J. M.; Baum, J. K.; Garren, M.; Vega Thurber, R. L. Responses of Coral-Associated Bacterial Communities to Local and Global Stressors. *Front. Mar. Sci.* **2017,** *4,* 262.

McSorley, R. Adaptations of Nematodes to Environmental Extremes. *Florida Entomol.* **2003,** *86,* 138–142.

Mcdvcdkova, K. A.; Khmelenina, V. N.; Baskunov, B. P.; Trotsenko, Y. A. Synthesis of Melanin by Moderately Thermophilic Methanotroph Methylocaldum Szegediense Depends on Cultivation Temperature. *Microbiology* **2008,** *77,* 112–114.

Meneghesso.; A.; Simionato.; D.; Gerotto.; C.; la Rocca.; N.; Finazzi.; G.; Morosinotto.; T. Photoacclimation of Photosynthesis in the Eustigmatophycean Nannochloropsis Gaditana. *Photosynth. Res.* **2016,** *129,* 1–15.

Microbiology Online. Microbes and Climate Change, 2015. http://www.microbiologyonline.org.uk/aboutmicrobiology/microbesandclimatechange. Accessed 15 Dec 2015.

Mojica, K. D.; Huisman, J.; Wilhelm, S. W.; Brussaard, C. P. Latitudinal Variation in Virus-Induced Mortality of Phytoplankton across the North Atlantic Ocean. *ISME J.* **2016,** *10* (2), 500–513.

Moore, S. K.; Mantua, N. J.; Hickey, B. M.; Yang, V. L. Recent Trends in Paralytic Shellfish Toxins in Puget Sound, Relationships to Climate, and Capacity for Prediction of Toxic Events. *Harmful Algae* **2009,** *8* (3), 463–477.

Moorhead, D. L.; Freckman, D. W.; Reynolds, J. F.; Whitford, W. G. A Simulation-Model of Soil Nematode Population-Dynamics–Effects of Moisture and Temperature. *Pedobiologia* (Jena) **1987,** *30,* 361–372.

Munteanu, R. The Effects of Changing Temperature and Precipitation on Free-Living Soil Nematoda in Norway, 2017; pp 1–30.

Murphy, G. E. P.; Romanuk, T. N.; Worm, B. Cascading Effects of Climate Change on Plankton Community Structure. *Ecol. Evol.* **2020,** *10,* 2170–2181. https://doi.org/10.1002/ece3.6055

Mykytczuk, N. C.; Trevors, J. T.; Twine, S. M.; Ferroni, G. D.; Leduc, L. G. Membrane Fluidity and Fatty Acid Comparisons in Psychrotrophic and Mesophilic Strains of Acidithiobacillus Ferrooxidans under Cold Growth Temperatures. *Arch. Microbiol.* **2010,** *192,* 1005–1018.

Nardo et al. Laccase and Peroxide Enzyme during Leaf Litter Decomposition of Quercusilex. *Soil Biol. Bio. Chem.* **2011,** *36,* 1538–1544.

Neher, D. A.; Weicht, T. R.; Moorhead, D. L.; Sinsabaugh, R. L. Elevated CO2 Alters Functional Attributes of Nematode Communities in Forest Soils. *Funct. Ecol.* **2004,** *18*: 584–591.

Nelson, K. E.; Clayton, R. A.; Gill, S. R.; Gwinn, M. L.; Dodson, R. J.; Haft, D. H.; Hickey, E. K.; Peterson, J. D.; Nelson, W. C.; Ketchum, K. A.; McDonald, L.; Utterback, T. R.; Malek, J. A.; Linher, K. D.; Garrett, M. M.; Stewart, A. M.; Cotton, M. D.; Pratt, M. S.; Phillips, C. A.; Richardson, D.; Heidelberg, J.; Sutton, G. G.; Fleischmann, R. D.; Eisen, J. A.; White, O.; Salzberg, S. L.; Smith, H. O.; Venter, J. C.; Fraser, C. M. Evidence for Lateral Gene Transfer between Archaea and Bacteria from Genome Sequence of Thermotoga Maritima. *Nature* **1999,** *399*: 323–329.

Niklaus, P. A.; Alphei, J.; Ebersberger, D.; Kampiichler, C.; Kandeler, E.; Tscherko, D. Six Years of in Situ CO2 Enrichment Evoke Changes in Soil Structure and Soil Biota of Nutrient-Poor Grassland. *Global Change Biol.* **2003,** *9,* 585–600.

Norici, A.; Bazzoni, A. M.; Pugnetti, A.; Raven, J. A.; Giordano, M. Impact of Irradiance on the C Allocation in the Coastal Marine Diatom Skeletonema Marinoi Sarno and Zingone. *PlantCell Environ.* **2011,** *34,* 1666–1677.

O'Brien, P. A.; Morrow, K. M.; Willis, B. L.; Bourne, D. G. Implications of Ocean Acidification for Marine Microorganisms from the Free-Living to the Host-Associated. *Front. Mar. Sci.* **2016,** *3,* 1–14.

Ostle, N. J.; Smith, P.; Fisher, R.; Ian Woodward, F. F.; Fisher, J. B.; Smith, J. U.; Galbraith, D.; levy, P.; levy, P.; Niall, P. N.; McNamara, P.; Bardgett, R. D. Integrating Plant-Soil Interactions into Global Carbon Cycle Models. *Br. Ecol. Soc.* **2009,** *97,* 851–863.

Paerl, H. W.; Huisman, J. Blooms Like It Hot. *Science* **2003,** *320,* 57–58. doi: 10.1126/science.1155398.

Panda, A. K.; Bisht, S. S.; Mandal, S. De.; Kumar, N. A. Microbial Diversity of Thermophiles Through the Lens of Next Generation Sequencing. In *Microbial Diversity in Genomic Era*; Das, S.; Dash, H. R., Eds.; Academic Press, 2019; pp 217–226.

Pandey, R. K.; Rana, A.; Sharma, P.; Pathak, R.; Rana, M.; Tewari, L. Thriving at High Temperatures: Molecular Adaptations in Hyperthermophic Microorganisms. *Int. J. Curr. Res. Aca. Rev.* **2016,** *4,* 45–51.

Paulucci, N. S.; Medeot, D. B.; Dardanelli, M. S.; de Lema, M. G. Growth Temperature and Salinity Impact Fatty Acid Composition and Degree of Unsaturation in Peanut-Nodulating Rhizobia. *Lipids* **2011,** *46,* 435–441.

Pecl, G. T., et al. Biodiversity Redistribution under Climate Change: Impacts on Ecosystems and Human Well-Being. *Science* **2017,** *355,* eaai9214.

Pennanen, T.; Liski, J.; Bååth, E.; Kitunen, V.; Uotila, J.; Westman, C. J., et al. Structure of the Microbial Communities in Coniferous Forest Soils in Relation to Site Fertility and Stand Development Stage. *Microb. Ecol.* **1999,** *38* (2), 168–179.

Petrou, K.; Kranz, S. A.; Trimborn, S.; Hassler, C. S.; Ameijeiras, S. B.; Sackett, O.; Ralph, P. J.; Davidson, A. T. Southern Ocean Phytoplankton Physiology in a Changing Climate. *J. Plant Physiol.* **2016,** *203,* 135–150. [CrossRef] [PubMed].

Philippot. L.; Raajimakers, J. M.; Lemanceau, P.; van der putten, W. L. The Microbial Ecology of Rhizosphere. *Nature.* **2013,** *11,* 789–799.

Piffaretti, J-C.; Schink, B.; Semenza, J. C. Editorial to the Thematic Issue Climate Change and Mmicrobiology. *FEMS Microbiol. Lett.* **2018** May**,** *365* (10), fny080. https://doi.org/10.1093/femsle/fny080

Piontek, J.; Lunau, M.; Handel, N.; Borchard, C.; Wurst, M.; Engel, A. Acidification Increases Microbial Polysaccharide Degradation in the Ocean. *Biogeosciences* **2010,** *7,* 1615–1624.

Pollo, S. M. J.; Zhaxybayeva, O.; Nesbø, C. L. Insights into Thermoadaptation and the Evolution of Mesophily from the Bacterial Phylum Thermotogae. *Can. J. Microbiol.* **2015,** *61,* 655–670.

Pörtner, H.-O., et al. *In Climate Change 2014—Impacts, Adaptation and Vulnerability*: *Part A*: *Global and Sectoral Aspects*: *Working Group II Contribution to the IPCC Fifth Assessment Report*; Field, C. B., et al., Eds.; Cambridge University Press, 2014; pp 411–484.

Pretikanen, J.; Peterson, M.; Baath, E. Comparison of Temperature Effects on Soil Respiration and Bacterial and Fungal Growth Rates. *FEMS Microbiol. Ecol.* **2005,** *52,* 49–58.

Prosser, J. I.; Rangel-Castro, J. I.; Killham, K. Studying Plant-Microbe Interactions Using Stable Isotope Technologies. *Curr. Opin. Biotechnol.* **2006,** *17,* 98–102.

Raich, J. W.; Potter, C. S. Global Patterns of Carbon Dioxide Emissions from Soils. *Adv. Earth Space Sci.* **1995,** *9* (1), 23–36.

Rampelotto, P. H. Resistance of Microorganisms to Extreme Environmental Conditions and Its Contribution to Astrobiology. *Sustainability* **2010,** *2,* 1602–1623.

Rapala, J.; Sivonen, K.; Lyra, C.; Niemelä, S. I. Variation of Microcystins, Cyanobacterial Hepatotoxins, in *Anabaena* spp. as a Function of Growth stimuli. *Appl. Environ. Microbiol.* **1997,** *63,* 2206–2212.

Reed, C. J.; Lewis, H.; Trejo, E.; Winston, V.; Evilia, C. *Protein Adaptations in Archaeal Extremophiles*; Hindawi Publishing Corporation: Archaea, 2013; pp 1–14.

Reinold, M.; Wong, H. L.; MacLeod, F. I.; Meltzer, J.; Thompson, A.; Burns, B. P. The Vulnerability of Microbial Ecosystems in a Changing Climate: Potential Impact in Shark Bay. *Life* **2019,** *9,* 71.

Rintoul, S. R., et al. Choosing the Future of Antarctica. *Nature* **2018,** *558,* 233–241. doi: 10.1038/s41586-018-0173-4.

Ripple, W. J., et al. World Scientists' Warning to Humanity: A Second Notice. *BioScience* *67,* 1026–1028.

Rousk, J.; Bååth, E.; Brookes, P. C.; Lauber, C. L.; Lozupone, C.; Caporaso, J. G., et al. Soil Bacterial and Fungal Communities across a pH Gradient in an Arable Soil. *ISME J.* **2010,** *4* (10), 1340–1351.

Ruan, W., et al.. The Response of Soil Nematode Community to Nitrogen, Water, and Grazing History in the Inner Mongolian Steppe, China. *Ecosystems* (*N. Y.*) **2012,** *15,* 1121–1133.

Ruess, L.; Michelsen, A.; Schmidt, I. K.; Jonasson, S. Simulated Climate Change Affecting Microorganisms, Nematode Density and Biodiversity in Subarctic Soils. *Plant Soil.* **1999**, *212*, 63–73.

Runion, G. B.; Curl, E. A.; Rogers, H. H.; Backman, P. A.; Rodriguez-Kabana, R.; Helms, B. E. Effects of Free-Air CO2 Enrichment on Microbial Populations in the Rhizosphere and Phyllosphere of Cotton. *Agric. Forest Meteorol.* **1994**, *70*, 117–130.

Rustad L.; Michelsen, A.; Schmidt, K. A Meta Analysis of the Response of Soil Respiration, Net N Mineralization and above Ground Plant Growth to Experimental Ecosystem Warming. *Oecology* **2001**, *126*, 543–562.

Ruuskanen, M. Lake Sediment Microbial Communities in the Anthropocene, 2019. doi: 10.20381/ruor-23892.

Salvucci, M. E.; Crafts-Brandner, S. J. Inhibition of Photosynthesis by Heat Stress: The Activation State of Rubisco as a Limiting Factor in Photosynthesis. *Physiol Plant* **2004**, *120* (2), 179–186.

Sarmento, H.; Montoya, J. M.; Vázquez-Domínguez, E.; Vaqué, D.; Gasol, J. M. Warming Effects on Marine Microbial Food Web Processes: How Far Can We Go When It Comes to Predictions? *Phil. Trans. R. Soc. B*: *Biol. Sci.* **2010**, *365* (1549), 2137–2149.

Schimel, D.; Braswell, B. H; Holland, B. H.; Mckeown, R. Climatic, Edaphic and Biotic Controls Over Storage and Turnover of Carbon in Soils. *Cycles* **1994**, *8*(3), 279–293.

Schimel, J.; Balser, T. C.; Wallenstein, M. Microbial Stress-Response Physiology and Its Implications for Ecosystem Function. *Ecology* **2007**, *88*, 1386–1394.

Schindlbacher, A.; Rodler, A.; Kuffner, M.; Zechmeister, B. Experimental Warming Effects on the Microbial Community of a Temperate Mountain Forest Soil. *Soil Bio. Biochem.* **2011**, *2* (3), 1417–1425.

Schindler, D. W. The Cumulative Effects of Climate Warming and Other Human Stresses on Fish. *Can. J. Aqua. Sci.* **2001**, *250*, 967–970.

Schratzberger, M.; Holterman, M.; Oevelen, D.; Helder, J. A Worm's World: Ecological Flexibility Pays Off for Free-Living Nematodes in Sediments and Soils. *BioScience* **2019**, *69* (11), 867–876.

Shade, A.; Peter, H.; Allisson, S. D., et al. Fundamentals of Microbial Community. *Resist. Resil.* **2001**, *3*, 417–430.

Shapiro, R. S.; Cowen, L. E. Thermal Control of Microbial Development and Virulence: Molecular Mechanisms of Microbial Temperature Sensing. *mBio* **2012**, *3* (5), 1–6.

Sheik, C. S.; Beasley, W. H.; Elshahed, M. S.; Zhou, H.; Luo, Y. Q.; Krumholz, L. R. Effect of Warming and Drought on Grassland Microbial Communities. *ISME J. 5*, 1692–1700.

Shih, J. M.; Platzer, E. G.; Thompson, S. N.; Carroll, E. J. Characterization of Key Glycolytic and Oxidative Enzymes in Steinernema Carpocapsae. *J. Nematol.* **1996**, *28*, 431–441.

Simmons, B. L., et al. Long-Term Experimental Warming Reduces Soil Nematode Populations in the McMurdo Dry Valleys, Antarctica. *Soil Biol. Biochem.* **2009**, *41*, 2052–2060.

Singh, A. U.; Prasad, D. Impact of Climate Change on Nematode Population. In *Dynamics of Crop Protection and Climate Change*; Chattopadhyay, C.; Prasad, D., Eds.; Studera Press, New Delhi. 2016; pp 371–377.

Singh, B. K.; Bardgett, R. D.; Smith, P.; Reay, D. S. Microorganisms and Climate Change: Terrestrial Feedbacks and Mitigation Options. *Nat. Rev. Microbiol.* **2010**, *8*, 779–790.

Singh, B. K.; Reay, D. Microorganisms and Climate Change: Terrestrial Feedbacks and Mitigation Options. *Nat. Rev. Microbiol.* **2010,** *8,* 779–790.

Solanki, A.; Gupta, D. Studies on Adaptations of Thermophilic Bacteria at Molecular Level. *Res. J. Chem. Environ.* **2013,** *1* (2), 29–31.

Somasekhar, N.; Prasad, J. S. Plant-Nematode Interactions: Consequences of Climate Change, 2012.

Song, M.; Li, X.; Jing, S.; Lei, L.; Wang, J.; Wan, S. Responses of Soil Nematodes to Water and Nitrogen Additions in an Old-Field Grassland. *Appl. Soil Ecol.* **2016,** *102,* 53–60.

Starosak, A. M.; Smith, A. C. Seasonal Patterns in Bacterioplankton Abundance and Production in Narragansett Bay. *Aqua. Microb. Ecol.* **2004,** *35,* 275–287.

Stetter, K. O. Hyperthermophilic Procaryotes. *FEMS Microbiol Rev.* **1996,** *18,* 149–58.

Stevnbak, K.; Maraldo, K.; Georgieva, S.; Bjornlund, L.; Beier, C.; Schmidt, I. K.; Christensen, S. Suppression of Soil Decomposers and Promotion of Long-Lived, Root Herbivorous Nematodes by Climate Change. *Eur. J. Soil Biol.* **2012,** *52,* 1–7.

Sulman, B. N.; Phillips, R. P.; Oishi, A. C.; Shevliakova, E.; Pacala, S. W. Microbe-Driven Turnover Offsets Mineral-Mediated Storage of Soil Carbon under Elevated CO2. *Nat. Clim. Change* **2014,** *4,* 1099–1102.

Sunagawa, S.; Coelho, L. P.; Chaffron, S.; Kultima, J. R.; Labadie, K.; Salazar, G.; Djahanschiri, B.; Zeller, G.; Mende, D. R.; Alberti, A., et al. Ocean Plankton. Structure and Function of the Global Ocean Microbiome. *Science* **2015,** *348* (6237), 1261359.

Suttle, C. A. Marine Viruses—Major Players in the Global Ecosystem. *Nature* **2007,** 5, 801–812.

Swart, A.; Marais, M.; Mouton, C.; du Preez, G. C. Non-Parasitic, Terrestrial and Aquatic Nematodes. In *Nematology in South Africa: A View from the 21st Century*; Fourie, H.; Spaull, V.; Jones, R.; Daneel, M.; De Waele, D., Eds.; Springer, 2017; pp 419–449.

Sylvain, Z. A.; et al. Soil Animal Responses to Moisture Availability Are Largely Scale, Not Ecosystem Dependent: Insight from a Cross-Site Study. *Global Change Biol.* **2014,** *20,* 2631–2643.

Taucher, J.; Jones, J.; James, A.; Brzezinski, M.A.; Carlson, C.A.; Riebesell, U.; Passow, U. Combined Effects of CO_2 and Temperature on Carbon Uptake and Partitioning by the Marine Diatoms Thalassiosira Weissflogii and Dactyliosolen Fragilissimus. *Limnol. Oceanogr.* **2015,** *60,* 901–919. [CrossRef].

Teoh, M. L.; Chu, W. L.; Marchant, H.; Phang, S. M. Influence of Culture Temperature on the Growth, Biochemical Composition and Fatty Acid Profiles of Six Antarctic Microalgae. *J. Appl. Phycol.* **2005,** *2,* 421–430.

Thakur, M. P.; Reich, P. B.; Fisichelli, N. A.; Stefanski, A.; Cesarz, S.; Rich, R. L.; Dobies, T.; Hobbie, S. E.; Eisenhauer, N. Nematode Community Shifts in Response to Experimental Warming and Canopy Conditions Are Associated with Plant Community Changes in the Temperate-Boreal Forest Ecotone. *Oecologia. 175,* 713–723.

Thomas, G. H. Evidence for Intensification of the Global Water Cycle: Review and Synthesis. *J. Hydrol.* **2006,** 83–95.

Thurber, R. V.; Willner-Hall, D.; Rodriguez Mueller, B.; Desnues, C.; Edwards, R A.; Angly, F.; Dinsdale, E.; Kelly, L.; Rohwer, F. Metagenomic Analysis of Stressed Coral Holobionts. *Environ. Microbiol.* **2009,** 11, 2148–2163.

Tikhonova, E. N.; Kravchenko, I. K. Activity and Diversity of Aerobic Methanotrophs in Thermal Springs of the Russian Far East. In *New and Future Developments in Microbial*

Biotechnology and Bioengineering: *Microbial biotechnology in Agro-environmental Sustainability*; Singh, J. S.; Singh, D. P., Eds.; Elsevier, 2019; pp 1–30.

Tobin, T. C.; Janzen, C. P. Microbial Communities in Fire-Affected Soils. In Dion, P.; Nautiyal, C. S., Eds.; *Soil Biol., Microbiol. Extreme Soils* **2008**, *13*, 299–316.

Todd, T.; Blair, J.; Milliken, G. Effects of Altered Soil-Water Availability on a Tallgrass Prairie Nematode Community. *Appl. Soil Ecol.* **1999**, *13*, 45–55.

Toussaint, et al. Primary Production along North South Transect of the Atlantic Ocean. *Nature* **2019**, *416*, 168–171.

Trainer, V. L.; Moore, S. K.; Hallegraeff, G.; Kudela, R. M.; Clement, A.; Mardones, J. I.; Cochlan, W. P. Pelagic Harmful Algal Blooms and Climate Change: Lessons from Nature's Experiments with Extremes. *Harmful Algae* **2019**. doi:10.1016/j.hal.2019.03.009

Treseder, K. A Meta-Analysis of Mycorrhizal Responses to Nitrogen, Phosphorus, and Atmospheric CO_2 in Field Studies. *New Phytol.* **2004**, *164*, 347–355.

Treseder, K. K.; Balser, T. C.; Bradford, M. A.; Brodie, E. L.; Dubinsky, E. A.; Eviner, V. T.; Hofmockel, K. S.; Lennon, J. T.; Levine, U. Y.; MacGregor, B. J.; Pett-Ridge, J.; Waldrop, M. P. *Biogeochemistry.* **2012**, *109*, 7–18.

Ugarte, C.; Zaborski, E. Soil Nematodes in Organic Farming Systems, 2020.

Ulrih, N. P.; Gmajner, D.; Raspor, P. Structural and Physicochemical Properties of Polar Lipids from Thermophilic Archaea. *Appl. Microbiol. Biotechnol.* **2009**, *84*, 249–260.

Uroz, S.; Oger, P.; Morin, E.; Frey-Klett, P. Distinct Ectomycorrhizospheres Share Similar Bacterial Communities as Revealed by Pyrosequencing-Based Analysis of 16S rRNA Genes. *Appl. Environ. Microbiol.* **2012**, *78* (8), 3020–3024.

US EPA. Climate Change: Greenhouse Gas Emissions: Greenhouse Gases Overview, 2016. https://www3.epa.gov/climatechange/ghgemissions/gases.html 2016. Accessed 20 March 2016.

van Elsas, J. D.; Chiurazzi, M.; Mallon, C A.; Elhottova, D.; Kristufek, V.; Salles, J. F. Microbial Diversity Determines the Invasion of Soil by a Bacterial Pathogen. *Proc. Natl. Acad. Sci. USA* **2012**, *109*, 1159–1164.

van Oppen, M. J. H.; Blackall, L. L. Coral Microbiome Dynamics, Functions and Design in a Changing World. *Nat. Rev. Microbiol.* **2019**, *17*, 557–567.

Vandegehuchte, M. L. et al. Responses of a Desert Nematode Community to Changes in Water Availability. *Ecosphere* **2015**, *6*, art44.

Vanwonterghem, I.; Webster, N. S. Coral Reef Microorganisms in a Changing Climate. *iScience* **2020**, *23* (4), 100972. ISSN 2589-0042. https://doi.org/10.1016/j.isci.2020.100972.

Wallace, H. R. *Nematode Ecology and Plant Disease*; Edward Arnold: London, 1973.

Wallenstein, M. D.; McMahon, S. K et al. Seasonal Variation in Enzyme Activities and Temperature Sensitivities in Artic Tundra Soil. *Global Change Biol.* **2009**, *15*, 1631–1639.

Walls, J. T.; Wyatt, K. H.; Doll, J. C.; Rubenstein, E. M.; Rober, A. R. Hot and Toxic: Temperature Regulates Microcystin Release from Cyanobacteria. *Sci. Total Environ.* **2018**, *610–611*, 786–795.

Walther, G. R.; Post, E.; Convey, P.; Menzel, A.; Parmesan, C.; Beebee, T. J. C.; Fromentin, J. M.; Hoegh-Guldberg, O.; Bairlein, F. Ecological Responses to Recent Climate Change. *Nature* **2002**, *416*, 389–395.

Wang, G.; Post, W. M. Consequence of Climate Change on Microbial Life in Ocean. *Microbial Today* **2012**, *81*, 610–617.

Wang, Q.; Cen, Z.; Zhao, J. The Survival Mechanisms of Thermophiles at High Temperatures: An Angle of Omics. *Physiology* **2015**, *30* (2), 97–106.

Wei, Z.; Hu, X.; Li, X.; Zhang, Y.; Jiang, L., et al. The Rhizospheric Microbial Community Structure and Diversity of Deciduous and Evergreen Forests in Taihu Lake Area, China. *PLOS ONE* **2017**, *12* (4), e 0174411.

Weiman, S. Microbes Help to Drive Global Carbon Cycling and Climate Change. *Microb. Mag.* **2015**, *10* (6), 233–238.

Weise, L.; Ulrich, A.; Morcano, M.; Gessler, A. Water Level Change Affect Carbon Turnover and Microbial Community Composition in Lake Sediments. *FEMS Microbiol. Ecol.* **2016**, *92*, 1–14.

Whitaker, J.; Ostle, N.; Nottingham, A. T.; Ccahuana, A.; Salinas, N.; Bardgett, R. D.; Meir, P.; McNamara, N. P. Microbial Community Composition Explains Soil Respiration Responses to Changing Carbon Inputs along an Andes-to-Amazon Elevation Gradient. *J. Ecol.* **2014**, *102*, 1058–1071.

Whitman, W. B.; Coleman, D. C.; Wiebe, W. J. Prokaryotes: The Unseen Majority. *Proc. Natl. Acad. Sci. USA* **1998**, *95*, 6578–6583. doi: 10.1073/pnas.95.12.6578; pmid: 961845.

Wilkening, A.; Rüb, C.; Sylvester, M.; Voos, W. Analysis of Heat-Induced Protein Aggregation in Human Mitochondria. *J. Biol. Chem.* **2018**, *293* (29), 11537–11552.

Wu, H. Y.; Zou, D. H.; Gao, K. S. Impacts of Increased Atmospheric CO_2 Concentration on Photosynthesis and Growth of Micro- and Macro-Algae. *Sci. China C: Life Sci.* **2008**, *51*, 1144–1150.

Wu, Z.; Dijistra, P., et al. Responses of Terrestrial Ecosystem to Temperature and Precipitation Change : A Meta Analysis of Experimental Manipulation. *Global Change Biol.* **2011**, *17*, 1927–1942.

Xenopoulous, M. A.; Lodge, D.; Alcamo, J.; Marker, M.; Schulze, K.; Van, D. P. Scenario's of Fresh Water Fish Extinction from Water. *Global Change Biol.* **2005**, *11*, 1557–1564.

Yadav, S.; Patil, J.; Kanwar, R. S. The Role of Free Living Nematode Population in the Organic Matter Recycling. *Int. J. Curr. Microbiol. Appl. Sci.* **2018**, *7* (06), 2726–2734.

Yamada, N.; Suzumura, M. Effects of Seawater Acidification on Hydrolytic Enzyme Activities. *J. Oceanogr.* **2010**, *66*, 233–241.

Yan, X.; Wang, K.; Song, L. et al. Daytime Warming Has Stronger Negative Effects on Soil Nematodes Than Night-Time Warming. *Sci. Rep.* **2017**, *7*, 44888.

Yang, Y.; Cai, L.; Zhang, R. Wei Sheng Wu Xue Bao. Acta Microbiol. Sinica **2015**, *55* (9), 1097–1104.

Yeates, G. W.; Newton, P. C. D.; Ross, D. J. Significant Changes in Soil Microfauna in Grazed Pasture under Elevated Carbon Dioxide. *Biol. Fertility Soils* **2003**, *38*, 319–326.

Yeates, G. W.; Orchard, V. A. Response of Pasture Soil Faunal Populations and Decomposition Processes to Elevated Carbon Dioxide and Temperature: A Climate Chamber Experiment. *Aust. Grassland Invert. Ecol. Conf.* **1993**, *6*, 148–154.

Yeates, G. W.; Tate, K. R.; Newton, P. C. D. Response of the Fauna of a Grassland Soil to Doubling of Atmospheric Carbon Dioxide Concentration. *Biol. Fertility Soils* **1997**, *25*, 307–315.

Yong, W. K. Tan, Y. H.; Sze-Wan, P.; Lim, P. E. Response of Microalgae in a Changing Climate and Environment. *Malaysian J. Sci.* **2016**, *35*, 167–187. doi: 10.22452/mjs. vol35no2.7

Zhanbin, L.; Jing, M.; Fu, C.; Xiaxiai, L.; Hupinh, H.; Shooliang, Z. Cracks Reinforce the interactions among Soil Bacterial Communities in the Coal Mining Area of Loess Plateau, China. *Int. J. Environ. Res. Public Health* **2019,** *16* (24), 4892.

Zidarova, R.; Pouneva, I. Physiological and Biochemical Characterization of Antarctic Isolate Choricystis Minor during Oxidative Stress at Different Temperatures and Light Intensities. *Gen. Appl. Plant Physiol.* **2006,** Special Issue, 109–115.

Zimmerman, Z.; Labonte, B. Climate Change and the Microbial Methane Banquet. *Clim. Instit.* **2015,** *27*.

CHAPTER 5

Climate Change and Its Impact on Plant–Microbe Interaction

S. S. KHANDARE[1*] and M. G. INGALE[2]

[1]*Department of Microbiology, Bajaj College of Science, Wardha, India*

[2]*Department of Microbiology, Bajaj College of Science, Wardha, India*

Corresponding author. E-mail: ksuhas21@gmail.com

ABSTRACT

The major challenge of the current century is global warming and climate change. If the policy is not changed, combustion of fossil fuels would lead to an increase in the concentration of carbon dioxide (CO_2) in the atmosphere and result in climate warming that may increase the temperature by 1.30°C by the end of the current century. Global warming is a severe issue that leads to swings in the average temperature of the world and causes significant shifts in weather and climate. Greenhouse gases (GHGs) concentrations increased in the atmosphere that exerts a warming effect like increased temperature and drought on positive plant–microbe interactions. As microorganisms are a significant component of carbon and nitrogen cycles, they play an essential role in the discharge and removal of GHGs and hold great significance, which in turn are responsible for global warming. Climate change is disturbing the propagation of species and associations between organisms. Microbes live in combination with many other species. Some are beneficial, some pathogenic, and some of which have little or no effect in complex communities. Natural communities are made of organisms that have varying attributes, dispersal capacity. Therefore, it is doubtful that they will all react to climatic change in the same manner. This presently causes it conceivable to test whether some general patterns occur and whether various gatherings of plant-related microorganisms react diversely or similarly to climate change. Soil microorganisms control the conversion of nutrients, provide supplements to plants, permit conjunction among neighbors, and changes in soil microorganism–plant communications, which could have massive repercussion for the composition of plant community and function of the ecosystem. Disjuncts in plant–pollinator and plant–herbivore collaborations have been relatively very much depicted. However, the plant–microbes relationship has received less attention. According to most of the studies, elevated CO_2 had a positive impact on arbuscular and ectomycorrhizal fungi and plant-associated microorganisms, whereas plant growth-promoting bacteria, endophytic fungi have variable effects.

Similarly, the effects of increased temperature were more variable on positive plant-related microorganisms. The microorganisms are critical players of carbon, nitrogen, and other biogeochemical cycles, and their job as for climate change requires consideration. Notwithstanding, micro-organisms are neglected from most conversations of climate change. It is because of the absence of satisfactory comprehension of microbial action which has not been considered appropriately in most climate change models. In this chapter, we shed light on how global climate change influence soil microbe–plant interactions directly and indirectly, and discuss the exciting questions and areas for future research. It was observed that microorganisms assume important role regarding global climate change, so microorganisms ought to never be ignored of their due significance in environmental change models as well as debates on this issue.

5.1 INTRODUCTION TO GLOBAL WARMING

Earth's climate changes frequently. It is only because of such natural process the environment of the world became favorable for the evolution of humans. In Earth's system, the planets are either hot or bitterly cold while the surface of the Earth is comparatively gentle with stable temperature. The Earth's atmosphere is having a skinny layer of gases which covers and protects the world. According to 97% of climate scientists, humans have altered Earth's atmosphere in dramatic ways over the past two centuries, thus leading to global warming. It is necessary to become accustomed to the atmospheric phenomenon. Massive amounts of radiation coming from the sun continually bombards with the Earth. These radiations hit the atmosphere of the Earth in the form of visible light, infrared (IR), ultraviolet (UV) and alternative types of radiation that are not visible to the human eye. UV radiation incorporates a short wavelength and a greater energy level than visible light, whereas IR radiation includes a greater wavelength and a weaker energy level.

Approximately, 30% of the radiation that hangs Earth's atmosphere is instantly mirrored back, bent on space by clouds, snow, ice, sand, and different reflective surfaces. Ocean and land absorb 70% of the remaining incoming radiation. As they heat up, the oceans, atmosphere, and land unharnessed heat within the kind of I.R. thermal radiation that passes out into space. This incoming and outgoing radiation equilibrium creates the liveable world, with a mean temperature of concerning 59°F (150°C). Earth would be as cold and inert as its Moon, or as bursting hot as Venus if this atmospherical balance is not maintained. The Moon that has about no environment is concerning less 243°F (less 153°C) on its dim position. Venus, on the inverse, fuses a thick environment that traps star radiation; the commonplace temperature on Venus concerns 864°F (46°C)("Global Climate Change Vital Signs").

The atmosphere of earth is affected by regular changes that mark the extent of sun-oriented vitality that arrives the Earth. These progressions grasp among the sun and changes in the Earth's circle. Changes inside the Earth's orbit likewise, because tilt and position of Earth's axis can even affect the amount of light reaching Earth's surface. The balance between energy getting into and leaving the planct's framework plays a significant role in determining Earth's temperature. When the world structure consumes power approaching from the sun, Earth warms. When the sun's energy is reflected into space, Earth evades from warming. A few factors, every common, will cause changes in Earth's energy balance, including:

- Changes within the atmospheric phenomenon that affects the quantity of warmth preserved by Earth's atmosphere.
- Variations within the sun's energy reaching Earth surface.
- Changes within the reflectivity of Earth's atmosphere and surface.

These factors have made Earth's atmosphere to change several times (Melillo et al., 2014). Most of the information on warming is subject to contradictions that bring about varieties of conclusion among what and who is right. As all reports on warming are not valid (some are thought of as enthusiastic, while some are viewed as slanted), exchange of ideas, perspectives and expectations within the view of global warming gets meatier and surprising, not just countries and their economy but conjointly affect the geological, sociology, and psychological frameworks and structures that exist nowadays. Effects of overall warming stand aplenty. The World Health Organization, United Nations Environment Program calculates a measurable 7,000,000 people die untimely every year from associated pollution diseases, just as strokes and cardiovascular sickness, respiratory ailment, and cancer. Pollution in most significant urban areas surpasses WHO air quality norms. Some pollutants that harm our well-being conjointly hurt the environment and are responsible for the temperature change. Dark carbon from diesel motors, cooking stoves, burning of waste, and ground level ozone are harmful, even if lived for short, are unsafe in the atmosphere. It is measurable that decreases in ephemeral atmospheric waste emissions from sources like traffic, farming, and industry may encourage the speed of overall warming by about 0.50°C by 2050. WHO conjointly states that consistently more than 1, 50,000 people succumb on grounds related or coupled to worldwide warming? ("WHO") Nature witnessed numerous catastrophes, beginning from storms in the United States of America to backwoods fires in different districts of Australia, Greece, Kingdom of Spain and the Portuguese Republic, from softening of Arctic icy masses and Antarctica ice-sheets to dissolving of permafrost floods in Yugoslavia, Romania, a geological region of Balkan state to dry seasons in the Sahel in landmass ("Aspects of India's Economy"). All such calamities leave a significant effect on different parts of the earth. Loss of species, relocation of species, reproduction season of animals, unfold of sicknesses like chikungunya, protozoal disease and phrenitis, infringement of human rights, and so forth are some of the outflows of such catastrophes.

5.2   GLOBAL WARMING—RELATIONSHIP WITH CLIMATE CHANGE

Global warming is one in all the foremost talked concerning term in biological science. The outside of the Earth consumes the solar energy that passes through the world's climate, whereas the majority of it is reflected into space. This strategy whenever continued usually, would not have troubled the climate of the world and in this manner, would not turn into a clarification for concern. But with industrialization, overspill, and land-use alteration, deforestation, and later in ways of life, the convergence of the different gases inside the environment has experienced change. Several investigators have expressed for effective pollution management measures to combat the increasing concentrations of such gases ("Kulkarni").

In 2014, Intergovernmental Panel on climate change revealed the Synthesis Report of the Fourth Assessment Report on temperature change and sets up the fact that our world is warming at a horrendous alarming rate that is discernible from the rise within the common air temperature and sea temperature, the far dissolving of ice-sheets, and the ascent within the ocean levels. They also revealed the very fact of temperature change and declared that it is on the far side the normal. Temperature change refers to the changes led to within the Earth's atmosphere system over a large amount of time going down primarily by human intervention or additional, human-made iatrogenic heating. In its Article 1, United Nations Framework Convention's on Climate Change characterizes global climate change as "a change that is credited directly or indirectly to human activity that alters the composition of the world atmosphere, and that is added to natural climate variability determined over comparable periods" ("Green Facts").

Understanding of such definition infers the following:

a. That man-made iatrogenic exercises are fundamentally to blame for climate change.
b. That global climate change brings a few changes within the structure of the world atmosphere, that is, an amendment within the permissible concentrations of gases within the atmosphere.
c. That it is not naturally iatrogenic.

According to the recent reports of Green Facts Express: currently the global mean temperature is about 0.8°C which is above pre-industrial levels. Seas have warmed by 0.09°C since the 1950s and are acidifying. Sea levels rose by concerning 20 cm since pre-industrial era and as of now ascending at 3.2 cm every decade; An extraordinary range of maximum heat waves occurred within the most recent decade. Significant food crop developing regions are progressively littered with water scarcity. A worldwide mean action of 4°C is getting ready to the distinction between the temperatures of the current day and those of the last geological period, when much of the area of central Europe and also the northern United States were coated with kilometers of ice, and also the current human-induced change is occurring at a far quicker rate, over a century, not millennia. If the presently planned actions are not enforced, a 4°C warming might happen by 2060s. Such an alarming level by 2100 would not be the tipping point:

additional warming to levels over 6°C would seemingly happen over the coming centuries ("Green Facts"). Humans have aggravated the problem by their involvement in destroying important resources through a number of means. The commitment of the agricultural sector toward the incidence of worldwide warming cannot be unnoticed. Fashionable agribusiness, food production, and delivery are significant contributors to greenhouse emissions: 14% of total greenhouse gas (GHG) discharges incorporated by agriculture, and more extensive rustic land-use choices have a far more substantial effect. Deforestation directly represents an additional 18% of emissions ("World Future Council"). During this specific situation, a recorded point of view must contemplate of Professor of Soil Science Dr Rattan Lal, from Ohio State University who has determined that in the course of the last 150 years, 476 billions of vast amounts of carbon have been emitted from agricultural soils because of improper cultivating and grazing practices, compared with just 270 gigatonnes (GT), discharged from the ignition of fossil fuels. An additional quoted figure is that within the most recent 300 years, around 200–250 Gt of carbon is lost. Regardless of accurate model, these decreases of living carbon potential have come about because of: deforestation, loss of diverseness, increased erosion, vanishing of soil organic matter, salinization of soils, water contamination at coastal sites and activity of the seas. Changes in land use can even considerably contribute to temperature change. Massive-scale changes like deforestation, disintegration, or machine cultivating ways could all add to swelled carbon focuses inside the air. It is relevant to say during this circumstance that the amendment within the land utilization incorporates a benefaction of 18% to the increase of GHGs due to deforestation, peat decay, and the burning and decomposition of biomass happening worldwide ("World Future Council").

5.3 GREEN HOUSE EFFECT

The switch of entering and outgoing radiation that warms the world is commonly referred to as the greenhouse effect. Climatologist accepts that expanding barometric centralization of carbon dioxide (CO_2) and different "greenhouse gasses" (GHGs) discharged by human exercises, such as consuming of nonrenewable energy sources and deforestation, are warming the world. The process is usually called the "greenhouse effect" that makes the planet inhabitable. Such gases within the atmosphere act simply like the mirror of a greenhouse, giving the daylight access and keeping heat from escaping out. However, the human actions have changed the composition of the climate through the development of ozone-depleting substances principally CO_2, nitrous oxide (N_2O), (CH_4) and so forth. Gases like CO_2, CH_4 and N_2O produce a halfway cover over the Earth's atmosphere and act just like the glass of a greenhouse. They, thus, attract the IR radiations and duplicate them into the Earth's air. This process of saddler gases is commonly received in greenhouses to require the help of the warmth produced from such gases within the faster growth of plants. Increase in temperature and changes in associated processes are legitimately associated with expanding GHG outflows within the atmosphere (Latake et al., 2015). As an outcome of the atmospheric phenomenon, the world's average temperature will increase. Such increment in temperature proceeds over a prolonged amount of time and is the fundamental reason for warming of the Earth and therefore named as "Global warming." A warming Earth is climate change, and it happens due to changed lifestyle throughout the world. Thus "global warming" is employed to refer with Earth's step-by-step increasing temperature. Additional floods in components of the planet and further powerful droughts in others indicate temperature change; uncontrolled fires in some areas and odd snowfalls in others indicate temperature change. A season of intense tornados and other strong hurricanes means additional energy within the atmosphere, which is due to temperature change. The Intergovernmental Panel on temperature change reported that world is inflicting warming through the emission of GHGs also known as ozone-depleting substances significantly CO_2 and CH_4 ("AR5 Climate Change 2014"). CO_2 and CH_4 concentrations, have magnified by 40% and 150%, respectively since the pre-industrial era. However, such increment has gotten far speedy with the progression of science and technology, the rise within the consuming of non-renewable energy sources and land-use change (Huang et al., 2016). UV radiation simply goes through the glass screens of a greenhouse and is consumed by the plants and onerous surfaces within. More fragile IR radiation, in any case, has an issue going through the glass screens and is at bay within, so warming the greenhouse ("Live Science").

GHG emission is inflated dramatically in recent years. These gases collect within the environment and inflicting concentrations to extend at time intervals. The most significant GHGs are CO_2, N_2O, CH_4 and halo-carbons. It is human iatrogenic heating that holds genuine responsibility regarding the enormous increment within the number of GHGs in the environment. There are various factors accountable for the same, the most noteworthy being the growing overpopulation. A deeper understanding of these gases will aid us to grasp climate change in an enormously improved manner. It has been discovered within the Fourth Assessment Report of the Intergovernmental Panel on temperature change printed in 2014 that the atmosphere concentration of GHGs has risen to 379 ppm up in the year 2005 (Abatenh et al., 2018).

5.3.1   CARBON DIOXIDE

CO_2 comes back from fuel use in several sections like building, transportation, heating, cooling, and production of concrete and alternative merchandise. It is conjointly discharged from regular processes like plant matter decaying, respiration, and organic matter decomposition by microbes. Of all the GHGs, CO_2 is thought to be the foremost vital contributor to heating. It is the rise within the concentration of the same gas that results in extreme atmosphere fluctuation and temperature change. Out of the six significant recognized GHGs, CO_2 has the highest life expectancy of nearly a thousand years. In industrialized countries, coal, petroleum gas and oil are accountable for carrying out the significant segment of the industrial functions. Thus, the release of CO_2 is consistently on the rise. The concentration of CO_2 was 280 ppm during the pre-industrial era but until 1998, the same amount rose to above 350 ppm. It has been expected that with the current pattern of increment in the concentration of CO_2, the concentration of the said gas will be near to 500 ppm before the ending of the 21st century (Leggett, 2009).

5.3.2   METHANE

Production of CH_4 occurs due to the daily anthropogenic activities that resemble fossil fuels production, its distribution and combustion, landfills and waste, droppings of eutherian farming, burning of biomass, and rice agriculture. A natural phenomenon that occurs in ground termites and seas are distinctive sources for CH_4 discharge. CH_4 is the second most gas that contributes to warming. Being an organic compound gas and having over 100 years' time span, it is thought to be 21 times additional damaging than carbon dioxide by weight however its availableness within the atmosphere is a smaller amount as compared with CO_2. CH_4 is additionally a by-product of rice farming and is found once organic matters decay in swamps (Abatenh et al., 2018).

5.3.3 NITROUS OXIDE

N_2O or laughing gas production occurs due to the use of fertilizer and burning of fuel. It is also conjointly produced naturally in soil and oceans. The concentration of N_2O within the atmosphere during the pre-industrial era was 0.27 ppm that has risen up to 0.31 ppm in 1998. Having the atmospherical time period of more than 100 years, N_2O gas once generated has the world warming potential of 298 times for 100 -year time horizon (Abatenh et al., 2018).

5.3.4 CHLOROFLUOROCARBONS

Amount of halocarbon gases is raised primarily because of anthropogenic and natural processes. Halocarbons include chlorofluorocarbons (CFCs) (CFC-11 and CFC-12) that were utilized widely as refrigeration agents and in alternative industrial processes. CFCs are highly responsible for damaging the layer of the atmosphere thereby resulting in carcinoma and therefore their production and use nowadays is hugely regulated by international protocol and national legislations ("Science Daily").

5.3.5 PERFLUOROCARBONS

Metal and semiconductor chip production has prompted the status of perfluorocarbons (PFCs) as a significant GHG. Their atmospherical period is a sort of high and concerning climate change, the main PFCs embrace per fluoromethane and perfluoro-ethane. Soluble PFCs embrace the aerated PFCs that are powerfully present within the ocean-atmosphere. Even though the environmental concentration of trifluoromethane is around 100,000 times less than CO_2, the powerful radiative forcing of perfluoroethane is as much as 1500th of the radioactive forcing of CO_2 ("Science Daily").

5.3.6 HYDROFLUOROCARBONS

While PFCs and fluoride have atmospherically time length going to hundreds of years, the atmospherical time frame of hydrofluorocarbons (HFCs) is numerous decades. In any case, their capacity to absorb IR radiations is higher, and it is such quality that makes it perceived as a severe greenhouse emission. Over the last 20 years, scientists are keeping a detailed eye on the atmospherical concentration of a fluorocarbon (HFC) gas. Scientists are concerned, because HFC-23 could be a potent greenhouse emission, with one of its emanations being comparable with the discharge of more than 12,000 tons of CO_2. As the HFCs do not include ozone-destroying chlorine or bromine atoms which drain the ozone layer, their utilization as refrigerants and in air conditioners keep on rising. Having the positive nature of ensuring the ozone layer, its utilization as of late in the developing world has expanded enormously. Projections uncover that by the year 2050, the utilization of HFCs will be 800 times more in the developing nations like that in the developed countries and the said emissions are probably going to warm the planet like 5–9 billion tons of CO_2 ("Science Daily").

5.3.7 SULFUR HEXAFLUORIDE

Under the Intergovernmental Panel on temperature change, sulfur hexafluoride (SF6) could be a potent greenhouse emission whose warming potential is 23,900 times over that of carbon dioxide more than 100-year timespan. SF6 is employed for insulation in wattage transmission supplies and within the semiconductor manufacturing ventures. It is conjointly utilized as a tracer for gas spill recognition. The sinks accessible for SF6 are negligible, and subsequently, hotspots for its gathering include all human-made sources. Even though SF6 is itself not harmful, it decays under electric pressure and in this manner produces poisonous products ("United Nations Climate Change").

5.3.8 OZONE

Ozone is a greenhouse emission that is frequently made and destroyed within the atmosphere by chemical reactions. Within the layer, anthropogenic activities have raised gas through the discharge of gases like carbon monoxide, hydrocarbons, and nitrogen oxide that reacts with chemicals to provide gas. As mentioned earlier, halocarbons released by human activities destroy ozone in the stratosphere and have caused the ozone hole over Antarctica (Gupta et al., 2014).

5.3.9 AEROSOLS

Fuel and burning of biomass have raised aerosols containing sulphur compounds, organic compounds and black carbon. Surface mining and industrial processes have increased the dust concentration in the atmosphere. Mineral dust released from the surface, sea salt aerosols, biogenic emissions from the land and oceans, sulfate and dust aerosols produced by volcanic eruptions are the examples of natural aerosols (Gupta et al., 2014).

5.3.10 EFFECT OF GLOBAL CHANGING CONDITIONS ON PLANT-MICROBE INTERACTION

Humans and other life forms of the Earth hugely influence climate structure. Recent climate changes have had extensive impacts on human and natural systems, including plants and microbes. The Earth's surface has been sequentially warming in the last three decades than any preceding decade since 1850. The warmest 30-year period of the previous 1400 years was the period from 1983 to 2012. In this varying environment, various parameters will be affected due to severe change in global climate. This is the situation for barometrical CO_2 concentration, temperature, and drought that boost continuously. Facts of observed climate change impact like changing rainfall or melting snow and ice in many regions, and are maximum and most significant for environmental systems. Many marines, terrestrial, and freshwater species have moved their geographic reaches, seasonal behavior, and patterns of immigration, species abundances, and species interactions in light of continuous climate change. Many studies of climate change which covers a wide range of regions and crops show that negative impacts on crop yields are more common than positive effects ("AR5 Climate Change 2014").

Soils and plants are inseparably connected. Plants and their interactions affect soil characters, which display a variety of effects on each other. The relationships among plants and their soil biota lead to complex inputs that control ecosystem processes and plant community dynamics. These impacts of plants on themselves and other species of plant and abiotic conditions of soil are termed as plant–soil feedbacks (PSFs). The net outcome of PSFs relies upon the harmony among beneficial and antagonistic associations with the plant and soil microbial communities, which can differ contingent upon both biotic (e.g., functional traits of the plant) and abiotic (e.g., soil pH, physical structure, and availability of nutrient) factors (Bennett and Klironomos, 2019). Climate change may affect PSFs by altering the succession both on the primary and secondary level, which are significant drivers of plant species substitution. Soil populations contain numerous species composed of a trophic network of primary producers, consumers, and of secondary and higher-level consumers, and changing climate (Coyle et al., 2017) could modify that.

The impact of environmental change on agribusiness has explicitly become a worldwide issue due to the ever-growing need to give food security and end hunger. The effect is all in all agroecosystem from crop yields and animals to fisheries, which thus can influence economic steadiness. The ramification of climate change on agriculture is not just because of impacts on plants yet besides because of the effects on biogeochemical properties of the soil on which they develop. A significant outcome of changing climates is its effect on the microorganisms present in the agricultural soil. Not exclusively does the climate-mediated changes in microbiota and their practices lessen food productivity, yet it stretches out to food safety, human well-being and economy when soil microbiota turn toxigenic under ecological influence ("Food and Agriculture Organization").

Change in climate is also resulting in an alteration of species distributions and at the same time affecting interactions between organisms. Natural communities are complex and composed of organisms with very different life-history traits, dispersal ability, and thermal tolerances. Furthermore, interactions between community members can be beneficial, harmful or have less to no functional impact, and these interactions may change with environmental stress (Vandenkoornhuyse et al., 2015). Numerous research studies show that shifts in species interactions due to climate change cascade can modify biodiversity and the function of terrestrial ecosystems, but only fewer studies focus on soil communities. Organisms in soil interrelate with each other as well as with plants in numerous ways that keep up ecosystem properties. Soil microbial interactions can give shape to the landscape patterns of plant and animal abundance, variety, and composition. Because of the increased temperature, the water content of the soil is estimated to diminish in some territories leading to higher water scarcity in several areas of the world. These climate-changing parameters are known to affect terrestrial organisms, such as plants and microorganisms. Interactions of plants and microbes are considered harmful when the net impacts of all soil organisms including symbiotic mutualists, pathogens, and decomposers decrease the performance of the plant, while interactions are considered positive when the advantages brought about by the soil community enhance the performance of plants, such as the production of biomass and survival (Langley and Hungate, 2014). Therefore, their significance in characterizing ecosystem properties, understanding how soil microbe–microbe and soil microbe–plant interactions react to climate change is a research priority that will reveal insight into significant ecosystem functions, such as soil carbon storage and net primary productivity (Fischer et al., 2014).

Nearly, all land plant taxa studied have deep-rooted symbioses with a considerable assortment of microbes. Some of them strengthen the growth of the plant and expands tolerance of the plant to biotic and abiotic stresses, whereas others can be pathogenic or nonpathogenic for their hosts. Majority of such plant growth-promoting microorganisms colonize the rhizosphere, the part of soil appended to the root surface and impacted by root exudates and by microbes. Some organisms can also enter into the root system of their hosts and enhance their beneficial impacts with an endophytic lifestyle. This is true for plant growth-promoting fungi (PGPF), such as arbuscular mycorrhizae, endophytic fungi, and other ectomycorrhizae, as well as plant growth-promoting rhizobacteria or plants growth-promoting bacteria. Due to change in climate, modified environmental conditions are probably going to induce changes in root exudation plant physiology, and increased concentration of CO_2, in many cases, leads to higher carbon allocation to the root zone and possibly also leads to altered root exudates composition. Changes in climate also influence the diversity and structure of microbial communities directly (e.g., temperature and seasonality) or indirectly (e.g., plant litter, plant composition, and root exudates). Microbial diversity of soil influences plant variety and is essential for the functions of the ecosystem, including cycling of carbon

(Jing et al., 2105). The plant growth-promoting microorganisms could play indispensable roles in the plant fitness maintenance and health of soil under stressful environments. Climate change-related factors may timely expand carbon allocation to the root zone, which will conceivably alter the root exudates composition, the C/N ratio or availability of the nutrients, with the subsequent impact on the abundance, composition, and/or activity of microbial communities that are associated with the plant (Vimal et al., 2017). As per the literature, warming, drought and increased CO_2 affect the plant beneficial microorganisms in different ways, the impacts being subject to the climate change factor studied, ecosystem type, plant species, soil type, and microbial genotype. Here, we review and discuss how the parameters of climate, including CO_2, temperature, and drought affect the plant–microbe interaction in soils. This shows that how plant–microbes interaction reacts to fluctuating climatic change.

5.4   PLANTS AND ASSOCIATED MICROBIOME IN RHIZOSPHERE

Arbuscular mycorrhizal fungi (AMF) are soil-borne fungi significantly imparts resistance to several abiotic stress factors and improves plant nutrient uptake. Majority of the AMF species belong to the subphylum Glomeromycotina of the Mucoromycota phylum. AMF has four orders, namely, Glomerales, Paraglomerales, Archaeosporales and Diversisporales, which have been identified and include 25 genera. They are obligate biotrophs and ingest photosynthetic products of plant and lipids to accomplish their life cycle. AMF-interceded growth advancement is not only by accelerating mineral, nutrient, and water uptake from the adjacent soil but also by protecting the plants from highly virulent fungal pathogens. Thus, arbuscular mycorhizal fungi are vibrant endosymbionts, which play an impactful role in plant efficiency, yield, and the ecosystem functioning. They are of significant value for sustainable crop development. However, due to the effects of climate change, such as expanded CO_2 concentrations in the atmosphere, warming of soil, or drought stress, many research investigations have reported indirect impacts of these parameters on this plant-associated fungi (Sun et al., 2018; Spatafora et al., 2016; Jiang et al., 2017).

With altered ecological conditions unlike mycorrhizal fungi that colonize plant roots and grow into the rhizosphere, there are microsymbionts such as certain endophytic fungi that live completely within plant tissues and may propagate inside the roots, stems, and emerges to sporulate on plant or host-tissue. All the fungal endophytes are well known to impart beneficial effects on their plant hosts. Fungal endophytes exist in a close, mutually positive relationship with their host plant, in that they offer ecological support to their host plants by permitting them to endure adverse stresses of biotic and abiotic type (T.S., S., 2017). Within fungal kingdom, they comprise a key component. These include the relief or mitigation of harm caused by pests or dangerous insects (Resquín-Romero et al., 2016). Moreover, reports demonstrate that plants colonized by these fungal endophytes are less vulnerable to the damaging effects of pests. The endophytes impart these advantages on their host plant by blocking the important stages of growth and development in the past; they also significantly influence the pattern of feeding and stages of reproduction within the pest, thereby affecting its overall survival (Vega, 2018). They have additionally been accounted for having a potential source of bioactive inoculants that could help accomplish sustainability in agriculture (Nath Yadav and Author, 2018). Chhipa et al. (2019) investigated the role of endophytic fungi in influencing the root exudations of plants in a positive manner, thereby drawing the attention of beneficial microbial flora of rhizosphere that encourages the mineral transportation from the soil that are required by plants (Chhipa et al., 2019).

Plants establish mutualistic associations with other microorganisms to effectively thrive in their natural niche, that is, advantageous to the other living entities in the ecological system. Plants are not simply just colonized by a wide range of fungi but also by various bacteria, and many of them impart beneficial effects to their hosts. Association between plant and microorganisms is an example of such positive interaction. Some tissue-colonizing bacteria establish closely linked association with their host plant, and indeed, award the benefits to the plant under both ideal and challenging situations (Santoyo et al., 2016). The bacteria may advance the growth of the plant, and its health and are of many different types, and involve the production of growth-promoting substances such as different growth-promoting hormones, easing of both biotic and abiotic stresses, antagonism of pathogens and generation of systemic responses (Miliute et al., 2015). As plant-associated bacteria depend on exudates of plant roots or metabolites of plant and are substantially impacted by environmental factors due to changes in plant physiology, it can be expected that environmental situations associated with climate change will affect these communities.

5.5   EFFECTS OF CHANGING ENVIRONMENTS ON PLANT–MICROBIOME INTERACTION

5.5.1   *EFFECTS OF ELEVATED CARBON DIOXIDE LEVELS*

Tremendous investigations have been done on the beneficial impacts of raised CO_2 concentrations on plant growth. The tropospheric concentration of CO_2 is expected to increased from 355 (v/v) to 710 ppm, by 2050. Elevated CO_2 concentration brings about advantages for plant development, although there may be differences among species to species. Considerably less is thought about the effect of CO_2 consequences on the rate and severity of biotic sicknesses of plants. Over the most recent years, around 3000 reports have been published on this matter. A few authors arrived at similar conclusions with various crops, forest species and natural ecosystems. Most soil fungi were found to tolerate more than 10–20-fold increase of CO_2 concentration in the atmosphere (Tyagi et al., 2014).

Elevated CO_2 concentrations may induce changes in colonization as well as community structures of AMF and ectomycorrhiza (ECM). According to the reports, production of mycelial biomass by *Hebeloma crustuliniforme* in *Pinus sylvestris* (L.) has increased by three-folds under raised CO_2 as compared with mycelial development under ambient CO_2 concentrations. *Pisolithus tinctorius* is a mycorrhizal fungus relies upon assimilates of *Pinus sylvestris* L., was found to grow much quicker and efficiently in the presence of raised CO_2, with three-fold higher mycorrhizal root clusters and with twofold higher biomass of extra mycorrhizal mycelia compared with ambient CO_2 levels. These reports exhibit that expanded CO_2 can build ECM colonization of host plants. Although, in the case of *Cenococcum geophilous* and *Suillus* sp. related with Scots pine (*P. sylvestris* L.), under these conditions, fungal biomass does not increase (Chanda et al., 2020). As per the reports, elevated ambient CO_2 levels were found to have effects on the growth of hyphae and root colonization ability of AMF in most of the studies. According to Zhu et al. (2016), higher CO_2 was found to accelerate plant roots colonization by mycorrhiza as an outcome of the improved carbon transfer to roots, which may bring about an expanded mineral take-up from soil however not really associated with nutrient transport to the host plant (Zhu et al., 2016). Concerning the community structure of AMF, raised CO_2 expanded the proportion of Glomeraceae to Gigasporaceae.

However, this impact might be disguised by the natural changes through time and by the reliance of various fungi on water accessibility and rainfall (Cotton et al., 2015; Veresoglou et al., 2016).

Few investigations have revealed the impact of expanded CO_2 levels on endophytic fungi other than AMF and ECM. The ericoids, dark septate endophytes, but not others, showed increased colonization in ericaceous dwarf shrubs under increased atmospheric CO_2 concentrations.

Mycorrhizal fungi that colonize the roots of plants developed into the rhizosphere, there are some other microsymbionts like endophytic fungi of genera *Atkinsonella, Balansia, Echinodothis, Balansiopsis, Myriogenospora, Epichloë, Neotyphodium*, and *Parepichloe* that exist in plant tissues and may develop inside roots, stems as well as leaves, advancing to sporulate on plant or host-tissue. On account of *Neotyphodium coenophialum* and its host, tall fescue Schedonorus phoenix, endophyte disease expressed to be higher under expanded CO_2 in contrast to ambient CO_2 level. In any case, CO_2 upgrade did not influence the interfaces between host grasses and purple top grass, *Tridens flavus* and their endophytic–fungal symbiont partner, *Acremonium lolii* and *Balansia epichloe*, respectively. The plant sugar content may furthermore change due to the effect of CO_2 on the host plant and its endosymbiont. This was affirmed with *N. lolii* and perennial ryegrass (*L. perenne*), where it was observed that contaminated plants had higher sugar level than the plants that are free from endophyte infection, and was higher than under ambient CO_2 levels. Raised CO_2 levels may prompt the expanded existence of endophyte contaminations, thus bringing in the general impacts on the ecosystem.

Additionally, under raised CO_2 conditions, the relationship between endophytic fungi and grasses guard plants against insect herbivores. Soil warming may have an impact on the favorable relationship among plants and fungal endophytes (Chanda et al., 2020). As an essential substrate for the metabolism of plants, atmospheric CO_2 have impacts on the distribution of carbon in belowground level and also on the chemistry of root exudation, conceivably influencing the interaction of rhizosphere with valuable soil microorganisms. Carbon and nitrogen content are essential markers for the quality of soil, which directly affects the function of plant growth-promoting rhizobacteria (PGPR) (Agbodjato et al., 2015).

Atmospheric CO_2 has an impact on microbial biomass as well as microbial diversity in the rhizosphere. The plant-intervened impacts of barometrical CO_2 on soil microbial communities are well characterized and recorded, showing a prevailing, plant-interceded mechanism. It is generalized, the variety in root-microorganism association under various climatic conditions occur because of changes in root exudates which are assessed to contain somewhere in the range of 5% and 40% of plant photosynthetically fixed carbon. Since carbon (C) rhizodeposition expands under raised CO_2 (eCO_2), it tends out that rhizosphere colonization by microorganisms depending on carbon from plant exudates will likewise be improved. While it is true that CO_2 can influence the overall composition of the microbial community over a range of various soil types, the degree to which eCO_2 influences microbial interactions in the rhizosphere remains arguable.

Both positive and negative effects of CO_2 have been reported in the past investigations. It equally remains challenging in how far eCO_2 is responsible for shifts between fungal or bacterial communities, and the resultant effect on the functioning of microorganisms present within the rhizosphere. Studies on plant responses and the nearness of specific rhizosphere organisms to CO_2 have proposed a potential connection between eCO_2, plant growth, and increase in colonization by PGPR. Plant growth-promoting rhizobacteria are normally strongly connected with plant roots and should, hence, be increasingly dependent on plant-derived carbon. Even though numerous examinations have addressed the impacts of eCO_2 on plant–rhizobia and plant–mycorrhiza associations, little is identified about the particular effects of eCO_2 on PGPR (Mohan et al., 2014). Williams et al. (2018) explored the effects of a pre-industrial concentration of sub-ambient ($saCO_2$) and a worst-case scenario projected concentration of eCO_2 on the colonization of rhizosphere of Arabidopsis roots by two well-studied soil bacteria *Pseudomonas simiae* WCS417, the colonizer of rhizosphere (Williams et al., 2018) (previously named *Pseudomonas fluorescens* WCS417 [Berendsen et al., 2015].) and *Pseudomonas putida* KT2440, the saprophytic soil colonizer. They proved that higher CO_2 levels accelerate the colonization of root by WCS417 in low carbon and nitrogen contained soil. To study whether rhizosphere colonization by WCS417 and KT2440 is altered by atmospheric CO_2, Williams et al. (2018) cultivated Arabidopsis in two different types of soil for 4 weeks at sub-ambient $saCO_2$ (200 ppm), ambient CO_2 (aCO2; 400 ppm) or eCO2(1200 ppm) before quantification of rhizosphere colonization. Extraordinarily, in soil with the poor nutrient condition, he found that titres of WCS417 bacteria within rhizosphere increased statistically from saCO$_2$ to eCO$_2$, on the other hand, this impact of CO$_2$ was missing in soil with the rich nutrient condition.

Moreover, the interaction between CO$_2$ and soil type, which was statistically significant, showed that the effects of CO$_2$ on the colonization of rhizosphere by WCS417 rely on the nutritional status of the soil. On the other hand, KT2440 titres within the rhizosphere were not statistically altered by CO$_2$, type of soil, or the interaction thereof, showing that the colonization by this saprophytic strain is not affected by the status of soil nutrition and atmospheric CO$_2$ (Williams et al., 2018). The stimulatory influence of atmospheric CO$_2$ on rhizosphere colonization by soil microorganisms depend on the quality of soil and species of bacteria. Yu et al. (2016) conducted an experiment on soybean and the data uncovered that there is a significant shifting of soil microbial flora in the soybean rhizosphere in Northeast China due to the presence of CO$_2$. The degree of variation was reliant on the cultivars of soybean. Their research showed that the CO$_2$-induced bacterial taxa might have an impact on soil nutrient cycling and productivity of ecosystem (Yu et al., 2016). Despite the fact that the biogeochemical characteristics of soil were not significantly associated with the whole microbial community, other factors should need to be further examined in regards to shifts in the microbial community. Many studies showed that increased CO$_2$ set of the interaction of legumes with rhizobia, which are notable for their nitrogen fixation in relationship with legumes as well as for their different plant growth-promoting traits. Many experiments prove that increase in atmospheric CO$_2$ levels might have differential effects on the growth and activity of beneficial plant bacteria, which might have further implications on their utility in agriculture or phytoremediation.

5.5.2 EFFECTS OF TEMPERATURE

Many scientists have examined AMF and ECM responses to a higher temperature. In most of the cases, it was found that higher temperature has a positive effect on the AMF colonization and length of the hypha. In other few cases, no impacts or negative impacts of elevated temperature on AMF were recorded. Environmental conditions influence the community and growth of AM fungi as well as the development of mycorrhizae. Many studies have investigated the diversity of AM fungi at different environmental conditions by using traditional morphological and molecular strategies. Nonetheless, to the extent we know, little data are existing about the impact of temperature on the structure of AM community (Botnen et al., 2015). High or low-temperature stress will affect plant morphological features, plant growth, and biomass production. The symbiosis of AMF has been exposed to increase the fitness of the host plants utilizing upgrading its development and biomass production. It has been accounted for that AMF plants developed better than non-AMF plants under low or high-temperature stress, which halfway credited to the upgraded photosynthesis and nutrient uptake, particularly for phosphate nutrition. A few scientists have explored the details that dry weights of shoot and root of AMF plants were higher than the non-AMF plants at low or high-temperature conditions (Liu et al., 2014). The incline in temperature positively affects arbuscular fungi colonization and hyphal length. It might likewise alter the structure of the AMF hypha and initiate a switch from vesicular hyphae responsible for storage in the more undisturbed soil to extra mycorrhizal hyphae required for development in hotter soils. This might be accountable for quicker distribution of carbon to the rhizosphere and for improved respiration of the extra mycorrhizal hyphae at an elevated soil temperature. AMF may improve plant growth and colonization for most of the strains of *Glomus* intraradices and Glomusmossae at a higher temperature. Respiration of ECM strains of *C. geophilum*, *Suillus intermedius* and *Lactarius pubescens* can be diminished under increasing temperatures (Chanda et al., 2020). Mohan et al. (2014) stated the outcomes acquired in research on the impact of rising temperatures on communities of mycorrhiza. In 17% of such studies, mycorrhizal profusion diminished, in 20% of them no major change was noticed, and the 63% of the works concluded that the mycorrhizal abundance could expand under raised temperature (Mohan et al., 2014). Augé et al. (2015) reported that AMF advancement was 10% higher when air temperatures were kept at or underneath 270°C than those that surpassed 270°C (Augé et al., 2014). Besides, Wilson et al. (2016) concluded that the immediate impact of expanding 30°C temperature diminishes AMF colonization, and this appeared to region-wise steady over the Mediterranean atmosphere gradient (Wilson et al., 2016). Despite the fact that the growth of external hyphae and the AMF associated diversity can accelerate at high temperatures (Zhang et al., 2016), the mycorrhizal activity generally decreases (Mohan et al., 2014). Thus, raised CO_2 as well as higher temperatures may activate changes in mycorrhizal communities.

In-plant–endophytic fungi associations it is known that soil warming may impact beneficial associations between plants and fungal endophytes. Temperature is a primary parameter that has a significant influence on the endophyte appearance in plant tissue. Soil warming may affect beneficial interaction between plants and fungal endophytes. Temperature is the main environmental parameter that plays a significant role in endophyte plant tissue; However, warming did not hinder the endophytic infection to the tail of fescue *S. phoenix* by *N. coenophialum*. Warming without influencing the composition and the richness of the community increased the density of different fungal endophyte genotypes within individual root sections of the plant. Any progressions in endophyte communities seen with changes in temperature are credited to the plant species variety being influenced by warming, leading to impacts on endophytic fungal communities. Endophytic fungi give an upper hand to their host plant by upgrading the protection from natural stresses (Chanda et al., 2020). At high soil temperatures endophytic fungus *Curvularia protuberata* has been associated with the survival of the grass *Dichanthelium lanuginosum* (Khare et al., 2018). In plant–microorganism relationship, and it is accepted that the impact of an ecological condition would be forced on both the plant and the microorganism. Velásquez et al. (2018) recently studied how natural conditions influence pathogenic organisms, as every pathogen has an ideal range of temperature for its growth and virulence. In *Agrobacterium* infection, the raised temperature appear to hinder type IV secretion associated pilus formation and expression of virulence (vir) genes (Velásquez et al., 2018). *Pectobacterium atrosepticum* a soft-rot bacterium demonstrated increased virulence at raised temperature, which is related to higher production of plant cell wall-degrading enzymes, quorum-sensing signals, and accelerated disease development. It is frequently not satisfactory whether conclusions made on in vitro information on temperature impacts on organisms consistently reflect what happens during a functioning in active plant interaction. This was particularly obvious in the investigation of the effect of temperature on type III secretion of *P. syringae*. While it has been very much recorded that raised temperature adversely influences type III secretion in vitro, increased type III translocation of effectors into host plants was identified during PstDC3000 disease in Arabidopsis at the raised temperature (Huot et al., 2017). Therefore, it would be desirable if future investigations evaluate environmental consequences on organisms and that should include more analysis in plant and utilized new procedures (e.g., double RNA sequencing [RNA-seq]) to uncover both host and micro-organism changes (Nobori and Tsuda, 2018).

Interaction between plants and microbe helps both the partners to adapt to the temperature challenges. The negative effect of temperature stress on plants could be alleviating by some rhizospheric bacteria and endophytes, they can also extend the capacity of host plants to grow at a various range of temperatures. A fascinating example is the association between tropical panic grass *Dichanthelium lanuginosum* and the fungus *Curvularia protuberata*, which permits both the life forms to develop at increased soil temperatures, while independently; neither the plant nor the fungi can survive at this condition. Moreover, *C. protuberata* requires infection by *Curvularia* thermal tolerance virus to confer heat tolerance to its host plant. Other than panic grass, C. Protuberate grant heat resistance that could be seen in tomato, proposing that the primary mechanism might be extensively relevant to assist different plants with coping with a raised temperature. A few microorganisms can even support plants to adapt to various stresses. An exciting example is a PsJN strain of *Burkholderia phytofirmans* which improves plant flexibility to heat in tomato, cold in grapevine, dry spell in wheat, and salt and freezing in Arabidopsis (Issa et al., 2018; Miotto-Vilanova et al., 2016).

5.5.3   *EFFECT OF DROUGHT STRESS*

Drought is one of the most studied abiotic stresses and an additional consequence of global warming factors with significant and adverse environmental and agronomic effects. The drought stress happens when soil water accessibility is decreased to crucial levels that stop the growth and expected performance of plants (Feller and Vaseva, 2014). Drought is also accountable for the decrease of plant growth and the roots, and aerial plant parts might be compressed. In light of the dry season, in any case, plants can show a set-up of physiological, biochemical, and morphological acclimatization to lessen the negative effect of the stress on fitness (Tátrai et al., 2016). As per the recent reports, symbiotic microorganisms might balance these phenotypic changes that permit plants to meet distressing circumstances (Dupont et al., 2015; Wani et al., 2015). AMF and ECM colonization and community structure can be significantly affected by drought. Arbuscular formation and hyphae enlargement of *Glomus* spp. (e.g., strain ZAC-19) was found to improve by drought, but at the same time, its colonization was found to be reduced (*Glomus fasciculatum*). This may lead to changes in the dissemination of photosynthesis in the rhizosphere just as in extra mycorrhizal mycelium.

Different plant species, mainly forest trees, communicate with ECM fungi which together with AMF provide organic nitrogen, phosphate, and various micronutrients to their plant host and upgrade seedling endurance. ECM colonization and community structure can be significantly affected by drought. Drought is also responsible for the diminishing of mycorrhizal fungal colonization of some plant species, for example, Norway spruce trees, a few Mediterranean bushes, some *Pinus* spp. (e.g., *P. oaxacana and P. muricata colonization with Rhizopogon* spp.), but not for *Pinus taeda* seedlings. Besides, a specific adaptation in colonization by some ECM species appears to happen, which may lead to community changes under diminished soil water accessibility. Henceforth, ECM strains carry beneficial impacts to plants considerably under drought conditions and subsequently may discover application in reforestation by the soothing of drought pressure in plants. *C. geophilumis* symbiosis gets revitalized under low water content as this parasite perseveres preferably under drought stress over others (Chanda et al., 2020). Different essential functions of plants straightforwardly influencing growth, development and tolerance to stress condition are modulated by endophytes (Wani et al., 2015). Secretion of various chemical metabolites factor (e.g., Growth hormones, other secondary metabolites), gene expression modulation, and secondary metabolic pathways of host plant etc. are the mechanisms that are used by endophytes while dealing with drought tolerance (Bacon and White, 2016; Deepika et al., 2016; Ramos et al., 2018). Plants that are inoculated with endophyte can show higher efficiency of photosynthesis as well as production of less amount of reactive oxygen species (ROS), that is under the control of endophyte-mediated upgradation of the plant antioxidant system (Azad and Kaminskyj, 2016). It was found that at molecular level, the inoculation of *Lolium perenne* plant with the leaf endophytic fungi *Epichloe festucae* (strain Fl1), resulted into changes in the expression of many host genes leading to the activation of secondary metabolism (Dupont et al., 2015). In harsh conditions which force extreme limitations for growth, the foundation of useful symbioses with microorganisms can assume a central role in adjustment of plants to higher solutes accumulation, diminished foliar conductance, diminished transpiration, or the thicker cuticles formation (Timmusk et al., 2014).

Ghaffari et al. (2016) revealed that the microbial presence contributes to improve the drought resistance in plants all through articulation of various osmoprotective pathways as well as differential gene expression (Ghaffari et al., 2016). Accumulation of greater amounts of solutes in tissues of vascular plants that permit osmotic alteration and helps to keep up physiological and biochemical processes is one of the mechanisms that is induced by endophytes in plants (Li et al., 2018). Experiment conducted by Hereme et al. (2020) demonstrates that the presence of endophytes assumes a positive role in the response modulation of *C. quitensis* to water stress at the biochemical (total soluble sugars [TSS], proline, and thiobarbituric acid-reactive substances [TBARS]), molecular (CqNCED1, CqABCG25, and CqRD22) and morphological (stomatal opening) levels (Hereme et al., 2020). This shows that their connection with endophytes is an effective strategy that initiates a superior response to water shortage in the plant. Plant-associated bacteria and endophytes specifically, may give increased osmotic (and likely also drought) stress resilience through balancing the expression of genes. Nonetheless, comprehensive knowledge of how to plant gene regulation is influenced by microbes and how modified ecological conditions due to climate change influence this communication is as yet lacking.

Plant–microbiome interaction involving plant growth-promoting bacteria is also affected by drought. Santos-Medellı'n et al. (2017) studied the drought influenced microbial community composition in soil, root endosphere, and rhizosphere and he found that if the community is more intimately associated with roots, then there will occur more significant shift in the composition of drought-stressed rice plants (Santos-Medellín et al., 2017). Also, in an investigation to look at the impacts of soil moisture on root microbiome of sorghum, Xu et al. (2018) found that in surrounding soil bacterial community, diversity is unaffected and unchanged for the most parts, drought fundamentally decreased diversity in the rhizosphere and the root endosphere (Xu et al., 2018).

Within the host, drought stress causes a shift in root metabolites. Regardless of whether and how this drought enhanced metabolite causes root microbiome composition to advance stress response in plants stays to be resolved. This intriguing connection recommends that under drought condition, there might be molecular communication among plants and related microbiome to restructure root microbiota to adapt to drought stress. Translating this molecular communication should propel our necessary information essential to utilize microbiota to upgrade drought resilience in crop plants. Drought stress influences agricultural productivity universally and is probably going to increase further. Interactions of rhizobacteria could assume a significant role in mitigating drought stress in plants. These rhizobacteria colonize the rhizosphere of plants and confer drought resistance by up-regulation or downregulation of genes that can respond to drought, for example, ascorbate peroxidase, S-adenosyl-methionine synthetase, and heat shock protein (Igiehon and Babalola, 2018).

According to Márquez-García et al. (2015), there is a formation of determinate nodules within the soybean plant as a result of symbiotic interaction between soybean and *Bradyrhizobium*. This symbiotic association is essential and satisfactory to meet the nitrogen needs of the soybean crop. It was noticed that when soybean plants are exposed to extreme drought conditions, number of nodules become diminished. Drought tolerance of nodules has been connected to the capacity to support the photosynthate supply to the root nodules during drought and to magnify the nodule biomass. The connections between the intensity and frequency of nodulation, root development, and architecture are still inadequately studied, especially those factors that control nodule thickness per unit root length in the absence and presence of stress (Marquez-Garcia et al., 2015).

5.6 EFFECT OF XENOBIOTICS ON PLANT MICROORGANISM INTERACTION

In many developing countries, recent farming ways follow unsustainable practices and have brought about a too enormous amount of unhealthful effluents being discharged into the air, soil, and water. Currently, various chemicals used in agriculture (i.e., pesticides, herbicides, fungicides, rodenticides) and synthetic manures getting utilized non-wisely that have unfavorably influenced beneficial soil microbiota. Applied science and nanomaterials have additionally confused the condition of inputs of soil and their degradation (Mishra et al., 2017). As per published reports, the indiscriminate use of xenobiotics ends up in ecological imbalances and environmental issues (Li and Jennings, 2017). Pesticides are xenobiotics that are significant agents in modern agriculture because of their effect on crop productivity and overall profit. The pesticide is used to control or kill the pests. Some examples of pesticides include herbicides, insecticides, and fungicides. These pesticides sometimes applied directly to the soil or can be involved in the form of sprays, which results in deposition of pesticide particles in nature.

Additionally, different complex chemical compounds used in the agriculture sector also contaminate the soil, and they can kill beneficial soil microflora. It is estimated that the majority of pesticides instead of attacking pests attack nontargeted healthy soil microbiota. In any case, the overuse of pesticides prompts the debasement of soil's microbial ecosystems. The emergence of recent pests is one among the main consequences of global climate change. Farmers select continual use of chemical pesticides for the management of pests (González-Alcaraz and van Gestel, 2016). Pesticides in the soil must undergo degradation, transport, and adsorption/desorption depending on the complexity of chemical or pesticides and soil properties. The random use of pesticides releases recalcitrant xenobiotic compounds like fuels, solvents, polycyclic aromatic hydrocarbons (PAHs), and alkanes into the environment. Within the soil system, pesticides may affect the structure and function of microbes and may change the physiological and biochemical conduct of soil microorganisms. Pesticides application may likewise hinder or destroy certain types of microorganisms and outnumber other types by discharging them from the competition.

Majority of plants form a symbiotic association with root-colonizing microorganisms, for example, bacteria and arbuscular mycorrhizal (AM) fungi, and this helps in plant growth promotion under normal as well as stressed condition. The functioning of these microbes could be affected by exogenous applications of pesticides. Some pesticides like chlorimuronethyl, chlorothalonil could diminish the wealth of genes that are engaged with nitrogen cycling. Pesticides could affect nodulation and biological nitrogen fixation in legumes by influencing virulence of attacking nodular microbes and the root fibers of the plants. Another significant reason for the deterioration of the environment is PAHs (Anthracene, Fluorene, Acenaphthene, Phenanthrene, etc.) (Abdel-Shafy and Mansour, 2016). Most PAHs are recalcitrant, hydrophobic. Thus, they accumulate as an environmental pollutant and persist in nature for a long time. Therefore, studying the response of native microbes and the strategy employed by them for degradation of PAHs has become essential. Various investigations have discovered that the PAHs is not the only issue that affects the microbial community. Sometimes, environmental factors like pH also matters as opposed to xenobiotics alone (Wu et al., 2016). To feed and supply the plentiful amounts of nutrients to an ever-increasing human population without disturbing the environment is the biggest challenge for the agriculture sector. Pesticides and their utilization are considered as enchantment projectiles in developing countries. Pesticides are very detrimental to the environment and humans because a ton of pesticides and their derivatives accumulates in the soil for a long period. The pesticides adversely influence biological functions, biochemistry, community structure, and diversity of microbes. The fertility of the soil is also greatly affected by pesticides. Controlled and sensible utilization of pesticides is needful because if their application is exceeding than the recommended level, then they can harm the ecosystem more destructively. The training of farmers, wholesalers, and industry policymakers for the controlled use of pesticide is essential to decrease the hazardous consequences of pesticides on humans and nature. Planned investigations are required for reducing the impact of pesticides on microbial communities, and their ecotoxicological impacts in the soil environment.

5.6.1 THE TERRESTRIAL AND AQUATIC ECOSYSTEM

Anthropogenic activities are responsible for climate change. In the case of the soil ecosystem, discharge of GHGs occurs through farmland, and it represents the combined impact of man-made as well as microbial activities. Plants are vital markers for the results of outcomes on soil organisms, while in the case of the aquatic ecosystem, no such different thing is concerned. In the emission and removal of GHGs, both terrestrial microorganisms and microorganisms from marine environments contribute a lot. The difficulty of CH_4 leaks occurs exclusively in seas and epitomizes how microorganisms apply a dominant impact on temperature change. By comparing the two ecosystems, it is found that each encompasses a variety of perspectives distinctive to themselves. The matter of permafrost is a kind of vital as far as the terrestrial ecosystem is concerned. It shows an endless cycle of adverse effects within which the world of microbes plays the essential role of catalyst and amazingly well builds up against them as drivers of temperature alteration. These successively have an impact on fluctuations of the microbial population and their activity. Presently, as plants are suffering via seasonal differences, the consequences of seasons are probably going to have on terrestrial microbes. This conjointly gives an extent of investigation on species of plant, which might be successfully used to configure the community of soil microorganism. In this manner, it is clearly understood that though the primary effect of temperature change is generally similar on all organisms, the power of resultant effects and the pathway of incidence differs between the two types of ecosystems. Each has its options that apply their products (Dutta and Dutta, 2016). Oceans additionally influence climate by discharging enormous amounts of dimethyl sulfide (DMS). It is removed because of degradation of dimethylsulfoniopropionate (DMSP) by bacteria and other eukaryotes by some microalgae through osmoprotectant. The dimethyl sulfide is transformed into a vaporized type of sulfate that later behaves as condensation nuclei in the cloud. DMS acts as a carrier of sulfur from the seas to the land, responsible for the precipitation of sulfuric acid with rainwater.

5.6.2   THE NECESSITY OF PROPER STUDY

In response to climate change, microorganisms have either positive or negative feedback. Feedbacks to climate change by benthic coral reef organisms can be taken as an example. A case of positive microbial feedback is the way that warming accelerates the decay of organic matter present in the soil and in this way will increase the flux of carbon from soil to the air. However, the effects of expanding air CO_2 levels may be cradle by microbes and subsequently hamper the change in temperature and apply a dominant impact. A right valuation for the propelled ecologies and response to worldwide temperature change and diverse anthropogenic components is straight away required, so these microbial ecologies can be utilized for the alleviation of climate change. Budzianowski (2015) has recommended declining CO_2 discharges from the environment as well as evacuating atmospheric CO_2 for the management of carbon (Budzianowski, 2015). Microorganisms have the capacity for each and subsequently may serve an instrument for battling temperature change. National Aeronautics and Space Administration classify ways to deal with environmental change into mitigation (lessening gas outflows and stabilizing their concentration) and adaptation (adjusting to the environmental change that is occurring). For all the earlier temperature change factors together with microorganisms that should be investigated, should be appropriately manipulated. Thus, there is a requirement to grasp a response of microbes to warming by building up a translucent system of microorganisms, from the hereditary and the physiological level up to composition and collaborations. However, the contradiction is that microorganisms have not been enough investigated, but even have been ignored in this matter. As an example, the majority of the investigations of temperature change on soil microorganisms have targeted around net parameters, similar to biomass, action of enzymes, or essential community profiles (Zheng et al., 2019).

Yet, climate change also depends on an under-appreciated factor known as "carbon-cycle feedbacks." The mode of action of microbes-mediated carbon cycle feedback with reference to temperature change remains baffling. Eventually, the net impact of environmental change on biological systems carbon budgets depends upon the harmony among photosynthesis and respiration (that is, autotrophic root respiration and heterotrophic respiration by soil microbes). While the facts about the assimilatory component (photosynthesis) of the carbon cycle and its response to environmental change are very much progressed, there is ambiguity in our comprehension of the response of soil respiration (Bardgett et al., 2008). Thorough analysis should be undertaken to evaluate the microbial response to the advancement of climate change. The identification and characterization of the microorganisms ought to be done on the premise of helpful and physiological roles in the vicinity of greenhouse emission. The data collected from this ought to be used to upgrade the comprehension of management of greenhouse outpouring to the air by microbes. Additionally, aboveground and underground communications nearby the nutrient athletics should be studied. This can be a result of plant association with communities within the soil, which is the most vital controller of soil nitrogen and carbon elements, has conjointly not been comprehended. There are no doubts that those changes in temperature affect micro soil flora, and thus the foundation and advancement of plant species. Subsequently, the responses of an ecosystem are fluctuating apparently. Another vital side is that the issue of permafrost has been recognized to be practically significant for mitigation of climate change (Budzianowski, 2015). However, defrosting of Arctic permafrost might be prevented if warming is limited to a worldwide average of 2°C. Around 200 nations have adopted climate agreement at the COP21 summit in Paris that intends to control the average global temperature rise of the planet within 20°C. Budzianowski (2015) has advised about the ideas for advanced carbon sequestration like soil carbon sequestration and GHG utilization to a helpful product like fertilizers and fuels (Budzianowski, 2015). On the other hand, to lessen N_2O discharge to the climate, the aim of the management should be to optimize the use of nitrogen fertilizer rather than merely decreasing its application. Emissions due to adding nitrogen to soils are the result of microbial soil activities. N_2O is released as a side product of nitrification and denitrification processes of the N_2 cycle (Winiwarter et al., 2020). By considering all the above realities, it is indeed known that the microbial side of temperature change might be a current issue of the present times. This reality must be appropriately examined and has the right to run due to significance. It should in this manner be fit in temperature change models, as any such model that neglects to incorporate microbial activities is insufficient. It is an issue of pressing demand as current human-made practices have made microorganisms to produce a tremendous amount of GHGs (Microbiology Online, 2015). In such a manner, many scientists have attempted to incorporate microbial information (biomass, growth kinetics, enzymes) into climate models. Nonetheless, for higher expectations, data associated with the diversity of microbes, their strength, community and structure alongside their physiological talents should be considered.

5.6.3   FUTURE OUTLOOK

Microorganisms can effectively complete the bioremediation of a few xenobiotic compounds present in soil with the utilization of modern high throughput molecular tools and through genetic manipulation techniques. For the detoxification of environmental pollutants, bioremediation is a promising technique. Future investigation dealing with the xenobiotic biodegradation should be centered on both the primary and applied territories. It is essential to understand the biochemistry and genetics of microbes for their fruitful application in the laboratory as well as in vivo conditions. Attempts are required to overcome any barrier between progress at laboratory and field scales. Environmental parameters, for example, the physiological states of microbes, temperature, pH, moisture, nutrient level, xenobiotic concentration should be a concern while continuing toward practical bioremediation of polluted sites. Emphasis should be given to the selection of perfect microorganisms with potentially high catabolic exercises, genes, plasmids, and catabolic enzymes for bio augmentation at the field scale (Li and Jennings, 2017).

The capability of efficient microorganisms could be executed in bioreactor systems at large scale with the goal of significant mitigation measures of pollutants. More significant trials are required to detect specific genes or enzymes that degrade specific recalcitrant compounds by utilizing genomics and proteomics-based investigations. Moreover, information regarding interaction among various microorganisms, for example, bacteria, fungi, actinomycetes, etc. are probably going to be valuable in the bioremediation of xenobiotics as synergistic associations could be liable for accelerated remediation. It is widely accepted that native microorganisms are the key players for keeping up plant well-being. In this manner, it is pivotal to promote useful microbial communities (González-Alcaraz and van Gestel, 2016).

The scientists should address the impact of climate change on plant pathogens. Till date, plant disease dynamics concerning models of climate change have been inadequately investigated. Therefore, climate models are required to understand better the areas with elevated levels of vulnerability, particularly regarding regional-scale changes. There is a need to adopt the interdisciplinary approach both by plant biotechnologists and plant pathologists for characterizing essential procedures and factors impacting crucial plant physiological procedures for controlling diverse plant infections to improve food security.

5.7 CONCLUSION

Microorganisms play a significant role in regulating the concentration of greenhouse gas within the atmosphere. However, it still cannot seem to be appropriately studied and accepted by the research community. Interactions of plant and soil communities could also be random while observing their responses toward natural fluctuations in climate. The potential output of these microorganisms could be detained by considering the immediate and indirect impacts of climate change on them. For controlling climate change microorganisms could be a significant regular asset. It is the right time to review this perspective, comprehend the acting mechanism more accurately, and subsequently appropriately use it for creating solutions. Combinations of observation, experimental testing and modeling of plants and soil microbial communications in response to climate are essential to foresee the function of the future ecosystem.

KEYWORDS

- **climate change**
- **global warming**
- **plant-microbe interaction**
- **carbon dioxide**
- **ectomycorrhiza**

REFERENCES

Abatenh, E.; Gizaw, B.; Tsegaye, Z.; Tefera, G. Microbial Function on Climate Change—A Review. *Environ. Pollut. Clim. Change* **2018**, *02* (01), 1–6. https://doi.org/10.4172/2573-458X.1000147.

Abdel-Shafy, H. I.; Mansour, M. S. M. A Review on Polycyclic Aromatic Hydrocarbons: Source, Environmental Impact, Effect on Human Health and Remediation. Egyptian Journal of Petroleum. Egyptian Petroleum Research Institute March 1, 2016, pp 107–123. https://doi.org/10.1016/j.ejpe.2015.03.011.

Agbodjato, N. A.; Noumavo, P. A.; Baba-Moussa, F.; Salami, H. A.; Sina, H.; Sèzan, A.; Bankolé, H.; Adjanohoun, A.; Baba-Moussa, L. Characterization of Potential Plant Growth Promoting Rhizobacteria Isolated from Maize (Zea Mays L.) in Central and Northern Benin (West Africa). *Appl. Environ. Soil Sci.* **2015**, 2015, 1–9. https://doi.org/10.1155/2015/901656.

AR5 Climate Change 2014: Impacts, Adaptation, and Vulnerability — IPCC. https://www.ipcc.ch/report/ar5/wg2/ (accessed Jun 7, 2020).

Aspects of India's Economy. https://www.rupe-india.org/about.html (accessed Jun 5, 2020).

Augé, R. M.; Toler, H. D.; Saxton, A. M. Arbuscular Mycorrhizal Symbiosis Alters Stomatal Conductance of Host Plants More under Drought than under Amply Watered Conditions: A Meta-Analysis. *Mycorrhiza* **2014**, *25* (1), 13–24. https://doi.org/10.1007/s00572-014-0585-4.

Azad, K.; Kaminskyj, S. A Fungal Endophyte Strategy for Mitigating the Effect of Salt and Drought Stress on Plant Growth. *Symbiosis* **2016**, *68* (1–3), 73–78. https://doi.org/10.1007/s13199-015-0370-y.

Bacon, C. W.; White, J. F. Functions, Mechanisms and Regulation of Endophytic and Epiphytic Microbial Communities of Plants. *Symbiosis*; Springer Netherlands, March 1, **2016**, 87–98. https://doi.org/10.1007/s13199-015-0350-2.

Bardgett, R. D.; Freeman, C.; Ostle, N. J. Microbial Contributions to Climate Change through Carbon Cycle Feedbacks. *ISME J.* Nature Publishing Group, August 10, **2008**, 805–814. https://doi.org/10.1038/ismej.2008.58.

Bennett, J. A.; Klironomos, J. Mechanisms of Plant–Soil Feedback: Interactions among Biotic and Abiotic Drivers. *New Phytol.* **2019**, *222* (1), 91–96. https://doi.org/10.1111/nph.15603.

Berendsen, R. L.; van Verk, M. C.; Stringlis, I. A.; Zamioudis, C.; Tommassen, J.; Pieterse, C. M. J.; Bakker, P. A. H. M. Unearthing the Genomes of Plant-Beneficial Pseudomonas Model Strains WCS358, WCS374 and WCS417. *BMC Genom.* **2015,** *16* (1). https://doi.org/10.1186/s12864-015-1632-z.

Botnen, S.; Kauserud, H.; Carlsen, T.; Blaalid, R.; Høiland, K. Mycorrhizal Fungal Communities in Coastal Sand Dunes and Heaths Investigated by Pyrosequencing Analyses. *Mycorrhiza* **2015,** *25* (6), 447–456. https://doi.org/10.1007/s00572-014-0624-1.

Budzianowski, W. M. Single Solvents, Solvent Blends, and Advanced Solvent Systems in CO_2 Capture by Absorption: A Review. Int. J. Global Warm. Inderscience Publishers, January 2, **2015,** 184–225. https://doi.org/10.1504/IJGW.2015.067749.

Chanda, A.; Maghrawy, H.; Sayour, H.; Gummadidala, P. M.; Gomaa, O. M. Impact of Climate Change on Plant-Associated Fungi. *Microbiol. Res.* **2020,** *183*, 83–96. https://doi.org/10.1007/978-3-030-41629-4_5.

Chhipa, H.; Deshmukh, S. K. Fungal Endophytes: Rising Tools in Sustainable Agriculture Production. In *Endophytes and Secondary Metabolites*, 2019; pp 1–24. https://doi.org/10.1007/978-3-319-76900-4_26-1.

Cotton, T. E. A.; Fitter, A. H.; Miller, R. M.; Dumbrell, A. J.; Helgason, T. Fungi in the Future: Interannual Variation and Effects of Atmospheric Change on Arbuscular Mycorrhizal Fungal Communities. *New Phytol.* **2015,** *205* (4), 1598–1607. https://doi.org/10.1111/nph.13224.

Coyle, D. R.; Nagendra, U. J.; Taylor, M. K.; Campbell, J. H.; Cunard, C. E.; Joslin, A. H.; Mundepi, A.; Phillips, C. A.; Callaham, M. A. Soil Fauna Responses to Natural Disturbances, Invasive Species, and Global Climate Change: Current State of the Science and a Call to Action. *Soil Biol. Biochem.* **2017,** *110*, 116–133. https://doi.org/10.1016/j.soilbio.2017.03.008.

Deepika, V. B.; Murali, T. S.; Satyamoorthy, K. Modulation of Genetic Clusters for Synthesis of Bioactive Molecules in Fungal Endophytes: A Review. *Microbiol. Res*. Elsevier GmbH, January 1, **2016,** 125–140. https://doi.org/10.1016/j.micres.2015.10.009.

Dupont, P. Y.; Eaton, C. J.; Wargent, J. J.; Fechtner, S.; Solomon, P.; Schmid, J.; Day, R. C.; Scott, B.; Cox, M. P. Fungal Endophyte Infection of Ryegrass Reprograms Host Metabolism and Alters Development. *New Phytol.* **2015,** *208* (4), 1227–1240. https://doi.org/10.1111/nph.13614.

Dutta, H.; Dutta, A. The Microbial Aspect of Climate Change. *Energ. Ecol. Environ.* **2016,** *1* (4), 209–232. https://doi.org/10.1007/s40974-016-0034-7.

Feller, U.; Vaseva, I. I. Extreme Climatic Events: Impacts of Drought and High Temperature on Physiological Processes in Agronomically important Plants. *Front. Environ. Sci*; Frontiers Media SA; October 6, **2014**. https://doi.org/10.3389/fenvs.2014.00039.

Fischer, D. G.; Chapman, S. K.; Classen, A. T.; Gehring, C. A.; Grady, K. C.; Schweitzer, J. A.; Whitham, T. G. *Plant Genetic Effects on Soils under Climate Change. Plant and Soil*; Kluwer Academic Publishers, November 29, 2014; pp 1–19. https://doi.org/10.1007/s11104-013-1972-x.

Food and Agriculture Organization of the United Nations: Climate Change, Agriculture and Food Security. http://www.fao.org/publications/sofa/2016/en/ (accessed Jun 7, 2020).

Ghaffari, M. R.; Ghabooli, M.; Khatabi, B.; Hajirezaei, M. R.; Schweizer, P.; Salekdeh, G. H. Metabolic and Transcriptional Response of Central Metabolism Affected by Root

Endophytic Fungus Piriformospora Indica under Salinity in Barley. *Plant Mol. Biol.* **2016,** *90* (6), 699–717. https://doi.org/10.1007/s11103-016-0461-z.

Global Climate Change Vital Signs of the Planet. Solutions Mitigation and Adaptation. https://climate.nasa.gov/solutions/adaptation-mitigation/ (accessed Jun 6, 2020).

González-Alcaraz, M. N.; van Gestel, C. A. M. Toxicity of a Metal(Loid)-Polluted Agricultural Soil to Enchytraeus Crypticus Changes under a Global Warming Perspective: Variations in Air Temperature and Soil Moisture Content. *Sci. Total Environ.* **2016,** *573*, 203–211. https://doi.org/10.1016/j.scitotenv.2016.08.061.

Green Facts. Impacts of a 4°C Global Warming. https://www.greenfacts.org/en/impacts-global-warming/l-2/index.htm (accessed Jun 5, 2020).

Gupta, C.; Dhan, P.; Gupta, S. Role of Microbes in Combating Global Warming. *Int. J. Pharm. Sci. Lett.* **2014,** *4* (2), 359–363.

Hereme, R.; Morales-Navarro, S.; Ballesteros, G.; Barrera, A.; Ramos, P.; Gundel, P. E.; Molina-Montenegro, M. A. Fungal Endophytes Exert Positive Effects on Colobanthus Quitensis under Water Stress but Neutral Under a Projected Climate Change Scenario in Antarctica. *Front. Microbiol.* **2020,** *11* (264), 2–12. https://doi.org/10.3389/fmicb.2020.00264.

Huang, J.; Yu, H.; Guan, X.; Wang, G.; Guo, R. Accelerated Dryland Expansion under Climate Change. *Nat. Clim. Change* **2016,** *6* (2), 166–171. https://doi.org/10.1038/nclimate2837.

Huot, B.; Castroverde, C. D. M.; Velásquez, A. C.; Hubbard, E.; Pulman, J. A.; Yao, J.; Childs, K. L.; Tsuda, K.; Montgomery, B. L.; He, S. Y. Dual Impact of Elevated Temperature on Plant Defence and Bacterial Virulence in Arabidopsis. *Nat. Commun.* **2017,** *8* (1). https://doi.org/10.1038/s41467-017-01674-2.

Igiehon, N. O.; Babalola, O. O. Below-Ground-above-Ground Plant-Microbial Interactions: Focusing on Soybean, Rhizobacteria and Mycorrhizal Fungi. *Open Microbiol. J.* **2018,** *12* (1), 261–279. https://doi.org/10.2174/1874285801812010261.

Issa, A.; Esmaeel, Q.; Sanchez, L.; Courteaux, B.; Guise, J. F.; Gibon, Y.; Ballias, P.; Clément, C.; Jacquard, C.; Vaillant-Gaveau, N.; Aït Barka, E. Impacts of Paraburkholderia Phytofirmans Strain PsJN on Tomato (Lycopersicon Esculentum L.) under High Temperature. *Front. Plant Sci.* **2018,** *871*, 1397–1397. https://doi.org/10.3389/fpls.2018.01397.

Jiang, Y.; Wang, W.; Xie, Q.; Liu, N.; Liu, L.; Wang, D.; Zhang, X.; Yang, C.; Chen, X.; Tang, D.; Wang, E. Plants Transfer Lipids to Sustain Colonization by Mutualistic Mycorrhizal and Parasitic Fungi. *Science* (80-.). **2017,** *356* (6343), 1172–1173. https://doi.org/10.1126/science.aam9970.

Jing, X.; Sanders, N. J.; Shi, Y.; Chu, H.; Classen, A. T.; Zhao, K.; Chen, L.; Shi, Y.; Jiang, Y.; He, J.-S. The Links between Ecosystem Multifunctionality and Above- and Belowground Biodiversity are Mediated by Climate. *Nat. Commun.* **2015,** *6* (1), 8159. https://doi.org/10.1038/ncomms9159.

Khare, E.; Mishra, J.; Arora, N. K. Multifaceted Interactions between Endophytes and Plant: Developments and Prospects. *Front. Microbiol*; Frontiers Media SA; November 15, **2018**. https://doi.org/10.3389/fmicb.2018.02732.

Kulkarni, S. *Petroleum and Refining Sector: Towards Cost Effective and Sustainable Practices*; Google Books.

Langley, J. A.; Hungate, B. A. Plant Community Feedbacks and Long-Term Ecosystem Responses to Multi-Factored Global Change. *AoB Plants* **2014,** *6*, plu035–plu035. https://doi.org/10.1093/aobpla/plu035.

Latake, P. T.; Pawar, P.; Ranveer, A. C. The Greenhouse Effect and Its Impacts on Environment. *IJIRCT* **2015,** *1* (3), 333–337.

Leggett, J. A. *Climate Change: Science Highlights: Congressional Research Service*, 2009.

Li, X.; He, X.; Hou, L.; Ren, Y.; Wang, S.; Su, F. Dark Septate Endophytes Isolated from a Xerophyte Plant Promote the Growth of Ammopiptanthus Mongolicus under Drought Condition. *Sci. Rep.* **2018,** *8* (1), 1–11. https://doi.org/10.1038/s41598-018-26183-0.

Li, Z.; Jennings, A. Worldwide Regulations of Standard Values of Pesticides for Human Health Risk Control: A Review. *Int. J. Environ. Res. Public Health*. MDPI AG; July 22, **2017,** 826. https://doi.org/10.3390/ijerph14070826.

Liu, A.; Chen, S.; Chang, R.; Liu, D.; Chen, H.; Ahammed, G. J.; Lin, X.; He, C. Arbuscular Mycorrhizae Improve Low Temperature Tolerance in Cucumber via Alterations in H2O2 Accumulation and ATPase Activity. *J. Plant Res.* **2014,** *127* (6), 775–785. https://doi.org/10.1007/s10265-014-0657-8.

Live Science: What Is the Greenhouse Effect? https://www.livescience.com/37743-greenhouse-effect.html (accessed Jun 5, 2020).

Marquez-Garcia, B.; Shaw, D; Cooper, J. W.; Karpinska, B; Quain, M. D.; Makgopa, E. M.; Kunert, K.; Foyer, C. H. Redox Markers for Drought-Induced Nodule Senescence, a Process Occurring after Drought-Induced Senescence of the Lowest Leaves in Soybean (Glycine Max). (4), 497–510 | 10.1093/Aob/Mcv030. *Ann. Bot.* **2015,** *116* (4), 497–510.

Melillo, J. M.;Terese, R.;Yohe, G. W. Overview and Report Findings Climate Change Impacts in the United States, 2014. https://doi.org/10.7930/J0Z31WJ2

Microbiology Online. Microbes and Climate Change, 2015.

Miliute, I.; Buzaite, O.; Baniulis, D.; Stanys, V. Bacterial Endophytes in Agricultural Crops and Their Role in Stress Tolerance: A Review. *Zemdirbyste-Agric.* **2015,** *102* (4), 465–478. https://doi.org/10.13080/z-a.2015.102.060.

Miotto-Vilanova, L.; Jacquard, C.; Courteaux, B.; Wortham, L.; Michel, J.; Clément, C.; Barka, E. A.; Sanchez, L. Burkholderia Phytofirmans PsJN Confers Grapevine Resistance against Botrytis Cinerea via a Direct Antimicrobial Effect Combined with a Better Resource Mobilization. *Front. Plant Sci.* **2016,** *7* (AUG2016), 1236. https://doi.org/10.3389/fpls.2016.01236.

Mishra, P. K.; Gregor, T.; Wimmer, R. BSG for Cellulose Nanofibres; **2017,** *12*.

Mohan, J. E.; Cowden, C. C.; Baas, P.; Dawadi, A.; Frankson, P. T.; Helmick, K.; Hughes, E.; Khan, S.; Lang, A.; Machmuller, M.; Taylor, M.; Witt, A.; Hobbie, E.; Mohan, J. E. Mycorrhizal Fungi Mediation of Terrestrial Ecosystem Responses to Global Change: Mini-Review Science Direct. *Fungal Ecol.* **2014,** 1–17. https://doi.org/10.1016/j.funeco.2014.01.005.

Nath Yadav, A.; Author, C. Endophytic Fungi for Plant Growth Promotion and Adaptation under Abiotic Stress Conditions. *Acta Sci.* **2018**.

Nobori, T.; Tsuda, K. In Planta Transcriptome Analysis of Pseudomonas Syringae. Bio-Protocol **2018,** *8* (17). https://doi.org/10.21769/bioprotoc.2987.

Ramos, P.; Rivas, N.; Pollmann, S.; Casati, P.; Molina-Montenegro, M. A. Hormonal and Physiological Changes Driven by Fungal Endophytes Increase Antarctic Plant Performance under UV-B Radiation. *Fungal Ecol.* **2018,** *34*, 76–82. https://doi.org/ 10.1016/j.funeco.2018.05.006.

Resquín-Romero, G.; Garrido-Jurado, I.; Delso, C.; Ríos-Moreno, A.; Quesada-Moraga, E. Transient Endophytic Colonizations of Plants Improve the Outcome of Foliar

Applications of Mycoinsecticides against Chewing Insects. *J. Invertebr. Pathol.* **2016,** *136,* 23–31. https://doi.org/10.1016/j.jip.2016.03.003.

Santos-Medellín, C.; Edwards, J.; Liechty, Z.; Nguyen, B.; Sundaresan, V. Drought Stress Results in a Compartment-Specific Restructuring of the Rice Root-Associated Microbiomes. *MBio* **2017,** *8* (4). https://doi.org/10.1128/mBio.00764-17.

Santoyo, G.; Moreno-Hagelsieb, G.; del Carmen Orozco-Mosqueda, M.; Glick, B. R. Plant Growth-Promoting Bacterial Endophytes. *Microbiol. Res.* Elsevier GmbH; February 1, **2016,** 92–99. https://doi.org/10.1016/j.micres.2015.11.008.

Science Daily: Emissions of Potent Greenhouse Gas Have Grown, Contradicting Reports of Huge Reductions. https://www.sciencedaily.com/releases/2020/01/200121113039. htm (accessed Jun 7, 2020).

Spatafora, J. W.; Chang, Y.; Benny, G. L.; Lazarus, K.; Smith, M. E.; Berbee, M. L.; Bonito, G.; Corradi, N.; Grigoriev, I.; Gryganskyi, A.; James, T. Y.; O'Donnell, K.; Roberson, R. W.; Taylor, T. N.; Uehling, J.; Vilgalys, R.; White, M. M.; Stajich, J. E. A Phylum-Level Phylogenetic Classification of Zygomycete Fungi Based on Genome-Scale Data. *Mycologia* **2016,** *108* (5), 1028–1046. https://doi.org/10.3852/16-042.

Sun, Z.; Song, J.; Xin, X.; Xie, X.; Zhao, B. Arbuscular Mycorrhizal Fungal 14-3-3 Proteins Are Involved in Arbuscule Formation and Responses to Abiotic Stresses during AM Symbiosis. *Front. Microbiol.* Mar, **2018,** *9.* https://doi.org/10.3389/fmicb.2018.00091.

T. S., S. Fungal Endophytes: An Elective Review. *Kavaka* **2017,** *48* (1), 1–9.

Tátrai, Z. A.; Sanoubar, R.; Pluhár, Z.; Mancarella, S.; Orsini, F.; Gianquinto, G. Morphological and Physiological Plant Responses to Drought Stress in Thymus Citriodorus. *Int. J. Agron.* **2016,** *2016,* 1–8. https://doi.org/10.1155/2016/4165750.

Timmusk, S.; Abd El-Daim, I. A.; Copolovici, L.; Tanilas, T.; Kännaste, A.; Behers, L.; Nevo, E.; Seisenbaeva, G.; Stenström, E.; Niinemets, Ü. Drought-Tolerance of Wheat Improved by Rhizosphere Bacteria from Harsh Environments: Enhanced Biomass Production and Reduced Emissions of Stress Volatiles. *PLoS One* **2014,** *9* (5), e96086. https://doi.org/10.1371/journal.pone.0096086.

Tyagi, S.; Singh, R.; Javeria, S. Effect of Climate Change on Plant-Microbe Interaction: An Overview. *Eur. J. Mol. Biotechnol.* **2014,** *5* (3), 149–156. https://doi.org/10.13187/ejmb.2014.5.149.

UnitedNationsClimateChange:GlobalWarmingPotentials(IPCCSecondAssessmentReport). https://unfccc.int/process/transparency-and-reporting/greenhouse-gas-data/green-house-gas-data-unfccc/global-warming-potentials (accessed Jun 7, 2020).

Vandenkoornhuyse, P.; Quaiser, A.; Duhamel, M.; Le Van, A.; Dufresne, A. The Importance of the Microbiome of the Plant Holobiont. *New Phytol.* **2015,** *206* (4), 1196–1206. https://doi.org/10.1111/nph.13312.

Vega, F. E. The Use of Fungal Entomopathogens as Endophytes in Biological Control: A Review. *Mycologia.* NLM (Medline); January 1, **2018,** 4–30. https://doi.org/10.1080/00275514.2017.1418578.

Velásquez, A. C.; Castroverde, C. D. M.; He, S. Y. Plant–Pathogen Warfare under Changing Climate Conditions. *Curr. Biol*; Cell Press, May 21, **2018,** R619–R634. https://doi.org/10.1016/j.cub.2018.03.054.

Veresoglou, S. D.; Anderson, I. C.; de Sousa, N. M. F.; Hempel, S.; Rillig, M. C. Resilience of Fungal Communities to Elevated CO_2. *Microb. Ecol.* **2016,** *72* (2), 493–495. https://doi.org/10.1007/s00248-016-0795-8.

Vimal, S. R.; Singh, J. S.; Arora, N. K.; Singh, S. Soil-Plant-Microbe Interactions in Stressed Agriculture Management: A Review. *Pedosphere* **2017,** *27* (2), 177–192. https://doi.org/10.1016/S1002-0160(17)60309-6.

Wani, Z. A.; Ashraf, N.; Mohiuddin, T.; Riyaz-Ul-Hassan, S. Plant-Endophyte Symbiosis, an Ecological Perspective. *Appl. Microbiol. Biotechnol.* **2015,** *99* (7), 2955–2965. https://doi.org/10.1007/s00253-015-6487-3.

WHO. Health, Environment and Climate Change Coalition (HECC). https://www.who.int/globalchange/coalition/en/ (accessed Jun 5, 2020).

Williams, A.; Pétriacq, P.; Beerling, D. J.; Cotton, T. E. A.; Ton, J. Impacts of Atmospheric CO_2 and Soil Nutritional Value on Plant Responses to Rhizosphere Colonization by Soil Bacteria. *Front. Plant Sci.* **2018,** *9*, 1493. https://doi.org/10.3389/fpls.2018.01493.

Wilson, H.; Johnson, B. R.; Bohannan, B.; Pfeifer-Meister, L.; Mueller, R.; Bridgham, S. D. Experimental Warming Decreases Arbuscular Mycorrhizal Fungal Colonization in Prairie Plants along a Mediterranean Climate Gradient. *Peer J* **2016,** *2016* (6). https://doi.org/10.7717/peerj.2083.

Winiwarter, W.; Mohankumar, S. E. P. Reducing Nitrous Oxide Emissions from Agriculture: Review on Options and Costs; International Institute for Applied Systems Analysis. 2015, 1-19.

World Future Council: How Does Agriculture Contribute to Climate Change? https://www.worldfuturecouncil.org/how-does-agriculture-contribute-to-climate-change/ (accessed Jun 7, 2020).

Wu, M.; Dick, W. A.; Li, W.; Wang, X.; Yang, Q.; Wang, T.; Xu, L.; Zhang, M.; Chen, L. Bioaugmentation and Biostimulation of Hydrocarbon Degradation and the Microbial Community in a Petroleum-Contaminated Soil. *Int. Biodeterior. Biodegrad.* **2016,** *107*, 158–164. https://doi.org/10.1016/j.ibiod.2015.11.019.

Xu, L.; Naylor, D.; Dong, Z.; Simmons, T.; Pierroz, G.; Hixson, K. K.; Kim, Y.-M.; Zink, E. M.; Engbrecht, K. M.; Wang, Y.; Gao, C.; Degraaf, S.; Madera, M. A.; Sievert, J. A.; Hollingsworth, J.; Birdseye, D.; Scheller, H. V; Hutmacher, R.; Dahlberg, J.; Jansson, C.; Taylor, J. W.; Lemaux, P. G.; Coleman-Derr, D. Drought Delays Development of the Sorghum Root Microbiome and Enriches for Monoderm Bacteria. https://doi.org/10.1073/pnas.1717308115.

Yu, Z.; Li, Y.; Wang, G.; Liu, J.; Liu, J.; Liu, X.; Herbert, S. J.; Jin, J. Effectiveness of Elevated CO_2 Mediating Bacterial Communities in the Soybean Rhizosphere Depends on Genotypes. *Agric. Ecosyst. Environ.* **2016,** *231*, 229–232. https://doi.org/10.1016/j.agee.2016.06.043.

Zhang, T.; Yang, X.; Guo, R.; Guo, J. Response of AM Fungi Spore Population to Elevated Temperature and Nitrogen Addition and Their Influence on the Plant Community Composition and Productivity. *Sci. Rep.* **2016,** *6* (1), 1–12. https://doi.org/10.1038/srep24749.

Zheng, Q.; Hu, Y.; Zhang, S.; Noll, L.; Böckle, T.; Dietrich, M.; Herbold, C. W.; Eichorst, S. A.; Woebken, D.; Richter, A.; Wanek, W. Soil Multifunctionality Is Affected by the Soil Environment and by Microbial Community Composition and Diversity. *Soil Biol. Biochem.* **2019,** *136*, 107521. https://doi.org/10.1016/j.soilbio.2019.107521.

Zhu, X.; Song, F.; Liu, S.; Liu, F. Role of Arbuscular Mycorrhiza in Alleviating Salinity Stress in Wheat (Triticum Aestivum L.) Grown under Ambient and Elevated CO_2. *J. Agron. Crop Sci.* **2016,** *202* (6), 486–496. https://doi.org/10.1111/jac.12175.

CHAPTER 6

Climate Change and Microbial Aquatic Life

TANZEELA REHMAN[1*], SHAHNAZ BASHIR[1], MOHAZIB NABI[1], and SUHAIB A. BANDH[2]

[1]*Sri Pratap College Campus, Cluster University, Srinagar, India*

[2]*Government Degree College DH Pora Kulgam, Jammu and Kashmir, India*

Corresponding author. E-mail: tanzeelashah5@gmail.com

ABSTRACT

Frequently, the microbial world is left out of discussions concerning climate change and global warming. Perhaps, because these little creatures arc not as visible to us as polar bears or blue whales or are not as commercially significant as salmon or tuna. However, as key players in the carbon and other biogeochemical cycles, their responses to climate change need our attention. It is hard to anticipate how such fundamental changes to the Earth would influence microbial food webs and biogeochemical cycles at this stage. Numerous studies are now showing that the aquatic microorganisms may function as predictors of change and can even act as sentinels for monitoring the global climate change. But if we are to accurately forecast the aquatic ecosystem changes, then long-term biological observations of the aquatic environments is critical. In this context, the present chapter attempts to determine the influence of climate change on microbial aquatic life in order to better understand it.

6.1 INTRODUCTION

Global warming and climate change are the most significant issues of the current environmental scenarios, created by the higher levels of greenhouse gases in the air. Although the human sources and impacts of greenhouse gases have received considerable attention, the importance and impacts of microorganisms remain neglected. For over 12,000 years, the global climate has remained stable, and that stability is of a greater significance for humanity (NASA, 2015). Throughout Earth's history, microorganisms have been changing the environment and vice versa. With the unprecedented impact of climate change on the ground, microorganisms respond, adapt and evolve in their environment because they take only a few hours as their generation time (Jonathan, 2016). Climate change is likely to have significant effects on freshwater biodiversity across the Arctic and may lead to different adaptive responses. On a local scale, the expected shifts in mean values and variability of climate-related abiotic forces may alter local interactions (Stenseth et al., 2002). While, on a large scale, climate variables like temperature and evapotranspiration are robust predictors of aquatic and terrestrial diversity. The 1990s was the warmest decade recorded in modern times (IPCC, 2001), and there were rapid changes in the biodiversity patterns during this period (Belgrano et al., 2004; Straile, 2005). In various aquatic communities with effects manifested by variations in the use of different types of inorganic Nitrogen, amino acids, enzyme activity, cell breathing, and ATP response (Rose, 1967; Crawford et al., 1974; Reay et al., 1999).

The abundance of microbes is inversely linked to the quantity of temperature in aquatic ecosystems (Brown et al., 2004). The warming causes profound lake oxygen depletion to impede air exchange so that these sites do not support life (Union of Scientists concerned, 2011). Viscosity is another most important characteristic of water, based on temperature, whose variation has a significant impact on consumer carriage capacity, growth rates, and average predator density (Beveridge et al., 2010). The density of water is also affected by fluctuations of temperature, which affects the streaming and transportation of microbial nutrients (Glöckner et al., 2012). However, smaller phytoplanktons in aquatic ecosystems benefit from global warming, as these tiny microbes are of greater temperature tolerance (Huertas et al., 2011).

6.2 GLOBAL OCEANIC MICROBIOME: STRUCTURAL AND FUNCTIONAL ASPECTS

The total microbial richness of oceans that is estimated to be 37,470 operational taxonomic units (OUTs) (Sunagawa et al., 2015) generally increases from the surface to mesopelagic zones (Sunagawa et al., 2015), attributed to the particle-associated (PA) environment (Stocker, 2012; Sunagawa et al., 2015), besides low productivity (Pernthaler, 2005; Sunagawa et al., 2015). Members of proteobacteria such as SAR11 and SAR86 dominate oceans both in relative abundance and taxonomic richness (Morris et al., 2002; Dupont et al., 2012; Sunagawa et al., 2015). Cyanobacteria, Deferribacteres and Thaumarchaeota, are found in abundance, although their taxonomic richness is small. The presence of *Prochlorococcus* and *Synechococcus* in mesopelagic waters, along with their contribution of about 1% of abundance, suggests their role in sinking particle flux (Lochte and Turley, 1988; Sunagawa et al., 2015). Microbiota of polar and temperate sea regions may differ from each other at the surface level, but at the deep sea level is found similar owing to the oceanic circulations (Salazar et al., 2016; Cao et al., 2020). However, the similarity is found among the microbiota of Arctic and Antarctic waters given to identical environmental filtering and microbial dispersal mechanisms (Cao et al., 2020). Furthermore, 78% of OTUs are native to the Southern Ocean, and 70% of OTUs are unique to the Arctic ocean (Zhang et al., 2019; Cao et al., 2020). But only 27.69% orthologs are specific to the Antarctic microbiome suggesting that Arctic and Antarctic microbiota possess greater similarities at the functional level than the taxonomic level (Meie et al., 2014; Cao et al., 2020). More than 35,000 microbial species have been estimated in the oceans, and that 73% of the variation in composition is explained by depth (Sunagawa et al., 2015; Cao et al., 2020). Furthermore, temperature and salinity determine the bacterial and archaeal community structure, whereas N/P/Si is confirmed to shape eukaryotic communities. Surface seawater samples have shown SAR11, SAR86, *Rhodobacteraceae*, *Rhodospiriaceae*as prominent bacterial members in the Mediterranean Sea, *Prochlorococcus* in the Atlantic Ocean, and *Synechococcus* and SAR11 in the Brazilian coastal sea while as from Archaea, Atlantic has shown the predominance of *Euryarchaeota* and *Thaumarchaeota* in the Mediterranean Sea (Zhou et al., 2018). Nitrogen fixation is a regulatory mechanism in such ecosystems as a strong relationship exists between nitrogen fixation rates and plankton abundances. Microbial community structure is shaped by bottom-up processes in response to the shift in the abundance of high nucleic acid bacteria (Bock et al., 2018). Bacterial richness in the Northwest Atlantic Ocean has been found highest at the shelf break where the waters from both on-shelf and off-shelf meet. Besides, seasons, depth, temperature, oxygen, and salinity also determine the community structures (Zorz et al., 2019). In tropical and subtropical surface waters, different microbiota such as prokaryotic and picoeukaryotic are shaped by distinct ecological mechanisms. The picoeukaryotes are shaped by dispersal-limitation, while the prokaryotes are structured both by dispersal-limitation, selection, and drift (Logares et al., 2020). Besides, vertical stratification of epipelagic community composition is structured by temperature (Sunagawa et al., 2015). Environmental factors like nutrient concentration, determine the community structure, especially in areas with low productivity, such as oligotrophic subtropical and tropical waters. However, particle-associated (PA) and free-living (FL) structures of bacteria are more distinct and diverse in bacteria than in the archaeal community (Suzuki et al., 2016). Oceanic zones such as epipelagic and mesopelagic, which mostly include viruses, prokaryotes, and picoeukaryotes, have >40 million non-redundant genes. Microbial symbionts comprising of approximately 40% of the biomass of their sponge hosts (de Goeij et al., 2013; Wilkins et al., 2019) convert dissolved organic carbon released by reef organisms into particulate organic carbon, which is then consumed by heterotrophs (Webster and Thomas, 2016; Wilkins et al., 2019). In fact, through coral-symbiodiniaceae mutualism and sponge-bacterial symbiosis, Darwin's Paradox of the existence of a highly productive coral ecosystem within the oligotrophic tropical sea environment is explained (Colman, 2015; Wilkins et al., 2019). Photosynthetic cyanobacteria and single-celled eukaryotic phytoplankton are known for half of the primary production on the Earth (Field et al., 1998; Bunse and Pinhassi, 2017), and almost 50% of organic carbon is channeled by heterotrophic bacteria as they are the only organisms with the capability of assimilating and transforming the dissolved organic matter (DOM) in marine ecosystems (Ducklow, 2000; Bunse and Pinhassi, 2017). It is because of the biological carbon pump system that both organic and inorganic carbon fixed by phytoplankton in the euphotic zone are transferred to the ocean interior and eventually to sediments (Hulse et al., 2017; Chisholm, 1995; Basu and Mackey, 2018). Hence take carbon out of the atmosphere for several thousand years and help in lowering the atmospheric carbon levels (Hutchins and Fu, 2017; Basu and Mackey, 2018). As far as carbon sequestration is concerned, the microbial carbon pump (MCP) is responsible for the conversion of huge quantities of labile dissolved organic carbon into recalcitrant dissolved organic carbon, which remains in the ocean for several thousand years

(Jiao et al., 2010; Jiao and Zheng, 2011). In addition to being capable of transporting carbon and silicon to the ocean interiors, diatoms play a vital role in petroleum formation as often silica, and fossil fuels are found together in petroleum sediments and rocks (Cermeño, 2016; Benoiston et al., 2017). Further, microbes assist in the sulfate redox reactions known for their significance in marine sulfur cycling (Bowles et al., 2014; Hu et al., 2018). Elemental sulfur is reduced by *Desulfocapsa thiozymogenes*, *Desulfobulbus propionicus*, and *Desulfocapsa sulfoexigens*sp. nov. (Kai et al., 1998; Hu et al., 2018). Prokaryotes and certain lower eukaryotes like *E. coli, Pseudomonas* sp., and *Candida maltosa* (yeast) have the ability of remineralization of phosphonate compounds (Kononova and Nesmeyanova, 2002; Paytan and McLaughlin, 2007). Besides, certain sediment-dwelling bacteria are known to regulate the phosphorus across the sediment/water interface and responsible for the formation of refractory organic phosphorus compounds and biogenic apatite (Gachter and Meyer, 1993; Paytan and McLaughlin, 2007). Nitrogen in the oceans is fixed by long known *Trichodesmium* (Capone et al., 2005; Voss et al., 2013), photo-heterotroph group-A (UCYN-A) (Zehr et al., 2008; Voss et al., 2013), and *Richelia intracellularis* (symbiotic association with diatoms) (Foster et al., 2011; Bombar et al., 2011; Voss et al., 2013). Cyanobacteria were the earliest organisms capable of performing oxygenic photosynthesis (Lyons et al., 2014; Benoiston et al., 2017), and it was only after the evolution of photosynthetic eukaryotic algae that the oxygen concentration of Earth increased from 1 to 5% to the current levels of 20% (Falkowski, 2015; Benoiston et al., 2017).

6.3 CLIMATE CHANGE: THE CATALYST FOR HARMFUL ALGAL BLOOMS

Algae are a diverse group of oxygen-evolving photosynthetic aquatic organisms. Some of its species produce toxins that negatively impact water supplies used for drinking, irrigation, fishing, and recreational purposes. Besides the anthropogenically induced nutrient enrichment, climatic change supports various species of harmful algal blooms and increases their dominance, growth, geographical distribution, and persistence (Paerl and Huisman, 2009). Climate change escalates the toxicity and prevalence of harmful algal blooms (HABs) (Hennon and Dyhrman, 2020). Owing to the temperature, carbon dioxide, and other greenhouse gases, the marine and freshwater habitats are being heated, acidified, and deoxygenated (Griffith and Gobler, 2020). The world over the oceanic surface temperature is supposed to increase in the mid-21st century by 0.4–1.4°C causing higher growth rates of tropical and subtropical harmful dinoflagellates Fuluyoa, Gambierdiscus, and Ostreopsis (Tester et al., 2020). Also, ambient carbon dioxide buildup, which acidifies the surface of the ocean when it is in contact, promotes the development of individual organisms such as toxic dinoflagellates and tropical benthic dinoflagellates (Doney et al., 2009; Hallegraeff, 2010). Freshwater harmful algal blooms showed the most induced intensity, from nearly every continent, and are universally higher than those of non-harmful eukaryotic algae (Paerl and Huisman, 2008; Paerl and Huisman, 2009). Most freshwater harmful algal blooms that may be dangerous to aquatic life occur during the peak summer temperatures (Pörtner and Farrell, 2008; Doney et al., 2012). Accumulation of carbon dioxde causes hypoxia and acidification due to the collection of organic matter in coastal waters, particularly in temperate zones. It also leads to the toxicity, severity, and prevalence of algae (Brandenburg et al., 2019; Sunda et al., 2006; Wallace et al., 2014), thus leading to depression in the pH and saturation of calcium carbonate in the oceans (Doney et al., 2009). These changes risk both calcifying and non-calcifying living organisms (Gobler and Bauman, 2016). In some ecosystems, the over optimum temperature causes a substantial physiological pressure to the aquatic organisms (Portner et al., 2008). Harmful algal blooms cause fish kills, seafood contamination with toxins, and alter ecosystem functions (Glibert et al., 2014). Due to climate change, many eutrophic habitats host HABs, experience thermal extremes, low dissolved oxygen, and low pH, making these locations more potential to harmful algae (Griffith and Gobler, 2020; Breitburg et al., 2018). The expected increase in global warming of 1.5°C by mid-century (2050) and increased nutrient pollution due to population explosion (Glibert, 2020) may affect the HABs.

Further, the alteration patterns of precipitation due to climate change increase the growth of HABs (Sinha et al., 2017). For example, the three significant and most extensive HAB events of this century including the Pseudo-nitzschia bloom from Alaska to Mexico, the *Alexandrium-catenella* blooms across south-eastern Australia, the "Godzilla red tide" of *A. catenella,* and *Pseudo chattonella verruculosa* in Chile was supported by the climate change effects (Trainer et al., 2020). The altering climate changes and effects the existence of benthic dinoflagellate HABs formed by genera Ostreopsis and Gambierdiscus (Tester et al., 2020). Studies show how climate change allows the intensification of such happenings within some regions and their poleward spread. While drifting to new environments, HABs may create a substantial risk to such environments and the humans living nearby. Native species would be the first to experience the drastic decline when exposed to HABs and its harmful effects (Colin and Dam, 2002; Bricelj et al., 2005).

6.4 SUSCEPTIBILITY OF COASTLINE ECOSYSTEMS TO HAB'S IN REACTION TO CLIMATE CHANGE'

Harmful algal blooms, which have adverse effects on fish, and other seafoods form a dismal scum or alter the functions of the ecosystem to a damaging extent and duration. Primarily due to eutrophication and introduction of exotic species and changes in the environmental conditions resulting from climate change (Anderson et al., 2002). The effect of climate change on the frequency and abundance of harmful algal blooms would be substantial because of the complexity of factors and their combined impact on habitat or growth of harmful blooms (Fu et al., 2012). In addition to temperature, altered salinity, increased precipitation and runoff, increased stratification and changes in the nutrient and light regimes affect the harmful blooms of the algae (Boyd and Doney, 2003; Hutchin et al., 2009; Fu et al., 2012). With the increasing incidence of the harmful algal blooms in the world's waters (Glibert et al., 2005), the public health effects of these blooms need urgently to be understood, predicted, and mitigated. Dangerous algal flora can have significant ecological and economic consequences through the impact on coastal marine resources and other marine life (Nixon, 1995). The long-term changes in phytoplankton assemblages and the assemblage of biomass (Ried et al., 1998; Edwards et al., 2002; Richardson and Schoeman, 2004) seem to be influenced by climate variability and regional warming. In recent decades, the incidence of blooms that are either toxic or otherwise harmful has been increasing in coastal regions worldwide. Dangerous algal blooms are now threatening almost any coastal state, covering larger areas of our coastline and involving many species (Anderson, 1995). The harmful blooms of algae are familiar to fish kills, bird kills, and occasional invertebrate kills. During a prolonged red tide in 1996, 150 manatees died due to toxin exposure due to harmful algal blooms along the southwest of Florida (Steidinger et al., 1996). The algae and bacteria secreted toxins can make both fish and people unhealthy. Conditions that support harmful algal blooms can increase in frequency and duration as precipitation patterns and temperature shifts due to climate change (Well et al., 2015; Glibert et al., 2014). As a result of continuing nutrient input from human activity, coastal dead zones are likely to expand (Diaz and Rozenberg, 2008). Oceanic changes may extend or contract oxygen-deficient or minimum oxygen-related areas in the oceans (Deutsch et al., 2014; Ullah et al., 2012). Also, changes in climatic wind patterns can influence ocean microbial productivity.

6.5 PROMOTION OF HARMFUL CYANOBACTERIAL BLOOMS TO SUMMER HEATWAVES

In recent decades, besides the rising air and water temperatures, frequent and extreme weather events, including heatwaves, have been witnessed worldwide (Urrutia-Cordero, 2020). In 2003, the hottest summer of the past 500 years was recorded in Europe (Levinson and Waple, 2004; Luterbacher et al., 2004; Jöhnk et al., 2007). Its effect was observed on the harmful Cyanobacterium such as *Mirocystis* in a eutrophic lake, Nieuwe Meer of Sweden. Here, temperature favored cyanobacteria over diatoms and green algae directly through their high-temperature optimum and buoyant nature. Besides, reduced cloudiness and decreased wind speed by increasing the stability of a water body and decreasing vertical mixing enhanced the cyanobacterial bloom formation (Jöhnk et al., 2007). Results from studies on Lake Krankesjönin southern Sweden revealed that the heat waves not only increase species-specific cyanobacterial bloom biomass, but also enhance their recruitment rates from the sediments (Urrutia-Cordero, 2020). In 2011, Lake Erie, because of phosphorus loading through long-term land use and agricultural practices to its western basin, led to the most massive harmful Algal bloom in history. Further, certain meteorological conditions such as precipitation events in spring and warm and stable late spring and summer, allowed algae to remain at the top of the water column.

Moreover, in late August, as nitrogen started depleting in the water body, nitrogen-fixing form, *Anabaena* replaced non-nitrogen fixing form

Microcystis in its western basin to increase the bio-available nitrogen in the water body (Michalak et al., 2013). Further, Lake Mondsee in Austria in July 2015 showed a massive increase in the biomass of *Cyanophyceae* in comparison to previous years over other algal classes (Bergkemper and Weisse, 2017).

Furthermore, model simulations incorporating weather data of northern Europe from May to July, 2018 simulating extreme temperature events such as heat waves on a shallow lake, Bryrup in Denmark, have shown an increase in phytoplanktons. In this lake, the average summer chlorophyll content and cyanobacterial biomass were found to be 39 and 58% higher than the normal years, respectively, along with the low-level decrease in nitrates (NO_3) and phosphates (PO_4), and thus changing the total nitrogen and total phosphorus concentrations to at least 13 and 3.5%, respectively (Chen et al., 2019). The most significant concern of these blooms is the increase in toxic forms such as *Microcystis* sp. and *Anabaena* sp. that produce hepatotoxin microcystin and neurotoxin anatoxin, respectively (Codd et al., 2005; Urrutia-Cordero, 2020). Moreover, their non-toxic forms are responsible for affecting the structure, function, and aesthetics of the lake ecosystems (Michalak et al., 2013). Nitrogen-fixing toxin-producing harmful cyanobacteria include *Anabaena*, *Aphanizomenon*, *Cylindrospermopsis*, *Lyngbya*, *Nodularia*, *Oscillatoria*, and *Trichodesmium* while as non-nitrogen fixing ones include *Microcystis* and *Planktothrix*. Several of them thrive well in fresh, estuarine and in marine systems such as in lakes like Victoria, Erie, Michigan, Okeechobee, Ponchartrain, Taihu; coastal waters like the Baltic Sea, the Caspian Sea, tributaries of Chesapeake Bay, North Carolina's Albemarle-Pamlico Sound, Florida Bay, the Swan River estuary in Australia, the Patos, and other coastal lagoon estuaries in Brazil (Paerl et al., 2011; Paerl and Otten, 2012). However, to prevent cyanobacterial growth in lakes and other reservoirs, artificial mixing is used to increase the dissolved oxygen content in water, increase the temperature in deeper layers. Simultaneously, decreasing the same in surface layers besides making nutrients available to phytoplankton from sediments and hypolimnion. The ratio of mixing depth to the euphotic area (Zm/Zeu) in an artificially mixed lake is increasing, which in turn increases the photon irradiance variations that cyanobacteria less adjust to when contrasting with algae and diatoms (Flameling and Kromkamp, 1997; Nicklisch, 1998; Litchman, 2000; Visser et al., 2015). Also confirmed by remote sensing data (Kahru et al., 1993; Paerl and Otten, 2012), these blooms increase the surface temperatures of waterbodies due to light absorption by certain photosynthetic and photoprotective pigments (Paerl et al., 1983; Paerl et al., 1985; Paerl and Otten, 2012). Therefore, through this positive feedback mechanism, cyanobacteria can maintain their growth rates by optimizing temperature and remaining dominant over other forms (Paerl and Otten, 2012). Moreover, these blooms can last even after the nutrients have depleted in waterbodies (Paerl, 1988; Paerl and Otten, 2012).

6.6   EFFECT OF TEMPERATURE ON AQUATIC MICROBIAL ECOSYSTEM DYNAMICS

As per the fifth Intergovernmental Panel on Climate Change (IPCC), the global average temperature is going to increase by 2.6–4.8°C (average 3.7°C) by the end of this century, which is relative to temperatures from 1986 to 2005 (IPCC, 2013; Wei et al., 2018). Therefore, this increased temperature is going to have a warming effect on the oceanic surfaces, and thus also causing vertical stratification. This stratification will then decrease the mixing power of the surface and deep waters and hence, increase the exposure of microorganisms to high temperatures and solar irradiation (Danovaro et al., 2011; Wei et al., 2018). Temperature plays a key role in determining the taxonomic and functional microbial community structure in the photic zone of the oceans (Sunagawa et al., 2015). An increase in microbial richness has been found from 4 to 12°C followed by the decrease up to the sampled temperature range (30°C). Temperature and dissolved oxygen have a strong effect on both taxonomic and functional composition on the surface layer of oceans (Sunagawa et al., 2015). Warming is anticipated to lead to temperate dinoflagellates and coccolithophores in high-latitude subpolar regions (Hallegraeff, 2010; Winter et al., 2014; Hutchins and Fu, 2017). An increase in total global biomass of tropical picocyanobacteria by 14–29% over the next century (Flombaum et al., 2013; Hutchins and Fu, 2017), shift in nitrogen-fixing bacteria to higher latitudes (Breitbarth et al., 2007; Boyd et al., 2013; Hutchins and Fu, 2017) along with the disappearance of *Trichodesmium* and *Crocosphaera* from their present tropical locations is also expected (Thomas et al., 2012; Fu et al., 2014; Hutchins and Fu, 2017). It may lead to the appearance of thermally tolerant ecotypes or replacement by already existing alternative ones (Hutchins and Fu, 2017). Furthermore, viral communities of oligotrophic marine waters of the Western Pacific Ocean have shown a difference in rates of their decay. Although viruses showed decaying effect due to increased temperature and photosynthetically active radiations; however, low-fluorescence viruses were found more susceptible to the warming effect than their high-fluorescence counterparts, which is supposed to be because of the difference in their nucleic acids, proteins, and lipids (Mojica and Brussard, 2014; Uedaira et al., 1998; Wei et al., 2018). Therefore, the enhanced viral decay in the future is going to decrease the infection rates of their hosts along with the increased supply of viral decay matter into marine DOM pool, which in turn is going to increase the bacterioplankton population and thus is going to have crucial effects on biological, ecological, and biogeochemical aspects of the marine microbiome (Wei et al., 2018) and are hence also known to play an important role in marine biogeochemical cycling and global climate regulation (Danovaro et al., 2011; Wei et al., 2018). Furthermore, temperature records of 10 years in five alpine lakes of the Austrian Alps were observed through reduced ice cover duration and increased average water temperature showing temperature as a limiting factor for the growth of the bacterial community in the early growing season. In contrast, only nutrients were found influential in the later season, explained by the fact that colder places show more effect of climate change than warmer ones. Hence, during early spring, the temperature of water shows a steep increase in a lesser period. Still, as soon as the threshold for temperature is crossed, nutrients and certain biotic factors become more influential (Jiang et al., 2019). Further, because of the effects of increased warming on dissolved oxygen in surface waters and stratification-driven isolation from atmospheric ventilation in deep waters, hypoxic water will have huge, harmful impacts on microbial metabolism and diversity along with the biogeochemical cycling of nutrients like that of nitrogen (Gruber, 2011; Hutchins and Fu, 2017). Experiments have shown a positive correlation between warming and bacterial population, respiration and biomass, along with an occasional decrease in bacterial growth efficiency (Hoppe et al., 2008; Sarmento et al., 2010; Lara et al., 2013; von Scheibner et al., 2014; Hutchins and Fu, 2017) which could either be explained by direct thermal effects on bacterial metabolism (Sarmento et al., 2010; Hutchins and Fu, 2017) or indirectly through enhanced organic carbon releases by plankton at rising temperatures (Engel et al., 2011; Thornton, 2014; Hutchins and Fu, 2017) along with the increase in protozoan grazing (Sarmento et al., 2010; Lara et al., 2013; Hutchins and Fu, 2017), thus increasing predation of bacteria (Hutchins and Fu, 2017). Moreover, Crocosphaera is favored by warming over Trichodesmium as the former adapts to a higher temperature (Fu et al., 2014; Hutchins and Fu, 2017). Indirectly temperature, through ocean stratification, is found to affect nitrogen fixers such as cyanobacteria due to ultraviolet radiations (UVR) (Gao et al., 2012; Hutchins and Fu, 2017) and hence, Trichodesmium is predicted to be more vulnerable to UVR because of its daytime nitrogenase function along with its habit of surface bloom formation (Sohm et al., 2011; Hutchins and Fu, 2017). An increase in temperature in the future may increase the N/P ratio of marine communities as warming is predicted to affect the cellular core-metabolism, which is expected to increase nitrogen limitation in oceans (Toseland et al., 2013; Boscolo-Galazzo et al., 2018). Furthermore, through a change in C/P ration of exported organic matter, latitudinal changes in plankton stoichiometry might be transferred to the deep ocean nutrient pool despite circulation (Teng et al., 2014; Boscolo-Galazzo et al., 2018). Thus, the increased C/N/P ratio of marine planktons may negatively affect the weakening of the biological pump efficiency. Besides, this variable stoichiometry might have various effects on ocean nutrient cycling by effecting the relative contribution of the MCP (Jiao et al., 2010; Boscolo-Galazzo et al., 2018).

6.7 TEMPERATURE CONTROL OF BACTERIOPLANKTON GROWTH IN LAKES

Bacterioplankton plays a significant role in the conversion and mineralization of organic matter in all aquatic ecosystems due to their short life duration and access to dissolved nutrients and organic carbon (Cho and Azam, 1990; Hall and Cotner, 2007). They are supposed to depend on inputs of autochthonous and allochthonous organic carbon fixed by phytoplankton in productive lakes and unproductive Clearwater lakes (Jones, 1992). Bacterioplankton is usually sensitive to climate change and an increase in nutrient concentrations or to pollution, due to their fast turnover rates and environmental interactions (Robarts and Carr, 2009). In the temperate climate zone, the water temperature is likely to increase due to climate change, particularly in the winter and spring season, thereby affecting the biology and hydrology of lakes (Hoppe et al., 2008; Boer et al., 1990), which are expected to have a significant impact on bacterioplankton community compositions, bacterial growth, and abundance (Xiong et al., 2016). The interactive effects of temperature on the metabolism of bacterioplankton communities and its response depend on the resource treatment. Increasing temperatures promote bacterial respiration (Hoppe et al., 2008; Scheibner et al., 2014; Wohlers et al., 2009), and therefore, show redundancy in bacterial growth in response to warming. While other observations support the fact that the growth rate of bacterioplankton is limited to the colder temperature and well adapted to optimum temperature (Felip et al., 1996; Shiah and Ducklow, 1994; Simon and Wiinsch, 1998). It is well understood that temperature plays a vital role in regulating bacterioplankton communities (Madigan et al., 2003; Apple et al., 2006).

6.8 CLIMATE CHANGE AND WATER-BORNE MICROBIAL DISEASES

Climate shifts increase the frequency and intensity of droughts and massive rainfall occurrences (IPCC, 2013), which promote water-borne diseases involving many types of water-borne infections, including pathogens across several taxa like viruses, protozoa, diarrhea, and flu-like helminths. These pathogens can cause liver, neurological, and other damage (Levy et al., 2018). The distribution of water-borne diseases is different from country to country, for example, Cholerae, Hepatitis E, and Schistosomiasis pathogens are limited in some tropical countries (Hunter P., 2003) while other pathogens are widespread (cryptosporidiosis and campylobacteriosis). The diffusion of pathogens depends on the hydrodynamics of surface waterbodies and is further mobilized by floods and heavy rainfall; temperatures also alter pathogen survival, replication, and virulence (Funari et al., 2012; Levy et al., 2016). Heavy rain leads the stormwater runoff into surface waters, where both potential pathogens and high indicator bacteria are observed (O'Shea and Field, 1992). The temperature usually increases the risk of cholera—a water-borne disease, both geographically and temporally. The rise in temperature accelerates the growth of plankton, and in the adjacent areas of bloom, more cases of cholera are recorded (Lipp et al., 2002; Colwell, 1996). There is a threat of altering water quality to increase in cholera incidences, particularly in Asia and South America, due to an increase in temperature.

Furthermore, increased temperatures may add to the continued rise in problems due to algal blooms in Europe (Hunter P., 2003; Lobitz et al., 2000; Pascual et al., 2000; Speelmon et al., 2000)). The naturally occurring *Vibrio cholerae* cause cholera, also diffused by floods and hurricanes, thus change the water distribution system. The disease is one of the most severe forms of water-borne disease, especially for developing countries, because of poverty and the use of poor sanitation and unsafe water (WHO, 2012). Higher temperatures usually neutralize the pathogens when entering the water environment. However, it differs insensitively, for instance, Giardia cysts and enteroviruses are less rapidly inactivated when compared with *Cryptosporidium oocysts* (Schijven and de Roda, 2005).

Moreover, viruses show a large variation in temperature sensitivity (Schijven and Hassanizadeh, 2000). The increased washing of wild animal and livestock feces into water further expands the zoonotic infections, posing a threat of increased strains of viruses entering the water bodies. These strains are resistant to treatment in sewage treatment plants, which provides water for bathing (Maalouf et al., 2010).

A study examined that much of the water-borne diseases are associated with excessive rainfall (Howe et al., 2002; Patz et al., 2000). For instance, an outbreak of *Escherichia coli* O157:H7 reported in Walkerton, Canada, was associated with flooding and heavy rainfall, which affected over 1000 people (Anon, 2000; O'Connor, 2002). Another outbreak of *E. coli* O157:H7 occurred in New York in September 1999 and was linked to contaminated well water and resulted due to excessive rainfall (Patzetal, 2000). In the United States, 230 water-borne outbreaks reportedly affecting an estimated 443,000 people from 1991 to 1998, among which 51% outbreaks were related to extreme rainfall events (Curriero et al., 2001; Craun et al., 2002). In the UK, 65 outbreaks were recorded from 1991 to 2000, affecting 4112 people (Stanwell-Smith et al., 2002).

Cryptosporidium cysts were reported to be the cause of the most massive water-borne disease outbreaks in the USA, UK, and Japan. In England and Wales, the cryptosporidiosis cases were associated with maximum river flow (Lake et al., 2005), while in Japan and Oregon its outbreaks happened due to intense rainfall after very long and unusual drought periods (Leland et al., 1993; Yamamoto et al., 2000). In 2007, various microorganisms (*Giardia, Cryptosporidium, Campylobacter*, and *norovirus*) were involved in water-borne outbreaks in the European Union (EFSA 2009).

KEYWORDS

- oceanic microbiome
- harmful algal blooms
- coastline ecosystems
- climate change
- cyanobacterial blooms

REFERENCES

Anderson, D. M.; Glibert, P. M.; Burkholder, J. M. Harmful Algal Blooms and Eutrophication: Nutrient Sources, Composition, and Consequences. *Estuaries* **2002**, *25*, 704–726.

Anon. Water-Borne Outbreak of Gastroenteritis Associated with a Contaminated Municipal Water Supply, Walkerton, Ontario, May–June 2000. *Can. Communicable Dis. Rep.* **2001**, *26*, 170–173.

Apple, J. K.; del Giorgio, P. A.; Kemp, W. M. Temperature Regulation of Bacterioplankton Production, Respiration, and Growth Efficiency in a Salt-Marsh Estuary. *Aquat. Microb. Ecol.* **2006**, *43*, 243–254.

Basu, S.; Mackey, K. R. M. Phytoplankton as Key Mediators of the Biological Carbon Pump: Their Responses to a Changing Climate. *Sustainability* **2018**, *10*, 869. doi:10.3390/su10030869

Belgrano, A.; Lima, M.; Stenseth, N. C.; Lindahl, O. Responses of Phytoplankton Communities to Climate Variability. In *Marine Ecosystems and Climate Variation*; Stenseth, N. C.; Ottersen, G.; Hurrell, J. W.; Belgrano, A., Eds.; Oxford University Press Oxford, 2004; pp 109–114.

Benoiston, A-.S.; Ibarbalz, F. M.; Bittner, L.; Guidi, L.; Jahn, O.; Dutkiewicz, S.; Bowler, C. The Evolution of Diatoms and Their Biogeochemical Functions. *Phil. Trans. R. Soc. B.* **2017**, *372*, 20160397. http://dx.doi.org/10.1098/rstb.2016.0397

Bergkemper, V.; Weisse, T. Phytoplankton Response to the Summer 2015 Heat Wave—A Case Study From Prealpine Lake Mondsee, Austria. *Inland Waters* **2017**, *7* (1), 88–99. doi: 10.1080/20442041.2017.1294352

Beveridge, O. S.; Petchey, O. L.; Humphries, S. Direct and Indirect Effects of Temperature on the Population Dynamics and Ecosystem Functioning of Aquatic Microbial Ecosystems. *J. Anim. Ecol.* **2010**, *79*, 1324–1331.

Bock, N.; VanWambeke, F.; Moïra Dion, M.; Duhame, S. Microbial Community Structure in the Western Tropical South Pacific. *Biogeosciences* **2018**, *15*, 3909–3925. https://doi.org/10.5194/bg-15-3909-2018

Boer, M. M.; Koster, E. A.; Lundberg, H. Greenhouse Impact in Fennoscandia-Preliminary Findings of a European Workshop on the Effects of Climate Change. *Ambio* **1990**, *19*, 2–10.

Bombar, D.; Moisander, P. H.; Dippner, J. W.; Foster, R. A.;Voss, M.; Karfeld, B.; Zehr, J. P. Distribution of Diazotrophic Microorganisms and nif H Gene Expression in the Mekong River Plume during Intermonsoon. *Mar. Ecol. Prog. Ser.* **2011**, *424*, 39–52. doi:10.3354/meps08976

Boscolo-Galazzo, F.; Crichton, K. A.; Barker, S.; Pearson, P. N. Temperature Dependency of Metabolic Rates in the Upper Ocean: A Positive Feedback to Global Climate Change? *Global Planet. Change* **2018**, *170*, 201–212.

Bowles, M. W.; Mogollón, J. M.; Kasten, S.; Zabel, M.; Hinrichs, K. U. Global Rates of Marine Sulfate Reduction and Implications for Sub-Seafloor Metabolic Activities. *Science* **2014**, *344*, 889–891.

Boyd, P. W., et al. Marine Phytoplankton Temperature versus Growth Responses from Polar to Tropical Waters—Outcome of a Scientific Community-Wide Study. *PLoS ONE* **2013**, *8*, e63091.

Boyd, P. W.; Doney, S. C. The Impact of Climate Change and Feedback Processes on the Ocean Carbon Cycle. In Ocean Biogeochemistry—the Role of the Ocean Carbon Cycle in Global Change; Fasham, M. J. R., Ed.; Springer-Verlag: Berlin, 2003; pp 157–193.

Brandenburg, K. M.; Velthuis, M.; Van de Waal, D. B. Meta-Analysis Reveals Enhanced Growth of Marine Harmful Algae from Temperate Regions with Warming and Elevated CO_2 Levels. *Global Change Biol.* **2019**. https://doi.org/10.1111/gcb.14678.

Breitbarth, E.; Oschlies, A.; LaRoche, L. Physiological Constraints on the Global Distribution of *Trichodesmium*-Effect of Temperature on Diazotrophy. *Biogeosciences* **2007**, *4*, 53–61.

Breitburg, D.; Levin, L. A.; Oschlies, A.; Grégoire, M.; Chavez, F. P.; Conley, D. J.; Jacinto, G. S. Declining Oxygen in the Global Ocean and Coastal Waters. *Science* **2018**, *359*, 7240.

Bricelj, V. M.; Connell, L.; Konoki, K.; MacQuarrie, S. P.; Scheuer, T.; Catterall, W. A.; Trainer, V. L. Sodium Channel Mutation Leading to Saxitoxin Resistance in Clams Increases Risk of PSP. *Nature* **2005**, *434*, 763.

Brown, J. H.; Gillooly, J. F.; Allen, A. P.; Savage, V. M.; West, G. B. Toward a Metabolic Theory of Ecology. *Ecology* **2004**, *85*, 1771–1789.

Bunse, C.; Pinhassi, J. Marine Bacterioplankton Seasonal Succession Dynamics. *Trends Microbiol.* **2017**, *25* (6), 494–505. http://dx.doi.org/10.1016/j.tim.2016.12.013.

Cao, S.; Zhang, W.; Ding, W.;Wang, M.; Fan, S.; Yang, B.; Mcminn, A.; Wang, M.; Bin-bin Xie, B.-b.; Qin, Q.-L.; Chen, X.-L.; He, J.; Zhang, Y.-Z. Structure and Function of the Arctic and Antarctic Marine Microbiota as Revealed by Metagenomics. *Microbiome* **2020**, *8*, 47. https://doi.org/10.1186/s40168-020-00826-9

Capone, D. G.; Burns, J. A.; Montoya, J. P.; Subramaniam, A.; Mahaffey, C.; Gundersoen, T.; Michaels, A. F.; Carpenter, E. J. Nitrogen Fixation by Trichodesmium spp.: An Important Source of New Nitrogen to the Tropical and Subtropical North Atlantic Ocean. *Global Biogeochem. Cycl.* **2005**, *19*, 1–17. doi:10.1029/2004GB002331

Cermeño, P. The Geological Story of Marine Diatoms and the Last Generation of Fossil Fuels. Perspect. *Phycol.* **2016**, *3*, 53–60. doi:10.1127/pip/2016/0050

Chen, W.; Nielsen, A.; Andersen, T. K.; Hu, F.; Chou, Q.; Søndergaard, M.; Jeppesen, E.; Trolle, D. Modeling the Ecological Response of a Temporarily Summer-Stratified Lake to Extreme Heatwaves. *Water* **2019**, *12*, 94. doi:10.3390/w12010094

Chisholm, S. W. The Iron Hypothesis: Basic Research Meets Environmental Policy. *Rev. Geophys.* **1995**, *33*, 1277–1286.

Cho, B. C.; Azam, F. Biogeochemical Significance of Bacterial Biomass in the Ocean's Euphotic Zone. *Marine Ecol. Prog. Ser.* **1990**, *63*, 253–359.

Codd, G. A.; Morrison, L. F.; Metcalf, J. S. Cyanobacterial Toxins: Risk Management for Health Protection.*Toxicol. Appl. Pharmacol.* **2005**, *203*, 264–272.

Colin, S. P.; Dam, H. G. Latitudinal Differentiation in the Effects of the Toxic Dinoflagellate Alexandrium spp. on the Feeding and Reproduction of Populations of the Copepod Acartia hudsonica. *Harmful Algae* **2002**, *1*, 113–125.

Colman, A. S. Sponge Symbionts and the Marine P Cycle. *Proc. Natl. Acad. Sci. USA* **2015**, *112*, 4191–4192. https://doi.org/10.1073/pnas.1502763112 PMID: 25825737.

Colwell, R. R. Global Climate Change and Infectious Disease: The Cholera Paradigm. *Science* **1996**, *274*, 2025–2031.

Craun, G. F.; Calderon, R. L.; Nwachuku, N. Causes of Water-Borne Outbreaks Reported in the United States, 1991–98. In *Drinking Water and Infectious Disease: Establishing the Link*; Hunter, P. R.; Waite, M.; Ronchi, E., Eds.; CRC Press: Boca Raton, 2002; pp. 105–117.

Crawford, C. C.; Hobbie, J. E.; Webb, K. L. The Utilization of Dissolved Free Amino Acids by Estuarine Microorganisms. *Ecol* **1974**, *55*, 551–563.

Curriero, F. C.; Patz, J. A.; Rose, J. B.; Lele, S. The Association between Extreme Precipitation and Water-Borne Disease Outbreaks in the United States, 1948–1994. *Am. J. Public Health* **2001**, *91*, 1194–1199.

Danovaro, R., et al. Marine Viruses and Global Climate Change. *FEMS Microbiol. Rev.* **2011**, *35* (6), 993–1034.

de Goeij, J. M.; van, Oevelen, D.; Vermeij, M. J. A, Osinga, R.; Middelburg, J. J.; de Goeij, A. F. P. M., et al. Surviving in a Marine Desert: The Sponge Loop Retains Resources within Coral Reefs. *Science* **2013**, *342*, 108–110. https://doi.org/10.1126/science.1241981 PMID: 24092742.

Diaz, S.; Grime, J. P.; Harris, J.; Mcpherson, E. Evidence of a Feedback Mechanism Limiting Plant Response to Elevated Carbon Dioxide. *Nature* **1993**, *364*, 616–617.

Doney, S. C.; Fabry, V. J.; Feely, R. A.; Kleypas, J. A. Ocean Acidification: The Other CO_2 Problem. *Annu. Rev. Mar. Sci.* **2009**, *1*, 169–192.

Doney, S. C.; Ruckelshaus, M.; Duffy, J. E.; Barry, J. P.; Chan, F.; English, C. A.; Galindo, H. M.; Grebmeier, J. M.; Hollowed, A. B.; Knowlton, N.; Polovina, J.; 2012. Climate Change Impacts on Marine Ecosystems. *Annu. Rev. Mar. Sci.* **2012**, *4*, 11–37.

Ducklow, H. W. Bacterial Production and Biomass in the Oceans. In *Microbial Ecology of the Oceans*; Kirchman, D. L., Ed.; John Wiley and Sons, 2000; pp 85–120.

Dupont, C. L., et al. Genomic Insights to SAR86, an Abundant and Uncultivated Marine Bacterial Lineage. *ISME J.* **2012**, *6*, 1186–1199.

Engel, A., et al. Effects of Sea Surface Warming on the Production and Composition of Dissolved Organic Matter during Phytoplankton Blooms: Results from a Mesocosm Study. *J. Plankton Res.* **2011**, *33*, 357–372.

European Food Safety Authority–European Centre for Disease Prevention and Control. *The Community Summary Report. Food-Borne Outbreaks in the European Union in 2007*; EFSA-ECDC: Brussels, Belgium, 2009.

Falkowski, P. G. *Life's Engines*: *How Microbes Made Earth Habitable*; Princeton University Press: Princeton, NJ, 2015.

Felip, M.; Pace, M. L.; Cole, J. J. Regulation of Planktonic Bacterial Growth Rates: The Effects of Temperature and Resources. *Microb. Ecol.* **1996**, *31*, 15–28.

Field, C. B., et al. Primary Production of the Biosphere: Integrating Terrestrial and Oceanic Components. *Science* **1998**, *281*, 237–240.

Flameling, I. A.; Kromkamp, J. Photoacclimation of Scenedesmus Protuberans (Chlorophyceae) to Fluctuating Irradiances Simulating Vertical Mixing. *J. Plankton. Res.* **1997**, *19*, 1011–1024.

Flombaum, P., et al. Present and Future Global Distributions of the Marine Cyanobacteria *Prochlorococcus* and *Synechococcus*. *Proc. Natl. Acad. Sci. USA* **2013**, *110*, 9824–9829.

Foster, R. A.; Kuypers, M. M.; Vagner, T.; Paerl, R. W.; Musat, N.; Zehr, J. P. Nitrogen Fixation and Transfer in Open Ocean Diatom–Cyanobacterial Symbioses. *ISME J.* **2011**, *5*, 1484–1493. doi:10.1038/ismej.2011.26

Fu, F. X.; Place, A. R.; Garcia, N. S.; Hutchins, D. A. CO_2 and Phosphate Availability Control the Toxicity of the Harmful Bloom Dinoflagellate *Karlodinium veneficum*. *Aquat. Microb. Ecol.* **2010**, 59, 55−65.

Fu, F.-X., et al. Differing Responses of Marine N_2 Fixers to Warming and Consequences for Future Diazotroph Community Structure. *Aquat. Microb. Ecol.* **2014**, *72*, 33–46.

Funari, E.; Manganelli, M.; Sinisi, L. Impact of Climate Change on Water-Borne Diseases. *Ann. Ist Super Sanità* **2012**, *48* (4), 473–487. doi: 10.4415/ANN_12_04_13

Gachter, R.; Meyer, J. S. The Role of Microorganisms in Mobilization and Fixation of Phosphorus in Sediments. Hydrobiologia **1993**, *253 (1)*, 103–121.

Gao, K.; Helbling, E. W.; Häder, D.-P.; Hutchins, D. A. 2012. Responses of Marine Primary Producers to Interactions between Ocean Acidification, Solar Radiation, and Warming. *Mar. Ecol. Prog. Ser.* **2012**, *470*, 167–189.

Glibert, P. M. Harmful Algae at the Complex Nexus of Eutrophication and Climate Change. *Harmful Algae* **2020**, *91*, 101583.

Glibert, P. M.; Allen, J. I.; Artioli, Y.; Beusen, A.; Bouwman, L.; Harle, J.; Holt, J. Vulnerability of Coastal Ecosystems to Changes in Harmful Algal Bloom Distribution in Response to Climate Change: Projections Based on Model Analysis. *Global Change Biol.* **2014**, *20*, 3845–3858.

Glibert, P.; Seitzinger, S.; Heil, C. A.; Burkholder, J. M.; Parrow, M. W.; Codispoti, L. A.; Kelly, V. The Role of Eutrophication in the Coastal Proliferation of Harmful Algal Blooms: New Perspectives and New Approaches. *Oceanography* **2005**, *18*, 198−209.

Glöckner, F. O.; Stal, L. J.; Sandaa, R. A.; Gasol, J. M.; O'Gara, F.; Hernandez, F.; Labrenz, M.; Stoica, E.; Varela, M. M.; Bordalo, A.; Pitta, P. Marine Microbial Diversity and Its Role in Ecosystem Functioning and Environmental Change. In *Marine Board Position Paper 17*; Calewaert, J. B.; McDonough, N., Eds.; Marine Board-ESF: Ostend, Belgium, 2012; pp 13–25.

Gobler, C. J.; Baumann, H. Hypoxia and Acidification in Ocean Ecosystems: Coupled Dynamics and Effects on Marine Life. *Biol. Lett.* **2016**, *12* (5), 20150976.

Griffith, A. W.; Gobler, C. J. Harmful Algal Blooms: A Climate Change Co-Stressor in Marine and Freshwater Ecosystems. *Harmful Algae* **2020**, *91*, 101590. https://doi.org/10.1016/j.hal.2019.03.008

Gruber, N. Warming Up, Turning Sour, Losing Breath: Ocean Biogeochemistry under Global Change. *Phil. Trans. R. Soc. A.* **2011**, *369*, 1980−1996.

Hall, E. K.; Cotner, J. B. Interactive Effect of Temperature and Resources on Carbon Cycling by Freshwater Bacterioplankton Communities. *Aquat. Microb. Ecol.* **2007**, *49* (1), 35–45. https://doi.org/10.3354/ame01124

Hallegraeff, G. M. Ocean Climate Change, Phytoplankton Community Responses, and Harmful Algal Blooms: A Formidable Predictive Challenge. *J. Phycol.* **2010,** *46*, 220–235.

Hallegraeff, G. M. Ocean Climate Change, Phytoplankton Community Responses, and Harmful Algal Blooms: A Formidable Predictive Challenge. *J. Phycol.* **2010,** *46* (2), 220–235.

Heisler, J.; Glibert, P. M.; Burkholder, J., et al. Eutrophication and Harmful Algal Blooms: A Scientific Consensus. *Harmful Algae* **2008,** *8*, 3–13.

Hennon, G. M. M.; Dyhrman, S. T. Progress and Promise of Omics for Predicting the Impacts of Climate Change on Harmful Algal Blooms. *Harmful Algae* **2020,** *91*, 101587. https://doi.org/10.1016/j.hal.2019.03.005.

Hoppe, H.; Sommer, U.; Jürgens, K. Climate Warming in Winter Affects the Coupling between Phytoplankton and Bacteria during the Spring Bloom: A Mesocosm Study. *Aquat. Microb. Ecol.* **2008** May, *51*, 105–115. doi: 10.3354/ame01998.

Howe, A. D.; Forster, S.; Morton, S.; Marshall, R.; Osborn, K. S.; Wright, P., et al. Cryptosporidium Oocysts in a Water Supply Associated with a Cryptosporidiosis Outbreak. *Emerg. Infect Dis.* **2002,** *8*, 619–624.

Hu, X.; Liu, J.; Liu, H.; Zhuang, G.; Xun, L. Sulfur Metabolism by Marine Heterotrophic Bacteria Involved in Sulfur Cycling in the Ocean. *Sci. China Earth Sci.* **2018,** *61* (10), 1369–1378. https://doi.org/10.1007/s11430-017-9234-x

Huertas, I. E.; Rouco, M.; Lo'pez-Rodas, V.; Costas, E. Warming Will Affect Phytoplankton Differently: Evidence through a Mechanistic Approach. *Proc. R. Soc. B.* **2011.** doi:10.1098/rspb.2011.0160

Hulse, D.; Arndt, S.; Wilson, J. D.; Munhoven, G.; Ridgwell, A. Understanding the Causes and Consequences of Past Marine Carbon Cycling Variability Through Models. *Earth-Sci. Rev.* **2017,** *171*, 349–382.

Hunter, P. Climate Change and Water-Borne and Vectorborne Disease. *J. Appl. Microbiol.* **2003,** *94*, 37–46. http://dx.doi.org/10.1046/j.1365-2672.94.s1.5.x

Hutchins, D. A.; Fu, F. Microorganisms and Ocean Global Change. *Nat. Microbiol.* **2017,** *2*, 17058.

Hutchins, D. A.; Mulholland, M. R.; Fu, F. X. Nutrient Cycles and Marine Microbes in a CO_2 Enriched Ocean. *Oceanography* **2009,** *22*, 128–145.

IPCC. Climate Change 2013: The Physical Science Basis Contribution of Working Group I to the Fifth Assessment Report of the Intergovernmental Panel on Climate Change. In Stocker, T. F.; Qin, D.; Plattner, G.-K.; Tignor, M.; Allen, S. K.; Boschung, J., et al., Eds.; Cambridge University Press: Cambridge and New York, NY, 2013; 1535 pp.

Jiang, Y.; Huang, H.; Ma, T.; Ru, J.; Blank, S.; Kurmayer, R.; Deng, L. Temperature Response of Planktonic Microbiota in Remote Alpine Lakes. *Front. Microbiol.* **2019,** *10*, 1714. doi: 10.3389/fmicb.2019.01714

Jiao, N.; Herndl, G. J.; Hansell, D. A.; Benner, R., et al. Microbial Production of Recalcitrant Dissolved Organic Matter: Long-Term Carbon Storage in the Global Ocean. *Nat. Rev. Microbiol.* **2010,** *8*, 593–599.

Jiao, N.; Zheng, Q. The Microbial Carbon Pump: From Genes to Ecosystems. *Appl. Environ. Microbiol.* **2011,** *77* (21), 439–7444. doi:10.1128/AEM.05640-11

Jöhnk, K. D.; Huisman, J.; Sharples, J.; Sommeijer, B.; Visser, P. M.; Stroom, J. M. Summer Heatwaves Promote Blooms of Harmful Cyanobacteria. *Global Change Biol.* **2007,** *14*, 495–512.

Jonathan, K. Microbes and Climate Change, 2016. Microbes and Climate Change Report on an American Academy of Microbiology and American Geophysical Union Colloquium, at Washington, DC, 2016.

Jones, R. I. The Influence of Humic Substances on Lacustrine Planktonic Food-Chains. *Hydrobiologia* **1992**, *229*, 73–91.

Kahru, M.; Leppänen, J-.M.; Rud, O. Cyanobacterial Blooms Cause Heating of the Sea Surface. *Marine Ecol. Prog. Ser.* **1993**, *101*, 1–7.

Kai, F.; Liesack, W.; Bo, T. Elemental Sulfur and Thiosulfate Disproportionationby *Desulfocapsa sulfoexigens* sp. nov. a New Anaerobic Bacterium Isolated from Marine Surface Sediment. *Appl. Environ. Microbiol.* **1998**, *64*, 119–125.

Kononova, S. V.; Nesmeyanova, M. A. *Biochemistry* **2002**, *67* (2), 184–195.

Lake, I. R.; Bentham, G.; Kovats, R. S.; Nichols, G. L. Effects of Weather and River Flow on Cryptosporidiosis. *J. Water Health* **2005**, *3* (4), 469–474.

Lara, E., et al. Experimental Evaluation of the Warming Effect on Viral, Bacterial and Protistan Communities in Two Contrasting Arctic Systems. *Aquat. Microb. Ecol.* **2013**, *70*, 17–32.

Leland, D.; McAnulty, J.; Keene, W.; Stevens, G. A Cryptosporidiosis Outbreak in a Filtered Water Supply. *J. Am. Water Works Assoc.* **1993**, *85* (6), 34–42.

Levinson, D. H.; Waple, A. M. State of Climate in 2003. *Bull. Am. Meteorol. Soc.* **2004**, *85*, S1–S72.

Levy, K.; Smith, S. M.; Carlton, E. J. Climate Change Impacts on Water-Borne Diseases: Moving toward Designing Interventions. *Curr. Environ. Health Rep.* **2018** June; *5* (2), 272–282. doi: 10.1007/s40572-018-0199-7

Levy, K.; Woster, A. P.; Goldstein, R. S.; Carlton, E. J. Untangling the Impacts of Climate Change on Waterborne Diseases: A Systematic Review of Relationships between Diarrheal Diseases and Temperature, Rainfall, Flooding, and Drought. *Environ. Sci. Technol.* **2016**, *50*, 4905–22. doi:10.1021/acs.est.5b06186. (PubMed: 27058059) This is a systematic review of the epidemiological literature that describes key areas of agreement and evaluates the biological plausibility of these associations.

Lipp, E. K.; Huq, A.; Colwell, R. R. Effects of Global Climate on Infectious Disease: The Cholera Model. *Clin. Microbiol. Rev.* **2002**, *15* (4), 757–770. http://dx.doi.org/10.1128/cmr.15.4.757-770.2002

Litchman, E. Growth Rates of Phytoplankton under Fluctuating Light. *Freshw. Biol.* **2000**, *44*, 223–235.

Lobitz, B.; Beck, L.; Huq, A.; Wood, B.; Fuchs, G.; Faruque, A. S. G.; Colwell, R. Climate and Infectious Disease: Use of Remote Sensing for Detection of Vibrio Cholerae by Indirect Measurement. *Proc. Natl. Acad. Sci.* **2000**, *97*, 1438–1443.

Lochte, K.; Turley, C. M. Bacteria and Cyanobacteria Associated with Phytodetritus in the Deep Sea. *Nature* **1988**, *333*, 67–69. doi: 10.1038/333067a0.

Logares, R., et al. Disentangling the Mechanisms Shaping the Surface Ocean Microbiota. *Microbiome* **2020**, *8*, 55. https://doi.org/10.1186/s40168-020-00827-8

Luterbacher, J.; Dietrich, D.; Xoplaki, E.; Grosjean, M.; Wanner, H. European Seasonal and Annual Temperature Variability, Trends, and Extremes since 1500. *Science* **2004**, *303*, 1499–1503.

Lyons, T. W.; Reinhard, C. T.; Planavsky, N. J. The Rise of Oxygen in Earth's Early Ocean and Atmosphere. *Nature* **2014**, *506*, 307–315. doi:10.1038/nature13068

Maalouf, H.; Pommepuy, M.; Le Guyader, F. Environmental Conditions Leading to Shellfish Contamination and Related Outbreaks. *Food Environ. Virol.* **2010,** *2* (3), 136–145. http://dx.doi.org/10.1007/s12560-010-9043-4

Madigan, M. T.; Martinko, J. M.; Parker, J. *Brock Biology of Microorganisms*; Prentice Hall, Upper Saddle River, NJ, 2003.

Meie, W. N., et al. Arctic Sea Ice in Transformation: A Review of Recent Observed Changes and Impacts on Biology and Human Activity. *Rev. Geophys.* **2014,** *52,* 185–217.

Michalak, A. M., et al. Record-Setting Algal Bloom in Lake Erie Caused by Agricultural and Meteorological Trends Consistent With Expected Future Conditions. *PNAS* **2013,** *110* (16), 6448–6452.

Mojica, K. D.; Brussaard, C. P. Factors Affecting Virus Dynamics and Microbial Host-Virus Interactions in Marine Environments. *FEMS Microbiol. Ecol.* **2014,** *89* (3), 495–515.

Morris, R. M., et al., SAR11 Clade Dominates Ocean Surface Bacterioplankton Communities. *Nature* **2002,** *420,* 806–810. doi: 10.1038/nature01240

NASA. http://climate.nasa.gov/solutions/adaptation-mitigation/. Accessed 15 Dec 2015.

Nicklisch, A. Growth and Light Absorption of Some Planktonic Cyanobacteria, Diatoms and Chlorophyceae under Simulated Natural Light Fluctuations. *J. Plankton. Res.* **1998,** *20,*105–119.

Nixon, S. W. Coastal Marine Eutrophication: A DefinitionS Cocial causes, and Future Concerns. *Ophelia.* **1995,** *41,* 199–219.

O'Connor, D. R. Report of the Walkerton Inquiry. Part 1. The Events of May 2000 and Related Issues. Toronto, 2002.

O'Shea, M. L.; Field, R. Detection and Disinfection of Pathogens in Storm-Generated Flows. *Can. J. Microbiol.* **1992,** *38,* 267–276.

Paerl, H. W. Nuisance Phytoplankton Blooms in Coastal, Estuarine, and Inland Waters. *Limnol. Oceanogr.* **1988,** *33,* 823–847.

Paerl, H. W.; Bland, P. T.; Bowles, N. D., et al. Adaptation to High Intensity, Low Wavelength Light among Surface Blooms of the Cyanobacterium Microcystis Aeruginosa. *Appl. Environ. Microbiol.* **1985,** *49,* 1046–1052.

Paerl, H. W.; Hall, N. S.; Calandrino, E. S. Controlling Harmful Cyanobacterial Blooms in a World Experiencing Anthropogenic and Climatic-Induced Change. *Sci. Total Environ.* **2011,** *409,* 739–1745.

Paerl, H. W.; Huisman, J. Blooms Like It Hot. *Science* **2008,** *320,* 57–58.

Paerl, H. W.; Huisman, J. Climate Change: A Catalyst for Global Expansion of Harmful Cyanobacterial Blooms. *Environ. Microbiol. Rep.* **2009** Feb; *1* (1):27–37. doi: 10.1111/j.1758-2229.2008.00004.x PMID: 23765717.

Paerl, H. W.; Otten, T. G. Harmful Cyanobacterial Blooms: Causes, Consequences, and Controls. *Microb. Ecol.* **2012**. doi: 10.1007/s00248-012-0159-y

Paerl, H. W.; Tucker, J.; Bland, P. T. Carotenoid Enhancement and Its Role in Maintaining Blue-Green Algal (Microcystis Aeruginosa) Surface Blooms. *Limnol Oceanogr.* **1983,** *28,* 847–857.

Pascual, M.; Rodo', X.; Ellner, S. P.; Colwell, R.; Bouma, M. J. Cholera Dynamics and El-Nin~o-Southern Oscillation. *Science* **2000,** *289,* 1766–1769.

Patz, J. A.; McGeehin, M. A.; Bernard, S. M. The Potential Health Impacts of Climate Variability and Change for the United States: Executive Summary of the Report of the Health Sector of the U.S. National Assessment. *Environ. Health Perspect* **2000,**, *108,* 367–376.

Paytan, A.; McLaughlin, K. The Oceanic Phosphorus Cycle. *Chem. Rev.* **2007**, *107* (2), 563–576.

Pernthaler, J. Predation on Prokaryotes in the Water Column and Its Ecological Implications. *Nat. Rev. Microbiol.* **2005**, *3*, 537–546. doi: 10.1038/nrmicro1180. ARCH Open Access.

Pörtner, H. O.; Farrell, A. P. Physiology and Climate Change. *Science* **2008**, *322* (5902), 690–692.

Reay, D. S.; Nedwell, D. B.; Priddle, J.; Ellis-Evans, J. C. Temperature Dependence of Inorganic Nitrogen Uptake: Reduced Affinity for Nitrate at Suboptimal Temperatures in Both Algae and Bacteria. *Appl. Environ. Microbiol.* **1999**, *65*, 2577–2584.

Robarts, R. D.; Carr, G. M. Bacteria, Bacterioplankton A2–Likens, Gene E. In *Encyclopedia of Inland Waters*; Likens, G. E., Ed.; Academic Press Oxford, 2009; pp 193–200. doi: 10.1016/B978-012370626-3.00124-1

Salazar, G.; Cornejo-Castillo, F. M.; Benítez-Barrios, V.; Fraile-Nuez, E.; Álvarez-Salgado, X. A.; Duarte, C. M.; Gasol, J. M.; Acinas, S. G. Global Diversity and Biogeography of Deep-Sea Pelagic Prokaryotes. *ISME J.* **2016**, *10*, 596–608.

Sarmento, H., et al. Warming Effects on Marine Microbial Food Web Processes: How Far Can We Go When It Comes to Predictions? *Phil. Trans. R. Soc. B* **2010**, *365*, 2137–2149.

Scheibner, M.; Dörge, P.; Biermann, A.; Sommer, U.; Hoppe, H.-G., Jürgens, K. Impact of Warming on Phyto-Bacterioplankton Coupling and Bacterial Community Composition in Experimental Mesocosms. *Environ Microbiol.* **2014**, *16* (3), 718–733.

Schijven, J. F.; de Roda Husman, A. M. Effect of Climate Changes on Waterborne Disease in The Netherlands. *Water Sci. Technol.* **2005**, *51* (5), 79–87.

Schijven, J. F.; Hassanizadeh, S. M. Removal of Viruses by Soil Passage: Overview of Modeling, Processes, and Parameters. *Crit. Rev. Environ. Sci. Technol.* **2000**, *30* (1), 49–127. http://dx.doi.org/10.1080/10643380091184174

Shiah, F. K.; Ducklow, H. W. Temperature Regulation of Heterotrophic Bacterioplankton Abundance, Production, and Specific Growth Rate in Chesapeake Bay. *Limnol. Oceanogr.* **1994**, *39*, 1243–1258.

Simon, M.; Wiinsch, C. Temperature Control of Bacterioplankton Growth in a Temperate Large Lake. *Aquat. Microb. Ecol.* **1998**, *16*, 119-130ʻ.

Sinha, E.; Michalak, A. M.; Balaji, V. Eutrophication Will Increase during the 21stᵗ Century as a Result of Precipitation Changes. *Science* **2017**, *357*, 405–408.

Sohm, J. A.; Webb, E. A.; Capone, D. A. Emerging Patterns of Marine Nitrogen Fixation. *Nat. Rev. Microbiol.* **2011**, *9*, 499–508.

Speelmon, E. C.; Checkley, W.; Gilman, R. H.; Patz, J.; Caleron, M.; Manga, S. Cholera Incidence and El-Ninˆo-related Higher Ambient Temperature. *J. Am. Med. Assoc.* **2000**, *283*, 3072–3074.

Stanwell-Smith, R.; Andersson, Y.; Levy, D. A. National Surveillance Systems. In *Drinking Water and Infectious Disease: Establishing the Link*; Hunter, P. R.; Waite, M.; Ronchi, E., Eds.; CRC Press: Bocata Raton, 2002; pp 25–40.

Steidinger, K. A.; Haddad, K. Biologic and Hydrographic Aspects of Red Tides. *Bioscience* **1981**, *31* (11), 814–819.

Stenseth, N. C.; Mysterud, A.; Ottersen, G.; Hurrell, J. W.; Chan, K-S.; Lima, M. Ecological Effects of Climate Fluctuations. *Science* **2002**, *297*, 1292–1296.

Stocker, R. Marine Microbes See a Sea of Gradients. *Science* **2012**, *338*, 628–633. doi: 10.1126/science.1208929

Stocker, T. F., et al. Intergovernmental Panel on Climate Change. Climate Change 2013: The Physical Science Basis. Contribution of Working Group I to the Fifth Assessment Report of the Intergovernmental Panel on Climate Change; Stocker, T. F., et al., Eds.; Cambridge University Press: Cambridge/New York, 2013; pp 953–1028.

Straile, D. Food Webs in Lakes—Seasonal Dynamics and the Impact of Climate Variability. In *Aquatic Food Webs—An Ecosystem Approach*; Belgrano, A.; Scharler, U. M.; Dunne, J. A.; Ulanowicz, R. E., Eds.; Oxford University Press: Oxford, 2005; pp 41–50.

Sunagawa, S., et al. Structure and Function of the Global Ocean Microbiome. *Science* **2015,** *348* (6237), 1–9.

Sunda, W. G.; Graneli, E.; Gobler, C. J. Positive Feedback and the Development and Persistence of Ecosystem Disruptive Algal Blooms. *J. Phycol.* **2006,** *42*, 963–974.

Suzuki, S.; Kaneko, R.; Kodama, T.; Hashihama, F.; Suwa, S.; Tanita, I.; Furuya, K.; Hamasaki, K. Comparison of Community Structures Between Particle-Associated and Free-Living Prokaryotes in Tropical and Subtropical Pacific Ocean Surface Waters. *J. Oceanogr.* **2016,** *73*, 383–395.

Teng, Y.; Primeau, F. W.; Moore, J. K.; Lomas, M. W.; Martiny, A. C. Global-Scale Variations of the Ratios of Carbon to Phosphorus in Exported Marine Organic Matter. *Nat. Geosci.* **2014**. https://doi.org/10.1038/NGEO2303

Tester, P.; Berdalet, E.; Litaker, R. W. Climate Change and Benthic Harmful Algae. *Harmful Algae* **2020,** *91*, 101655.

Thomas, M. K.; Kremer, C. T.; Klausmeier, C. A.; Litchman, E. A Global Pattern of Thermal Adaptation in Marine Phytoplankton. *Science* **2012,** *338*, 1085–1088.

Thornton, D. C. O. Dissolved Organic Matter (DOM) Release by Phytoplankton in the Contemporary and Future Ocean. *Eur. J. Phycol.* **2014,** *49*, 20–46.

Toseland, A.; Daines, S. J.; Clark, J. R.; Kirkham, A.; Strauss, J.; Uhlig, C., et al. The Impact of Temperature on Marine Phytoplankton Resource Allocation and Metabolism. *Nat Clim. Change.* **2013,** *3*, 979–984.

Trainer, V. L.; Moore, S. K.; Hallegraeff, G.; Kudela, R. M.; Clement, A.; Mardones, J. I.; Cochlan, W. P. Pelagic Harmful Algal Blooms and Climate Change: Lessons from Nature's Experiments with Extremes. *Harmful Algae* **2020,** *91*, 101591. https://doi.org/10.1016/j.hal.2019.03.009.

Uedaira, H.; Morii, H.; Ogata, K.; Ishii, S.; Sarai, A. Multi-State Thermal Transitions of Proteins-DNA-Binding Domain of the *c*-Myboncoprotein. *Pure Appl. Chem.* **1998,** *70* (3), 671–676.

Union of Concerned Scientists. 2011. http://www.climatehotmap.org/globalwarmingeffects/lakesandrivers.Html. Accessed 20 March 2016

Urrutia-Cordero, P.; Zhang, H.; Chaguaceda, F.; Geng, H.; Hansson, L.-A. Climate Warming and Heat Waves Alter Harmful Cyanobacterial Blooms Along the BenthicPelagic Interface. *Ecology* **2020,** *101* (7), e03025. 10.1002/ecy.3025

Visser, P. M.; Ibelings, B. W.; Bormans, M.; Huisman, J. Artificial Mixing to Control Cyanobacterial Blooms: A Review. *Aquat. Ecol.* **2015,** *50*, 423–441. doi: 10.1007/s10452-015-9537-0

von Scheibner, M., et al. Impact of Warming on Phytobacterioplankton Coupling and Bacterial Community Composition in Experimental Mesocosms. *Environ. Microbiol.* **2014,** *16*, 718–733.

Voss, M.; Bange, H. W.; Dippner, J. W.; Middelburg, J. J.; Montoya, J. P.; Ward, B. The Marine Nitrogen ycle: Recent Discoveries, Uncertainties and the Potential Relevance of Climate Change. *Phil. Trans. R. Soc. B.* **2013,** *368,* 20130121. http://dx.doi.org/10.1098/rstb.2013.0121

Wallace, R. B.; Baumann, H.; Grear, J. S.; Aller, R. C.; Gobler, C. J. Coastal Ocean Acidification: The Other Eutrophication Problem. Invited Feature Article in Estuar. *Coast. Shelf Sci.* **2014,** *148,* 1–13.

Webster, N. S.; Thomas, T. The Sponge Hologenome. *mBio* **2016,** *7* (2), e00135–16. https://doi.org/10.1128/mBio.00135-16

Wei, W.; Zhang, R.; Peng, L.; Liang, Y.; Jiao, N. Effects of Temperature and Photosynthetically Active Radiation on Virioplankton Decay in the Western Pacific Ocean. *Sci. Rep.* **2018,** *8,* 1525.

Wilkins, L. G. E.; Leray, M.; O'Dea, A.; Yuen, B.; Peixoto, R. S.; Pereira, T. J., et al. Hostassociated Microbiomes Drive Structure and Function of Marine Ecosystems. *PLoS Biol.* **2019,** *17* (11), e3000533. https://doi.org/10.1371/journal.pbio.3000533

Winter, A., et al. Poleward Expansion of the Coccolithophore *Emiliania Huxleyi. J. Plankton Res.* **2014,** *36,* 316–325.

Wohlers. J.; Engel, A.; Zöllner, E.; Breithaupt, P.; Jürgens, K.; Hoppe, H-G.; Sommer, U.; Riebesell, U. Changes in Biogenic Carbon Flow in Response to Sea Surface Warming. *Proc. Nat. Acad. Sci. USA* **2009,** *106* (17), 7067–772.

World Health Organization—World Meteorological Organization. *Atlas of Health and Climate*; WHO, WMO: Geneva, 2012. http://www.wmo.int/ebooks/ WHO/Atlas_EN_web.pdf.

Xiong, et al. Thermal Discharge-Created Increasing Temperatures Alter the Bacterioplankton Composition and Functional Redundancy. *AMB Expr.* **2016,** 6, 68. doi: 10.1186/s13568-016-0238-4

Yamamoto, N.; Urabe, K.; Takaoka, M.; Nakazawa, K.; Gotoh, A.; Haga, M., et al. Outbreak of Cryptosporidiosis after Contamination of the Public Water Supply in Saitama Prefecture, Japan in 1996. Kansenshogaku Zasshi. *J. Japan Assoc. Infect. Dis.* **2000,** *74 (6),* 518–526.

Zehr, J. P.; Bench, S. R.; Carter, B. J.; Hewson, I.; Niazi, F.; Shi, T.; Tripp, H. J.; Affourtit, J. P. Globally Distributed Uncultivated Oceanic N_2-Fixing Cyanobacteria Lack Oxygenic Photosystem II. *Science* **2008,** *322,* 1110–1112. doi:10.1126/science.1165340

Zhang, W., et al. Marine Biofilms Constitute a Bank of Hidden Microbial Diversity and Functional Potential. *Nat. Commun.* **2019,** *10,* 1–10.

Zhou, J.; Song, X.; Zhang, C.-Y.; Chen, G.-F.; Lao, Y.-M.; Jin, H.; Cai, Z.-H. Distribution Patterns of Microbial Community Structure Along a 7000-Mile Latitudinal Transect from the Mediterranean Sea across the Atlantic Ocean to the Brazilian Coastal Sea. *Microbial Ecol.* **2018,** *76,* 592–609. https://doi.org/10.1007/s00248-018-1150-z

Zorz, J.; Willis, C.; Comeau, A. M.; Langille, M. G. I.; Johnson, C. L.; Li, W. K. W.; LaRoche, J. Drivers of Regional Bacterial Community Structure and Diversity in the Northwest Atlantic Ocean. *Front. Microbiol.* **2019,** *10,* 281. doi: 10.3389/fmicb.2019.00281

CHAPTER 7

Microbial Food-borne Diseases Due to Climate Change

JOHN MOHD WAR[*], ANEES UN NISA, ABDUL HAMID WANI, and MOHD YAQUB BHAT

Section of Mycology and Plant Pathology, Department of Botany, University of Kashmir, Srinagar, Jammu and Kashmir, India

[*]*Corresponding author. E-mail: johnbotanyku@gmail.com*

ABSTRACT

Climate change is a global concern shadowing its impacts on the whole planet earth. The problem arises due to increased concentration of greenhouse gases in the atmosphere. The increased concentration of these gases causes an increase in the average temperature of earth's surface. The assessment report fifth of IPCC predicts that the average earth's surface temperature is likely to exceed $1.5°C–2.0°C$ by the year 2100. The changing climate is exerting significant impacts on all aspects of life on planet earth. Although much attention was given to its impacts on macrobiome, but now its impact on microbiome is also being studied. The changing climate affects the occurrence and persistence of all microorganisms and their vectors, leading to increased bacterial, viral, and pathogenic contamination of water and food that in turn is placing a threat to animal health by having direct and indirect impacts on food safety. There are sufficient evidences to support the interrelation between the survival, growth, and transmission factors (vectors) of causal organisms of food-borne illness, such as bacteria, fungi, and viruses and the changing climate patterns. The possible effect of changing climate on public health including its effect on microbial food-borne illness are now being studied and significant evidences support for its negative impacts on food safety especially in developing countries. Although food-borne diseases in maximum cases are a mild gastrointestinal problem, some cases are much more than gastroenteritis resulting in hospitalization and even death. These diseases are also posing a great thrust on the economy of a country due to treatment cost and many more factors, and hence are a major concern in most countries. Thus, food safety and security are the most significant climate change-related threats at global level. The current chapter discusses the increasing risks of microbial food-borne diseases due to climate change as well as the preventive measures that need to be addressed to the scientific community.

7.1   INTRODUCTION

Climate change is a change in worldwide or regional climate patterns that is mainly ascribed to the increased levels of atmospheric carbon dioxide that is a key greenhouse gas. The main reason of this increased carbon dioxide concentration is mainly anthropogenic such as burning of fossil fuels. Deforestation is also a major anthropogenic factor responsible for increased concentration of atmospheric carbon dioxide, it decreases the potential number of plants sequestering carbon dioxide and converting it into oxygen. Farming contributes to methane production which is also a greenhouse gas. Greenhouse gases are important components of the atmosphere that act as an envelope and trap the sun light, exerting a warming effect thereby maintaining the temperature of our planet capable of sustaining life. The increased greenhouse gas concentration causes increase in the earth's average temperature. The Fifth Assessment Report of IPCC predicts that there will be a rise of 1.5°C–2.0°C of average earth's surface temperature by the year 2100 (IPCC, 2018). The increased warming effects cause significant alteration in the rainfall patterns and rainfall amount. Climate models predict an increase in temperature from 3.2°C to 3.7°C and a change in the amount and pattern of rainfall for drylands globally by the late 21st century (Solomon et al., 2007). This is a very serious problem as even a minute fluctuation in global average temperature can cause severe shifts in weather and climate (US EPA). Climate change not only includes increase in the average earth's temperature but also encompasses thrilling weather events, shifting habitats and populations, rising sea levels, and a range of other effects. All these effects of changing climate are vital signs that our planet earth is displaying. Scientists are already familiar with its impacts on planet earth and have documented its impacts such as melting of glaciers leading to an increase in sea level, increased temperatures, wildlife shifts to cooler areas and extinction of many species, severe droughts in many regions, increased risk of wildfires, shortage of drinking waters, structural and functional alterations of ecosystem, and many more effects. The changing climate is at present among the utmost composite worldwide issues exerting significant impacts on all aspects (scientific, political, social, economic, ethical, and moral aspects) of life on planet earth (NASA). Climate change is linked to human health in a very complicated manner. Changes in the climatic variables are linked directly or indirectly to many human health issues ranging from mild stress to severe dreadful diseases. Climate change has led to the degradation of food quality, water, and air that resulted in food- water- and air-borne diseases (Fig. 7.1). If the climate change continues to happen, projections are there regarding the spread of diseases from the southern to the northern latitudes, especially the re-emergence of diseases that had already been eliminated, such as yellow fever, malaria, etc. Climate changes are also linked to the distribution of rodent-borne diseases, including Hantavirus disease and leptospirosis as well as vector-borne diseases, such as Lyme disease, leishmaniasis, tick-borne encephalitis, dengue, etc. (Gubler et al., 2001).

In summary, changing climate is affecting the whole life on planet earth. Although much attention was given to its impacts on macrobiome, its effects on microbiome is also being studied now. Microorganisms are an important part of the planet earth particularly to human life.They are present in all ecosystems and display an immense role in every type of habitat. The roles of microorganisms include: producers, decomposers, mutualists, biocontrol agents, pathogens, key contributors of biogeo-chemical cycling, role in sewage and wastewater treatment, some produce natural plant stimulants and many other important functions in maintaining ecological equilibrium of the planet earth. However, due to complexity of microbiomes, their response to changing climate is poorly understood (Bardgett et al., 2008; Zhou et al., 2012). To understand their response, it is necessary to identify the climate-sensitive microorganisms and understand their relationship with the factors responsible for the changing climate. So, various studies were conducted by many researchers to study their relations with the changing climatic variables, for example, Rui et al. (2015)

studied the impact of changing climatic factors on bacterial communities of meadow soils in alpine areas. Changes in the environmental conditions due to changing climate are likely to influence all aspects of microorganism's life. Soil microbial communities get affected by changing climatic variables either directly or indirectly through alterations in the plant physiological processes and composition of plant communities (Bardgett et al., 2013).

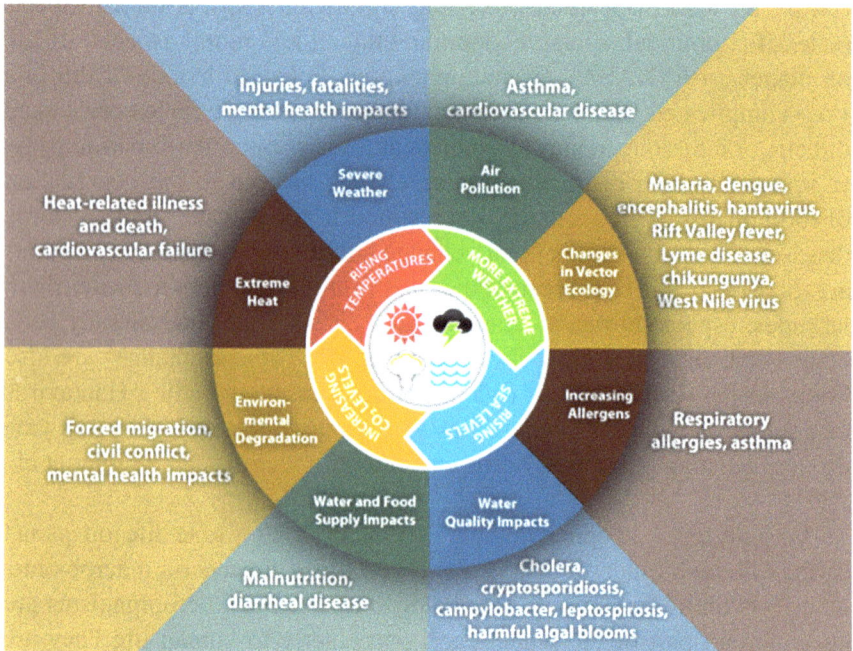

FIGURE 7.1 Impacts of climate change on human health.

Source: Centers for Disease Control and Prevention (CDC), 2020: https://www.cdc.gov/climateandhealth/effects/default.htm.

The changing climate affects the existence and persistence of all microorganisms and their vectors leading to increased pathogenic microbial contamination of water and food that in turn is placing a threat to animal health including humans by laying direct and indirect effects on food security. Food is a global commodity and its production is getting affected by changing climatic factors, even the increased contamination of water will affect the food output and may cause food-borne illness. There are sufficient evidences to support the interrelation between the survival, growth, and transmission factors (vectors) of causal organisms of food-borne diseases and the changing climate patterns. The causal organisms (such as bacteria, virus, mycotoxins, and phycotoxins) of diseases that are food-borne could emerge with changing climatic variables. Thus, populations are at risk of food-borne disease outbreaks, such as hepatitis A, typhoid fever, and diarrheal diseases, including dysentery, cholera, norovirus infections, and also to the exposure of toxic compounds through contaminated water and food. Food-borne illnesses are a key health issue in most of the countries. Food-orne disease in most cases is a mild gastro-intestinal problem, in some cases, it is much more than the gastroenteritis resulting in hospitalization and even death. According to World Health Organisation (WHO)report 2015, there were about 600 million food-borne illnesses and 420,000 associated deaths globally in 2010 (WHO, 2015). These diseases are putting a great thrust on country's economy. There are up to 823,000 enteric infections every year in New Zealand, posing a great pressure on country's economy (Snel et al., 2009). Generally, the common enteric infectious food-borne diseases in New Zealand is salmonellosis caused by *Salmonella* bacteria causing symptoms of nausea, diarrhea, headache, abdominal pain, and sometimes vomiting in humans (Heymann, 2004). The bacteria show rapid multiplication rate with changing climatic temperature variable (Britton et al., 2010). Therefore, climate change could upsurge salmonellosis infection in New Zealand posing a great threat to human health. The possible effects of changing climate on human health including its effect on microbial food-borne illness are now being studied and significant evidences support for its negative effects on food safety mostly in developing countries. The present chapter examines the effect of climate change on various microbial pathogens including bacteria and viruses and mycotoxin producing fungi, which are associated with food-borne illness. The impact of climate change and the related food safety issues will be discussed and the measures that are needed to respond to the negative impacts of climate change will be addressed to the scientific community through this chapter.

7.2 MICROBIAL FOOD-BORNE DISEASES AND CLIMATE CHANGE

Food-borne diseases are caused by taking contaminated food. The contamination can be an infectious organism, such as bacteria, parasite, virus, or the toxins such as mycotoxins (fungi), phycotoxins (algae), or chemicals. The food-borne illness is gastroenteritis in majority of the cases, however, the diseases may prove fatal at times. Gastroenteritis may result either directly wherein the microorganisms colonize the GI (gastrointestinal) tract and then grow and invade the host tissue or indirectly the pathogen contaminates the food by the secretion of exotoxins leading to food intoxication. The symptoms of food-borne illness include nausea, vomiting, abdominal pain, stomach cramps, fever, headaches, diarrhea, muscle aches, and in occasional cases may result in hospitalization and death (Public Health Agency of Canada, 2009). Food-borne illness is responsible for large epidemics in many countries, for example, *E. coli* O157 epidemic in the United States lasted from 2003 to2012 (Heiman et al., 2015). The advent of great epidemics linked to the consumption of food are rising because of globalization, international trade and travel, environmental contamination with feces in areas of poor sanitation, increased rate of natural catastrophes like floods. In the USA, there are approximately 3.5 million cases of bacterial food-borne illness per year which are caused by *Campylobacter, Clostridium, Bacillus cereus, Brucella, Escherichia coli, Yersinia enterocolitica, Vibrio, Salmonella, Staphylococcus aureus*, and *Listeria monocytogenes* (Scallan et al., 2011). The dispersal of such pathogens occurs by means of water, wind, humans, and animals.

Climate change is real and irrefutable, that is associated with rise in temperature, increase in sea level, and declining ice and snow cover (IPCC, 2018). The planet warming is occurring at unexpected rate due to anthropogenic activities (Neukom et al., 2019). The changing climate is affecting the whole life present on the biosphere. Humans are getting affected by the changing climatic variables either by direct mortalities due to heat waves (Robine et al., 2008) or indirectly by spread of infectious diseases because of insufficient or contaminated food and water (Parkinson and Butler, 2005). Climate change is negatively affecting human health by one or the other way, for example, flooding, one of the outcomes of changing climate, leads to contamination of drinking water sources followed by increased risk of infectious diseases mostly in developing countries (Hunter, 2003). Similarly, changing climatic variables like drought and warm weather conditions increases the chances of getting zoonotic diseases (diseases that spread from animals to people either by direct interaction with diseased animals or indirectly through their wastes and products). Climate change also increases the vulnerability of animals to diseases, alters their seasonality, influences the life cycle of their transmission vectors and thus increases the incidence of human infection (Naicker, 2011). With increase in temperature, the food preparation and consumption pattern may change as people will prefer foods from outdoor warmer environment, such as raw salads and barbecued foods that are easily contaminated, hence may increase the chances of getting infections (Kovats et al., 2004). In addition to changing climatic variables, the inadequate hygiene conditions may cause increased chances of epidemics due to unsafe food. The total climate change-related mortalities and disease burden has been estimated as 0.3% and 0.4%, respectively (Kendrovski and Gjorgjev, 2012). The Lancet countdown on health and climate change 2018 report mentions climate change as a "biggest global threat to 21st century" which if not tackled properly could lead to disasters that may interrupt with the fundamental public health infrastructure and may overpower health services (Watts et al., 2018). Hence, immediate attention should be given to develop and improve measures for examination of food availability, safe drinking water, response to outbreaks of food-borne illness, prevention of food-borne diseases, provisions for providing food safety educations, and collecting information of the affected people.

The microbial food-borne diseases are of great concern in majority of the countries as they not only cause population thinning but are also leaching country's economy. The World Bank report of 2018 on the economic load of food-borne diseases in middle- and low-income countries estimated the total annual productivity loss of US$ 95.2 billion and 15 billion US$ per year for the treatment of food-borne illness. The risk of infectious food and water-borne illnesses is increasing globally with changing climatic variables (Rose et al., 2001; Ebi et al., 2006). In Canada, every year, there were about 4 million cases of microbial food-borne illness from 2000 to 2010 (Thomas et al., 2013). The microbial food-borne pathogens are extremely affected by climatic variables and there are substantial evidences to support this (Park et al., 2018). Climate change impacts all the three aspects of a classical epidemiological triad, tht is, host, environment and agent, and hence, can have an inordinate effect on the infectious diseases (Kendrovski and Gjorgjev, 2012). Climate-driven factors such as increase in temperature and precipitation affect the food-borne illness pathogens by affecting their growth and abundance, increasing the prevalence of a pathogen in agricultural products, animals and the environment (Ebi, 2011) and alter the life cycles of their transmission vectors (Agunos et al., 2014). There will also be change in food handling, preferences, and cooking practices due to changing climatic factors (Ravel et al., 2010; Milazzo et al., 2017). Thus, from the above evidences, it is clear that climate-induced environmental changes will end up with net burden of food- and water-borne diseases. In the following sections, various microbial pathogens that are sensitive to change in one or more climatic variables and leading to food-borne illness are listed in (Table 7.1). These include two major groups, that is, bacteria (e.g., *Salmonella, Campylobacter, Listeria monocytogenes, Vibrio, Bacillus cereus, E. coli, Shigella*) and viruses (e.g., *Norovirus*, Hepatitis A virus, *Rotavirus*).

7.2.1 *CLIMATE-SENSITIVE BACTERIAL FOOD-BORNE ILLNESS PATHOGENS AND THEIR DISEASES*

Bacteria represent one of the key causal organisms of food-borne illnesses worldwide, they are unicellular microorganisms having ubiquitous distribution. They thrive almost in all habitats. They are responsible for many diseases in humans, such as salmonellosis, alimentary tract infections, cholera, typhoid, and tetanus, etc. The major diseases of concern in most countries is food-borne illness. The symptoms of bacterial food-borne diseases range from slight diarrhea to serious illness and death (e.g., *E. coli* O157: H7 strain can cause hemolytic uremic syndrome (HUS), a rare disorder especially in children resulting in kidney failure and death). Bacterial contamination of foods can occur in fields or during storage, processing, shipping, or during preparation of foods in the kitchen. They may already be present on purchased foods. Raw foods are not hygienic and are prone to more contamination than cooked foods. Cooked foods if kept for more than 2 hours at room temperature, provide a favorable atmosphere for bacterial growth.

Bacterial growth and survival is largely reliant on environmental circumstances, such as temperature, moisture, and pH (Montville et al., 2012). Any change in one of these climatic factors will alter bacterial growth, abundance, dispersal, and survival by direct or indirect means. Elevation in temperature favors the increased growth of bacterial pathogens creating stress in livestock to shed more enteric pathogens, for example, *Salmonella* from diary animals and farm environments, hence affecting its occurrence in agricultural crops and the environment (Pangloli et al., 2008). Similarly, intense and untimely precipitation could increase the dispersal of pathogens directly via flooding. Changing climate impacts bacterial pathogens indirectly by affecting the life cycle of the wildlife vectors. These vectors are associated with the increased contamination of food sources, for example, insects and rodents in the farm fields are linked with the increase in contamination of broiler chicken with *Campylobacter* (Agunos et al., 2014). In summary, bacterial food-borne illness pathogens are sensitive to changing climate (variations in temperature, precipitation, humidity, and extreme weather events, such as drought or floods etc.). The main climate-sensitive bacterial pathogens that are of major concern are discussed in the following sections.

7.2.1.1 SALMONELLA

It is a rod-shaped gram-negative bacterium accountable for the enteric fever and gastroenteritis (Salmonellosis) diseases. The bacterium has a single species and thousands of serotypes (serovars) that are responsible for different grades of infections. Globally, 93.8 million estimated cases of human illnesses are known annually, out of which 155,000 mortalities per year is due to *Salmonella* (Hoelzer et al., 2011). Salmonella gastroenteritis (or salmonellosis) occurs when the bacteria from the contaminated water or foods, such as meat, eggs, poultry, vegetables or their products enter the GI tract of host, multiply there (incubation period 8–48 h), invade the tissue and produce enterotoxins that disrupt the functioning of intestinal mucosa by disrupting its epithelial cover. The disease is mostly characterized by symptoms of nausea, diarrhea, vomiting, abdominal pain. Cramps that appear can last for few days and usually may dip off within a week. The primary inoculum source of the bacterium is animal and human intestines (Crump et al., 2015).

Salmonella typhoid fever occurs by taking contaminated food and water and by direct personal contacts. The primary source of inoculum is mainly human feces. The infected individual excretes the bacteria in large numbers which are then ingested by another host and later reach the small intestines and then pierce the epithelial lining and then get transported to lymphatic tissue, blood, liver, spleen and gallbladder. The basic symptoms of the diseases are fever, abdominal pain, constipation that may progress to intestinal hemorrhage, bowel perforation, and even death if untreated (Crump et al., 2015; Walia et al., 2005).

Salmonella is present almost in every environment and its transmission from environmental sources is directly linked to climatic factors, such as temperature, precipitation, and relative humidity (Kelly-Hope et al., 2008). These variables are affected by changing climate which by direct or indirect ways increases the spread of the bacteria and poses risk the public health.. Sufficient evidences are available to support the impacts of changing climate on *Salmonella* infections. In the coming decades in Ireland, 2% increase in the rate of *Salmonella* gastroenteritis is expected to occur because of the changing climate (Cullen, 2009). Global warming is increasing the temperature of earth's surface and the increased temperature is associated with increased salmonellosis infections. The temperature between 7°C and 30°C is favorable for the bacterium to replicate in foods and manure (Hocking et al., 1997). Salmonellosis shows seasonality and is reported generally in summer (D'Souza et al., 2004; Akil et al., 2014). Various investigators have studied the relationship between temperature and salmonellosis and their results showed a positive association between the two (Britton et al., 2010; D'Souza et 2004; Zhang et al., 2007). Time-series analysis studies carried out in 10 European countries revealed that salmonellosis cases for each 1°C rise of weekly temperature increases by 5%–10% when ambient temperatures are above 6°C (Kovats et al., 2004). In a study in Mississippi, southern state of the US, four new cases of salmonellosis are added for each of 1°F increase in temperature (Akil et al., 2014). Higher temperatures are the causes of large number of cases of salmonellosis infections, and the infection is quite constant among different cities and countries (Tirado et al., 2010). The increasing temperatures due to changing climate can be an alarm to community health authorities that there are increasing chances of such diseases. The WHO 2017 report on "Protecting health in Europe from climate change: 2017 update" mentions the case study of Kazakhstan wherein it is found that increase in 1°C in mean monthly temperature causes 5.5% increase in the incidence of *Salmonella* gastroenteritis (salmonellosis). Another outcome of changing climate is change in precipitation events that also has significant impacts on the food-borne illness pathogens. The bacteria have high survival rate in moist soils with poor drainage (Chandler and Craven, 1980; Holley et al., 2007). Variable relationship between *Salmonella* infections and rainfall have been stated. Various studies have revealed a positive relationship between *Salmonella* infections and rainfall in watersheds or coastal waters (Martinez-Urtaza et al., 2004; Haley et al., 2009; Levantesi et al., 2012).

However, no relationship between *Salmonella* infections and monthly average precipitation was detected (Akil et al., 2014). In contrast, some studies have reported negative relationships between incidence of *Salmonella* and the rainfall (Vereen et al., 2013). Worldwide, some efforts have been taken to model future changes in *Salmonella* infections. Studies investigated in Australia have shown that if the climate change remains unchecked, there will be around 48% and 56% increases in morbidity burden of *Salmonella* infections in subtropical regions by 2030 and 2050, respectively (Zhang et al., 2012).

7.2.1.2   CAMPYLOBACTER

Campylobacter, typically comma-shaped or S-shaped gram-negative motile bacteria responsible for the *Campylobacter* gastroenteritis (campylobacteriosis) is the most common basis of food poisoning in the US and Europe. The symptoms of this disease include diarrhea, abdominal pain, vomiting, nausea, and fever. However, sometimes preceded infections by *Campylobacter* may cause severe complications, such as arthritis, septicemia, and Guillain-Barre syndrome (GBS). Food-borne disease annual report in New Zealand mentions that there were reports of 117 per 100,000 hospitalized cases of GBS in 2014. According to WHO, the most common reason for gastroenteritic infections in humans throughout the biosphere is due to *Campylobacter*. *Campylobacter* is found almost in all natural environments, in animals, and birds' gut (Snelling et al., 2005; Humphrey et al., 2007). Infection occurs from contaminated foods and water, due to consumption of undercooked poultry, contact with poultry and pet animals, or due to travel (Melby et al., 1990; Rodrigues et al., 2001). Consumption of poultry in Denmark is considered as the major route of *Campylobacter* infection leading to large number of food-borne illnesses (Neimann et al., 2003). *Campylobacter jejuni* represents the most common organism for food poisoning in Europe and United States. In European regions in 2018, about 246,571 cases of *Campylobacter* infections were reported by the European Food Safety Authority (EFSA) and an estimated 9 million cases of human campylobacteriosis have been reported in European Union every year.

Impact of climatic change on *Campylobacter* has been studied by many workers. The campylobacteriosis incidences vary with the geography

(Kapperud and Aasen 1992) and season (D'Souza et al., 2004; Wallace et al., 1997; Patrick et al., 2004). Time-series analysis revealed the effects of changing climate on the *Campylobacter* gastroenteritis in 18 countries of three continents, that is, Europe, Australia, and North America, and the results showed that most countries in Europe showed the peak of infection in early spring. However, all the countries did not follow the same trend, Canada (North America) shows the peak of infection in late June and early July (Kovats et al., 2005). In Montreal, Canada, by 2055, there would be a 4.5°C increase in temperature which is predicted to increase campylo-bacteriosis cases by 23% (Allard et al., 2011). In temperate regions, the highest infection cases were observed in summers (Nylen et al., 2002; Tam et al., 2006).

Higher temperatures mainly in combination with several hours of daylight is related to higher campylobacteriosis infections (Patrick et al., 2004). In the coming decades, in Ireland, 3% increase in the incidence of disease is expected to occur due to changing climate (Cullen, 2009). In a 10-year study from 1999 to 2010, in Israel, an increase in 1°C above the threshold of 27°C resulted in increasing infections of *Campylobacter jejuni* and *Campylobacter coli* by 16.1% and 18.8%, respectively (Rosenberg et al., 2018). Various evidences are available to support the impacts of temperature and rainfall on the lifecycle of *Campylobacter* present in its natural environment or on food stuff (Patrick et al., 2004; Fleury et al., 2006; Bi et al., 2009). In Czech Republic, increased number of cases of *Campylobacter* infections were recorded after flooding (McMichael et al., 1996). Rainfall has variable effects on *Campylobacter* infection. Various researches have reported positive relations between rainfall and *Campylobacter* occurrence in watersheds (Vereen et al., 2007; Wilkes et al., 2009). In Canada, in the Solomon River, a negative relationship between the occurrence of *Campylobacter* spp. and rainfall was reported (Jokinen et al., 2001). Weisent et al. (2014) studied the importance of relationship of climatic factors with the sources, pools and communication routes of the *Campylobacter* and found that the results were in accordance with the previous studies that bacteria show seasonality and the infections are observed more in summers (2014). Additionally, their results also showed that the risk of infections is more during droughts (rainfall may well improve persistence and cause leaching of bacteria, droughts may concentrate the bacterial load (Patrick et al., 2004; Sinton et al., 2007) (Table 7.1). in Maryland, USA, extreme precipitations are linked with

TABLE 7.1 Climate-Sensitive Microbes and Their Transmission Routes to Humans.

Microbial agent	Host	Transmission route
Salmonella	Poultry and pigs	Fecal/oral
Campylobacter	Poultry	Fecal/oral
Listeria monocytogenes	Livestock	In the Northern Hemisphere, listeriosis has a distinct seasonal incidence in livestock probably related with feeding of silage
Vibrio	Shellfish, Fish	Fecal/oral
Escherichia coli O157	Cattle and other ruminants	Fecal/oral
Shigella	Poultry	Fecal/oral
Bacillus cereus	Plants, tracts of insects	Contaminated food
Anthrax, Clostridium	Livestock and wild birds	Ingestion of spores through contaminated food, water, soil, and environmental routes
Yersinia	Birds and rodents with regional differences in the species of animal infected Pigs are a major reservoir	Linked with outbreaks that occur after droughts Dealing with pigs at the slaughter house is also a great risk to humans
Leptospira	All farm animal species	Contaminated food and water
Norovirus	Pigs, cows, mice	Fecal/oral
Hepatitis A Virus	Shellfish	Fecal/oral, contaminated food and water, direct personal contact
Rotavirus	Chicken, pigs, cattle	Personal contact, contaminated food and water
Rift Valley fever virus	Multiple species of livestock and wildlife	Handling of animal tissue, unpasteurized milk of infected animals, mosquito, hematophagous flies.
Nipah virus	Bats and pigs	Directly from bats to humans through food in the consumption of date palm sap, infected pigs also a great threat to humans

TABLE 7.1 *(Continued)*

Microbial agent	Host	Transmission route
Hepatitis E Virus	Domestic and wild animals	Fecal/oral, contamination of water and Shellfishes by pig manure via irrigation
Encephalitis tick–borne virus	Sheep, goats	Uncooked milk
Cysticercus bovis	Cattle	Fecal/oral
Liver fluke	Sheep, cattle	Fecal
Toxoplasma gondii	Cats, pigs, sheep	Fecal(cats), contaminated meat, infected food handler
Cryptosporidium	Cattle, sheep	Fecal/oral, water-borne
Giardia	Cattle, cats, dogs	Fecal/oral, water-borne

Source: Adapted from Tirado et al. *Food Research International* **2010**, *43* (7), 1745, with permission from Elsevier).

increased number of *Campylobacter* infections (Soneja et al., 2016). The studies also revealed that climatic variables may also impact carriage rates and contamination in environmental reservoirs from one division to next.

7.2.1.3 LISTERIA MONOCYTOGENES

It is a facultative anaerobic, gram-positive, rod-shaped bacterium responsible for a high mortality rate food-borne illness called listeriosis. The largest ever globally recorded outbreak of listeriosis occurred in South Africa during early 2017 (Spies, 2018). The sources of outbreak were found in early march 2018 in food production facility (Polokwane, South Africa) where the pathogen was found in the food products containing meat (Motsoaledi, 2017). Foods are considered a major source of *Listeria* pathogen, and in the US, 99% of listeriosis cases are food-borne (Scallan et al., 2011). The pathogen strains have been found to survive for years and decades in food processing systems (Ferreira et al., 2014). Like other food-borne illness pathogens, it also causes diseases like fever, diarrhea, etc. but the infection is rarely diagnosed. The symptoms appear only when the bacteria reach to other parts beyond the gut. The symptoms include fever, muscle pain, convulsions, and balance loss. In expecting mothers, it causes preterm labor, fetal infections and also sometimes fetal deaths. The bacteria are extensively scattered in the biosphere and are present in soil, vegetation, water, and in some animal feces and contaminated foods (mostly foods that are ready to eat) which then become the sources of the infection. The contaminated foods include unpasteurized milk and its products, raw fruits, raw vegetables, and meat products (deli meat). Although the number of individuals getting infected by this disease is small (0.1–10 cases per million per year depending on regions and countries of the world) but the high death rate related to this infection is making it a serious concern for public health (WHO, 2018). In 2017, listeriosis outbreak caused 180 deaths in South Africa (Motsoaledi, 2017). To prevent such outbreaks, it is important to adapt preventive measures and also attention is to be paid toward the factors that may enhance the frequency of the diseases in the near future. One such factor that is nowadays receiving more attention toward its contribution to food-borne diseases is climate change.

Changing climate has a profound impact on human health, particularly, when infectious diseases are concerned. Like other food-borne illness pathogens, the hasty growth of *Listeria* on food products increases with increase in temperature which is due to the enhanced rate of bacterial division (Miettinen et al., 1999). There are various evidences in support of the positive correlation between temperature and the occurrence of listeriosis cases (Goulet et al., 2006; Musengimana et al., 2016; Hellberg and Chu, 2016). In addition to temperature, other environmental factors affected by global climate change including precipitation, rainfall, and more dry seasons affect *Listeria* transmission. Investigation on fish farms in Finland have been conducted and the study revealed that the number of *Listeria* spp., together with *L. monocytogenes* increased following rainfall periods, possibly due to runoff from contaminated rivers and nearby brooks (Miettinen an Wirtanen 2006). Duration of rainfall has varying effects on the transmission of the pathogens, short rainfall favors transmission of pathogens from soil into plants while rainfall for longer duration has washout effects (Hellberg and Chu, 2016). Moist soils have been found to favor extended persistence of the pathogen than dry soils (McLaughlin et al., 2011; Strawn et al., 2013). With the changing climate, there will be disasters like flooding and drought due to which scarcity of water will become evidently more. As a result of this, cleaning practices will be altered that in turn favor the pathogen in food processing systems (Chersich et al., 2018). The potable municipal water will be limited and farmers will rely on other sources for irrigation such as surface waters which naturally contains the pathogen (Odjadjare et al., 2010; Olaniran et al., 2015). The polluted water used in irrigating agricultural fields contaminates fruits and vegetables which then become the sources of infection (Markland et al., 2017). Thus, there is a known consequence of changing climate on the dispersal and survival of *Listeria* pathogens and by direct or indirect means it leads to contamination of food that affects human health due to food-borne illness.

7.2.1.4 VIBRIO

Vibrio is a comma-shaped uniflagellate genus of gram-negative bacteria that is the major culturable fraction of marine picoplankton (Farmer Farmer et al., 2005). The *Vibrio* spp. linked with human illness include *Vibrio vulnificus, Vibrio cholera*, and *Vibrio parahaemolyticus*. *Vibrio cholera* survives in both fresh and marine water and is responsible for water-borne illness, that is, cholera (Reidl an Klose, 20002). *Vibrio vulnificus* and *Vibrio parahaemolyticus* grow only in sea water and are responsible for food-borne illness, that is, gastroenteritis (Su et al., 2007;

Horseman and Surani, 2011). Worldwide, the prime cause of bacterial infections related to seafood is *Vibrio parahaemolyticus (*Abanto et al., 2020*)*. In the US, *Vibrio parahaemolyticus* is the major cause of gastro-enteritis and the infection reported in 2011 was 76% higher as compared with the 1996–1998 levels, causing an estimated 100 deaths and 80,000 illnesses every year (Food Net, 2012). The infections by *Vibrio vulnificus* is rare, but among food-borne pathogens, it has the highest death rate, with an annual estimate of 96 cases of food-borne diseases and 36 deaths in the US (Scallan et al., 2011). The disease is characterized by symptoms of diarrhea, vomiting, abdominal pain, fever, sudden chills, and shock in high-risk individuals. Individuals get infected by consuming raw and less-cooked sea foods (oyster).

Global warming is not only disturbing terrestrial ecosystem but it also causes significant warming of ocean waters, affecting whole marine life. All European seas have increased temperatures and have warmed at the global rate of four- to sevenfold during the previous few decades (Reid et al., 2011). Increase in temperature, heavy precipitation, elevation in sea levels, flooding, and change in salinity has its impact on aquatic flora including human pathogens such as *Vibrio* spp. (FAO, 2008c). Changing climate has its having profound effects on the environments and spreading of aquatic life and may result in more disease outbreaks by *Vibrio* spp. specially, where sea foods are consumed uncooked or undercooked (Morley et al., 2018). With the climate warming, the diseases associated with *Vibrio* are increasing worldwide (Harvell et al., 2002). Not only ocean warming but also the change in salinity that occurs due to changing climate affects *Vibrio* pathogen responsible for many infectious diseases in humans. Varying relationship exists between the *Vibrio* abundance and the salinity (Paz et al., 2007; Zimmerman et al., 2007; Hsieh et al., 2008; Caburlotto et al., 2012; Oberbeckmann et al., 2012). The increased occurrence of *Vibrio* is mostly evident in much warming regions of the biosphere, such as the US Atlantic coast (Newton et al., 2012) and Northern European countries (Baker-Austin et al., 2013). In recent decades, the number of *Vibrio* infections related to food-borne infections and recreational bathing has increased significantly coinciding specifically with heat waves. The possible reason for this increased incidence may be due to change in the patterns of disease transmission or increase in the number and growth of *Vibrio* spp. in the environment (Vezzulli et al., 2013). In the US, *Vibrio parahaemolyticus* and *Vibrio vulnificus* abundance was found to be positively linked with increased surface temperature of sea waters (Johnson et al., 2012) and the *Vibrio* infections show seasonal variability also, that is, infection rate is more in late springs and early summers, showing peak in august followed by a steady decrease in September onwards (Food Net, 2012). The leading food poisoning danger in Taiwan in the most recent 10 years was due to *Vibrio parahaemolyticus* (Taiwan Food and Drug Administration). Study using seasonal Autoregressive Integrated Moving Average (ARIMA) models to investigate the relation between the climatic aspects and the occurrence of *Vibrio parahaemolyticus* in Taiwan showed that temperature and salinity of the sea were associated positively with *Vibrio parahaemolyticus* outbreaks, but rainfall was inversely associated (Hsiao et al., 2016). In another study, it was found that the emergence of illness due to *Vibrio parahaemolyticus* in Peru was linked with ocean warming by the advent of El Nino conditions (Abanto et al., 2020). In the Republic of Korea, analysis of trends in food-borne illness by *Vibrio para-haemolyticus* shows that there is a solid positive association between the temperature and the pathogen (Park et al., 2019). Increased temperature actually affects the virulence genes of *Vibrio* sp. and enhance its virulence properties (Rodriguez-Castro et al., 2010).

7.2.1.5   *ESCHERICHIA COLI*

It is a gram-negative bacterium, belonging to family of microbes called coliforms. Majority of the strains of *E. coli* live in mutual relationship in the guts of humans but some are responsible for many diarrheal diseases, for example, *E. coli* O157:H7 is responsible for the outbreak of food-borne disease that may even lead to death. The transmission of the bacteria occurs by the intake of contaminated foods (raw or less-cooked meat products, raw milk, fruits and vegetables contaminated with feces of animals at any step of cultivation, handling, or processing), contaminated water or contacts with persons handling contaminated food. The symptoms of infection range from non-bloody to bloody diarrhea, dehydration, abdominal pain, fever, and weakness (Cody et al., 2019). In young children, it may cause HUS which is a serious complication with symptoms of renal failure, low red blood count and platelet count, and then ultimately death (Salvadori and Bertoni, 2013).

Change in weather events, length of season, and increased air tempera-tures has an impact on survival, growth, and dispersal of *E. coli* (Hellberg and Chu, 2016*)*. For example, heat-induced stress has been reported to be responsible for the secretion of Shiga toxin by *E. coli* from cattle herds in Michigan, USA (Venegas-Vargas et al., 2016). Soils having poor drainage conditions and high moisture content have been shown to provide favorable environment for the survival of *E. coli (*Chandler and Craven, 1980; Solo-Gabriele et al., 2000*)*. In another study carried out in Blackstone River watershed, Massachusetts, USA, it has been found that *E. coli* density generally increases during the periods of wet weather as compared with dry weather (Wu et al., 2009). During drought, soil salinity increases and the increased soil salinity in agriculture lands of the southwestern USA was found to have negative correlations with the survival rate of *E. coli* O157:H7 (Ma et al., 2012). Similar study in the Salinas valley of California showed that with decreased water content of the soil, the populations of *E. coli* also decreased (Gutierrez-Rodriguez et al., 2012), while as flooding has been found to increase its concentrations to highest water levels (Cooley et al., 2007). In a lagoon (South of France), it was observed that during the periods of higher rainfall, there was an increased concentration of *E. coli* bacteria in waters but decreased concentration of bacteria was found in oysters (Derolez et al., 2013). The decreased concentration in oysters possibly will be due to reduced filtration rates because of increased turbidity and decreased salinity that occurs due to rainfall. Overall, a positive correlation between rainfall and levels of *E. coli* bacteria has been found in rivers and estuaries (Cooley et al., 2007; Kim et al., 2013; Lipp et al., 2001; Vermeulen and Hofstra, 2013).

7.2.1.6　*SHIGELLA*

It is a nonmotile, rod-shaped, facultative anaerobic gram-negative bacterium present in every habitat and responsible for food-borne illness like shigellosis and bacillary dysentery. The causal organisms of shigellosis are one of the many types of *Shigella* bacteria (*Yang et al., 2005*). The bacillary dysentery, that is, severe shigellosis is caused by three species of *Shigella*, that is, *S. sonnei, S. Flexneri*, and *S. dysenteriae* (WHO 2018). *S. sonnei* was the sole pathogen causing bacillary dysentery in Yazd Province Iran (Aminharati et al., 2018). The most common and prime source of infection is through personal contact (transmission via oral or fecal route) and also by contaminated foods and water (Louise et al., 2020). The person develops symptoms within 1–7 days of infection. The disease is characterized with stomach cramps, fever, nausea, and diarrhea that contains mucus and may be bloody in severe shigellosis (bacillary dysentery). About 20% of the hospitalized patients suffering from shigellosis die, raising a concern for developing health strategies to manage this disease (Kotloff et al., 1999). Although any one can get infected with this disease but children between 2 and 4 years of age are more prone to such diseases.

The environmental factors that get influenced with global climate change evidently affect the epidemiology of food-borne illness pathogens, together with *Shigella* (Hellberg and Chu, 2016). Haung et al. (2008) investigated the impacts of climate variability on the incidence of bacillary dysentery in Northeast China and mentioned that the changing climatic variables (temperature, precipitation, and relative humidity) are showing positive correlation with the prevalence of bacillary dysentery (2008). An increase in 1°C causes an increase of 12 cases of *Shigella* dysentery in China's northern city if other factors remain unchanged (Zhang et al., 2007). Studies on enteric infections in Canada (New Brunswick) also revealed a positive association of incidence of *Shigella* infections with temperature and humidity. The infections were highest during summers and more humid conditions (Valcour et al., 2016; Wei et al., 2017). Humidity also has a significant impact on the rise in bacillary dysentery cases. Mild humid weather conditions favor the increased occurrence of bacillary dysentery whereas decreased humidity during drought conditions decreases the reproductive potential of the bacteria and hence decreases the incidence of dysentery infections (Liu et al., 2017a. Investigations in the Binyang country have revealed that both absolute and relative humidity have positive association with the occurrence of bacillary dysentery (Liu et al., 2017b). Studies in rural parts of China showed a significant interaction between mean monthly rainfall and reproductive capacity of *Shigella* (Wang et al., 2005). In Korea, the study estimated that increase in weakly average of daily temperatures and precipitation was associated with the increased rate of shigellosis (Song et al., 2018). Increasing precipitation due to hefty rainfall causes changes in the use of drinking water, thereby increasing the overall prevalence of dysentery infections (Hossain et al., 2014). A 100 mm increase in precipitation, 1% increase in humidity and 1°C rise in temperature causes 4%, 1%, and 6% or 7% increase in the occurrence of bacillary dysentery respectively, as studied in Kon Tum province Vietnam (Lee et al., 2017). All studies revealed that changing climate especially global warming affects the survival and transmission of the pathogen, thereby increasing the risk of shigellosis.

7.2.1.7 BACILLUS CEREUS

It is a motile spore-forming, gram-positive, rod-shaped bacterium that causes diarrheal diseases in humans. The bacterium is present in soil, vegetation, and on left over foods, particularly rice, soups, and sauces. The bacterium contaminates the foods that have been set aside for long periods at room temperatures by the secretion of enterotoxins in it. The contaminated food when eaten by the host causes symptoms of vomiting, nausea, diarrhea, abdominal cramps within 8–16 h of incubation period, and long-lasting health effects may result in hospitalization and even death (Crump et al., 2015). *Bacillus cereus* causes two forms of food-borne diseases depending upon the strain and toxin produced: one is emetic type and other diarrheal type. The diarrheal kind is associated with diverse number of foods, such as meat, vegetables, and dairy products, whereas the emetic type is primarily associated with foods containing starch, such as rice and pasta.

Climate change particularly the increasing temperature has significant impacts on *Bacillus cereus* food-borne illness. In Poland, it has been found that *Bacillus cereus* shows considerable seasonality in processed and raw milk, with highest number in late spring and summer and much lower in winter and autumn (Bartoszewicz et al., 2008). There are also various evidences in support of positive association of the bacteria with temperature (Park et al., 2019; Kim et al., 2015).

7.3 IMPACT OF CLIMATE CHANGE ON VIRAL FOOD-BORNE DISEASES

Viruses are obligate parasitic infectious agents made of deoxyribose nucleic acid (DNA) or ribose nucleic acid (RNA) as a genetic material, fenced by a protein covering called capsid. Capsid is made of subunits termed as capsomeres. They are positioned between living and nonliving environments. They cannot multiply in outside environment like bacteria; they proliferate exclusively in living cells of other organisms and are responsible for the large number of diseases that may even cause the death of an organism. The transmission of viruses is mainly due personal contact, contaminated foods, and water or contact with inanimate objects. Food-borne transmission of virus occurs either by infected food handler that is not hygienic, or by contact of foods with human and animal sewages, polluted water or by animals (for zoonotic viruses) upon consuming animal origin products (meat, fishes, etc.) that are contaminated with virus (Vasickova et al., 2010). The main source of food-borne viruses are the human and animal intestines. The feces shed by the infected person becomes the sources of infection. A number of viruses linked with food-borne illness include Hepatitis A, *Norovirus*, Hepatitis E virus, *Rotavirus*, Astroviruses, Aichi virus, Coxsackievirus A and B, parvoviruses, adenoviruses serotypes 40 and 41, Sapoviruses, and other picornaviruses and enteroviruses (D'Souza et al., 2007; Jaykus et al., 2013).

Viruses in their living environment are under the shadow of various climatic factors that will not only have influence on the abundance and dispersal of their vectors but will also impact the interaction of virus with its vector (Gould and Higgs, 2019). The main routes of viral infection of foods like sewage and polluted water, infected food handler, and animals for zoonotic viruses all may be under the influence of climate change. The epidemiological outcome, frequency, and death rate due to viral infections may be modified by climate change (Rohayem, 2009). Changing climate is predicted to influence the transmission ways of virus by changing the ecology of the host and by causing socioeconomic change that affects the host populations (Chan et al., 1999). For example, floods, the outcome of climate change, cause the overflow of untreated sewage into the water-bodies, increasing the chances of enteric virus contaminations during the production of agricultural foods and shellfish. Shellfishes are one among the number of commonly utilized foods that are infected easily by food-borne viruses (Lees, 2009; Greening, 2006). Changes in climatic variables like temperature and carbon dioxide concentration have a known impact on the multiplication and behavior of viral pathogens (Hepatitis A and Rotavirus). The increase in temperature and carbon dioxide concentration was shown to increase the mutation rate of viral genetic material (RNA), as an adaptation to changing climate (Tarek et al., 2019). Some of the climate-sensitive viral pathogens that cause diarrheal diseases include:

7.3.1 NOROVIRUS

It is a non-enveloped, highly infectious, single-stranded RNA virus of the family Caliciviridae. It has a wide genome diversity, classified into five genogroups; G1–GV. Globally, the most basic cause of viral food-borne illness (gastroenteritis) among all age groups is due to *Norovirus (*Patel et al., 2008*)*. In the US, 9.4 million cases of food-borne illnesses are reported yearly with 55,961 cases of hospitalization and 1351 mortalities due to 31 key pathogens. Most illnesses were because of *Norovirus* (58%), 11% by *Salmonella* species (non-typhoidal), 10% by *Clostridium perfringens*, and 9% by *Campylobacter* species (Scallan et al., 2011). The source of infection are foods contaminated with virus particles from vomit or feces of sick persons and by infected food handler. The infection develops within 12–48 h once the host comes in contact with the virus. The symptoms include vomiting, nausea, diarrhea, stomach pain, muscle aches, low fever, chills, fatigue, and headaches (Public Health Agency of Canada, 2019). In Europe, increased prevalence of *Norovirus* infection has been linked to low immunity among populations, cool and dry temperature, and the emergence of new genetic variants (Lopman et al., 2009).

Norovirus is also termed as winter vomiting bug because the virus exhibits winter seasonality that has been extensively recognized (Patel et al., 2009; Ahmed et al., 2013). Changing climate has the potential to influence the seasonality of *Norovirus* by changing its transmission ways, influencing its resistance to environmental conditions and host susceptibility (Rohayem, 2009). In England and Wales, long-term study has predicted temperature change as a greatest risk factor for *Norovirus* occurrence (Lopman et al., 2009). Rainfall has also been considered as a significant factor for its seasonality, probably due its water-borne transmission (Marshall and Bruggink, 2011). So, if the prevalence of *Norovirus* outbreak is associated with weather fluctuations, changing climate may influence the occurrence and the spread of the infection.

7.3.2   *HEPATITIS A VIRUS*

Hepatitis A virus (HAV) is one among the most recurrent cause of food-borne infections. The shape of the virus is spherical containing single-stranded RNA with an icosahedral protein coat about 30 nm in diameter. HAV is a member of the family Picornaviridae (Drexler et al., 2015; Lauber and Gorbalenya, 2012). The sources of infection are contaminated water or water products (undercooked or raw shellfishes and oysters), contaminated foods, or infected food handler (Greening, 2006; Robertson et al., 1992;

Hernandez et al., 1997). The virus is responsible for a severe medical condition hepatitis A disease which is characterized with inflammation of liver with symptoms like fever, diarrhea, nausea, dark urine, jaundice, light colored stools, upset stomach and pain, joint pain, loss of appetite, and rarely, it may cause renal failure and death of an individual. The WHO global hepatitis report *(2017)* for first time in 2015 provides the global estimate of viral hepatitis cases. As per the report, roughly 4% of the global population lives with hepatitis (viral) and about 1.34 million deaths per year occur due to this disease. Since 2000, the deaths due to viral hepatitis has increased by 22%. The World Health Assembly in May 2016 authorized the *global health sector strategy* on viral hepatitis 2016–2021 which calls for the eradication of viral hepatitis as a public health danger by 2030 (reducing morbidity by 90% and mortality by 65%).

There are various evidences to support that increase in the occurrence of viral food-borne disease depends on the season (Phillips et al., 1980; Miossec et al., 1998; Ehrnst and Eriksson, 1993). However, some pathogens like HAV do not exhibit seasonal variation as reported in Greece Saudi Arabia and in some parts of Brazil (Papaevangelou et al., 1982; Fathalla et al., 2000; Villar et al., 2002). In Rio de Janeiro, HAV infections occur throughout the year, but during warm seasons with more rains, its incidence increases which may be due to contamination of water during overflows that usually fill the water bodies. The contaminated water bodies could lead to excess viral infections (Villar et al., 2002). Changing climate alters seasonality of the virus by altering its routes of transmission. The outcome effects of changing climate, such as increased temperatures and carbon dioxide concentrations increase the rate of mutation of the HAV genome mainly due to induced errors in RNA polymerase that help the virus to adapt to such climate changes and consequently will lead to increased pathogenicity (Tarek et al., 2019).

7.3.3 ROTAVIRUS

It is the basic and common reason of diarrhea in children. Rotavirus infections are associated with 2 lakh deaths in children under the age of 5 years in 2013 (Lanata et al., 2013; Walker et al., 2013). Rotavirus is a wheel-shaped, non-enveloped highly infectious double-stranded RNA virus having three concentric capsid layers with a diameter of 100 nm.

It primarily spreads by direct contact and secondary transmission occurs by infected water and foods (Jothikumar et al., 2009; Logan et al., 2006). Food-borne illnesses may lead to secondary cases by inter-human transmission. The virus is present in all regions of the world. In temperate regions, it is mainly found in winter and in tropical regions, throughout the year. Rotavirus infection is characterized by watery diarrhea, abdominal pain, vomiting, fever, and in rare cases hospitalization of newborns (Data sheet on food-borne biological hazards, 2012). Rotavirus infection is treated with vaccine, given in two doses with first dose between the age of 6 weeks and 14 weeks. Vaccines can reduce the cases of hospitalization and number of deaths in developing countries (WHO, 2013).

The biology of virus is highly effected due to climate change. Climate change associated burden of diarrheal diseases is increasing generally (Kolstad and Johansson, 2011). Moreover, the change in climatic variables like temperature and CO_2 concentration are the major causes of appeared infections at the global level (Tarek et al., 2019). The prevalence of rotavirus is generally linked to cool and drier environmental conditions (Cook et al., 1990; Atchison et al., 2010). Variable relationship between temperature and rotavirus infections has been reported. In the tropical regions, meta-analysis has estimated that for each 1°C rise in temperature, there is a decrease of 4%–10% incidence of rotavirus-associated diarrheal diseases (Levy et al., 2009; Carlton et al., 2016). On the contrary, time-series analysis carried out in Dhaka (Bangladesh) has reported that rise in each 1°C above a threshold of 29°C is associated with a 40.2% increase in rotavirus disease incidence (Hashizume et al., 2008). In temperate areas, virus shows seasonality with drier conditions that encourage transmission rates (Moe and Shirley, 1982; Ansari et al., 1991). Several studies were carried out to predict the influence of increased temperatures on the pathogen growth, survival, virulence, and transmission factors (Levy et al., 2016), however, pathogens show a large heterogeneity (Carlton et al., 2016). In a study carried out for 10 different areas of the world, it has been shown that the increased temperature is increasing mutation rate in RNA polymerase of the pathogen that in turn enhances the pathogenicity of the virus and helps the virus to acclimatize to changing climate (Tarek et al., 2019). These adaptations are increasing the threats to human health mostly in developing countries. So, effective measures should be considered to monitor and develop strategies to prevent public health from various effects of changing climate.

7.4 CLIMATE CHANGE AND MYCOTOXIN PRODUCING FUNGI CAUSING FOOD-BORNE ILLNESS

Mycotoxins are the metabolites secreted by fungi that thrive on a number of crops, such as corns, cereals, oilseeds, spices, tree nuts, etc. Animals fed with infected feed also contain toxin in milk as aflatoxin M1. These are less common cause of food-borne illness that are likely to increase if the changing climatic trends remain unchecked (Smith and Fazil, 2019). It has been estimated that about 25% of global crop production is exposed and contaminated with mycotoxins (Food safety, 2019). Human population may get exposed to mycotoxin intoxication either directly by consumption of toxin-infected crops or indirectly by consuming foods derived from livestock animals that have used up the toxin-contaminated feed (Tirado et al., 2010). Mycotoxins are highly toxic chemical substances responsible for a number of diseases in humans including various forms of cancer and death. The five classes of fungal toxins that are agriculturally important and major concern for public health are ochratoxin A (OTA), aflatoxins (AFTs), trichothecenes (TCT), fumonisins (FUMs), and zearalenone (ZEA) secreted by *Penicillium, Aspergillus*, and *Fusarium* species (Lee and Rye, 2017; Milicevic et al., 2014). AFTs are among the most poisonous toxins secreted by *Aspergillus flavus* and *A. parasiticus*, which are the known contaminants of agricultural commodities (Reddy et al., 2009). These are most potent hepatocarcinogens and hence may prove fatal to both animal and human life (IARC, 1993, 2012). FUMs, ZEA and TCT are the mycotoxins produced by many *Fusarium* species that are associated with a number of life-threatening diseases including cancer. The exposure of FUMs (FB1) is responsible for hepatic and food pipe cancers (IARC, 2002) and neural tube defects in humans (Marasas et al., 2002). TCT are known for the suppression of immunity, exhibit renal and neuro toxicity. ZEA, a lactone with estrogen-like properties produces the most predominant form of the mycotoxicosis in domestic animals (Kos et al., 2017). Ochratoxin, secreted by *Aspergillus* and *Penicillium* species (Milicevic et al., 2009), is a potent nephrotoxin responsible for porcine nephropathy (Milicevic et al., 2008). Scientific Committee on Food (SCF) in 1998 concluded that it possesses nephrotoxic, carcinogenic, immunotoxic, teratogenic, and probably neurotoxic properties. In Africa, the prevalence of mycotoxin intoxication is common, with the International Food Safety Authorities Network (INFOSAN) detailing numerous food safety events yearly, each involving morbidities or mortalities among mostly vulnerable populations (Savelli et al., 2019).

The impacts of climatic factors on the amount of mycotoxin production has been studied by various researchers. The rise in temperature, humidity, and precipitation leads to increase in the production of mycotoxins by creating an optimal environmental condition for mold growth (Patriarca and Fernández Pinto, 2017). Changing climate has its influence on pests and insects also that indirectly trigger the colonization of fungi and mycotoxin production (Tirado et al., 2010). Mycotoxins can be produced in the crop before harvest and may increase after postharvest if environmental conditions favor the mold growth. The FAO stated that changing climate could result in harvesting grains with more than 12%–14% moisture level, thus increasing the chances of mycotoxin production (FAO, 2008b). Temperature and moisture content are the prime factors that influence the production of mycotoxins (Sanchıs and Magan, 2014). Mycotoxins are mainly found in areas with a warm and humid climate (Perracia et al., 1999). In 2012, in Serbia, continued periods of hot summers and extreme droughts resulted in increased incidence of *Aspergillus* spp. on maize grains than that occurred before 2008 (Levic et al., 2013). Owing to the greater corn contamination, higher concentrations of aflatoxin AFM1 were found in milk (Stefanović, 2014). Recent studies in Europe predicted that aflatoxin B1 in maize will turn out to be a big issue of food safety within next 100 years with a rise of 2°C temperature in Europe (Battilani et al., 2016). In Hungary, the increased levels of aflatoxin in maize has been ascribed to changing climatic factors (Dobolyi et al., 2013). Other important mycotoxins secreted by *Fusarium* and *Penicillium* genera are also influenced with changing climatic conditions. ZEA (a lactone), produced by *Fusarium* spp. occurs chiefly in temperate conditions with wet weather and high moisture (Moretti et al., 2018). In Central European countries, the most affected cereal crop that contains highest levels of this toxin is maize (Milicevic et al., 2019). The factors like heavy rainfall, mild winters, unrestrained temperature and relative humidity lead to increased growth of mold and hence increased levels of mycotoxin during the harvesting of maize (Kos et al., 2017). TCT are also produced by *Fusarium* spp. In a study carried out in Serbia, the concentration of this mycotoxin (i.e., TCT) was found to have been increased in 2010–2014 than from the previous report collected in 2005–2010 due to alteration in weather conditions in terms of temperature and precipitation (Jajić et al., 2017). As such, the increase in the concentration of CO_2 due to climate change has been reported to elevate the levels of deoxynivalenol in case of wheat (Bencze et al., 2017). In temperate areas, the prevalence of *Fusarium* is less and it is predicted that with changing climate, there will be increase in extreme weather events in temperate areas that in turn will enhance the frequency of *Fusarium* and thus the production of more toxins (Food safety, 2019; Milicevic et al., 2019). Overall, changing climate, mainly increased temperature will reduce the global yield of crops (Zhao et al., 2017). This decreased yield particularly in developing countries will compel small-scale farmers to sell and eat what they grow that in turn will lead to increased cases of food-borne illnesses and will put public health at risk (Food safety, 2019).

7.5   ADDRESSING MEASURES TO RESPOND TO THE NEGATIVE IMPACTS OF CHANGING CLIMATE ON FOOD SAFETY

Food safety issues can open up at any stage of the food chain from primary production to consumption. As the world is actively adapting sustainable agricultural practices of production and change in food systems in response to changing climate, attention toward food safety is extremely important so as to ensure nutritionally safe food throughout the food chain. The most affected theme with respect to the question of food security and safety is food production (Koluman, 2010; Koluman et al., 2012). The Codex Alimentarius Commission (CAC) has mentioned that *"food safety is the assurance that food will not cause harm to the consumer when it is prepared and/or eaten according to its intended use"* (Uçar et al., 2016). Food safety promising is a complicated job, however, if food safety-related regulations are followed from production to consumption stage, most of the food-borne diseases can be prevented (Medeiros et al., 2001). The improvement in the applicability of food safety needs effective food control systems (McMeekin et al., 2006). Presently, the PAS 220, HACCP, and ISO 22000 are some commonly applied food safety systems that have been internationally approved (Uçar et al., 2016).

   Climate change is likely to influence food safety directly or indirectly, posing a threat to public health (Food safety, 2019). The aspects of food like food security, food safety, and food system challenges, all are presumed to represent the most momentous climate change-related to public health globally (Springmann et al., 2016). The increase in severity of extreme weather events due to changing climate is affecting food security by increasing survival and growth of food-borne illness pathogens, increasing pest outbreaks and toxin amount in crops, and reducing the global yield of crops that leads to increased starvations (Food safety, 2019). This will pose a significant challenge on agriculture in the 21st century due to increased demand to food supply globally under decreased availability of natural resources and increasing threat from climate change (Kumar et al., 2016). The impact of changing climate on food safety is expected to contribute to increased malnutrition, mainly in developing countries (Miraglia et al., 2009; WHO, 2017). When the food supplies are limited, people consume unsafe foods and less healthy diets that pose threat to health and lead to increased malnutrition. While the direct impacts of changing climate on food safety is the pathogen load or contamination levels in food, it also has indirect impacts on food safety through the human attitude to changing climate (Moniruzzaman, 2015). With changing climate, the attitude of farming practices has also changed. While such practices helped in maintaining food production, the introduction of new crops and cultivation methods also increase the risk of introducing food-borne diseases that people and health systems are not familiar with. There are substantial evidences to support that the outcome effects of changing climate are putting burden on food safety.

Although the problem of controlling changing climate is a big and multi task issue but its impact on health-related problems can be mitigated. The outcome effects of changing climate should be tackled as per the guidelines of legislative framework, coordination and management, surveillance, monitoring, laboratorial and inspection services, and communication and education. These guidelines highlight the application of food safety management strategies that is in accordance with FAO/WHO guidance (Koluman et al., 2017). WHO must be well supportive to the authorities to prepare and respond to such changes, particularly in developing countries and most affected countries (Food safety, 2019). The various ways to address measures to respond to the negative impacts of climate change on food safety are as (FAO, 2008c). (Table 7.2.):

7.5.1 MONITORING AND SURVEILLANCE

The evolving problems and trends of food safety can be addressed and identified early by integrated monitoring and surveillance of threats for both the environment and food. The information collected may contribute

TABLE 7.2   Measures to Address the Impacts of Climate Change and Variability on Food Safety and the Existing International Initiatives

Adaptation strategies	Main target	Initiatives
Intersectoral coordination	Veterinary, public health, environmental health services and food safety	Biosecurity approach
Integrated monitoring and surveillance	Zoonosis, food-borne diseases, chemical and microbiological food contaminants, pesticide residues, veterinary drugs, etc.	One Health approach Global Environmental Monitoring System for chemical contaminants
Risk assessment and predictive modeling	Emerging microbiological risks and zoonosis, chemical risks, fungal growth and mycotoxins; Harmful Algal Bloom (HAB) and marine toxins	FAO/WHO Risk Assessment Work Intergovernmental Oceanic Commission (IOC) – HAB program
Improved detection methods	Research-based development of tools for rapid detection of pathogens, chemical and microbiological contaminants, Biotoxins (mycotoxins and marine toxins)	Application of Nanotechnology and Biotechnology
Good practices	Food safety and agriculture, livestock and aquaculture production sectors	Good agricultural practices Good hygiene practices Good veterinary practices Good aquaculture practices
Risk management guidance	Food contaminants, plant and animal health, etc. Mycotoxins	FAO Food-Chain Crisis management framework FAO Worldwide Mycotoxins
	HABs and fisheries safety	Regulations for food and feed
	Agriculture sector	FAO/IOC/WHO Biotoxin Risk management plans
	Human and veterinary health services	EMPRES-Food Safety
Emergency preparedness and response		INFOSAN

Source: Reprinted from Tirado et al. *Food Res. Int.* **2010**, *43* (7), 1745, with permission from Elsevier.

to predictive modeling and risk assessments (Smith et al., 2015). Countries with poor monitoring and surveillance systems are inept to detect chemical and environmental contamination thereby increasing the threat to human health through the serious and long-lasting exposure to contaminants (Food safety, 2019).

7.5.2 PREDICTIVE MODELING

The modeling approach could help in the development of adaptation measures necessary for facing the evolving threats related to climate changes. It is possible to guess the probability of global climate change outcomes, such as ecological systems and emerging hazards by using predictive modeling approach. In Canada, the development of risk modeling framework has provided an organized platform for productive discussion on changing climate impacts on food safety. This framework was used to forecast the various impacts of changing climate on human health, such as protozoa in drinking water, mycotoxins in wheat, and *Vibrio parahaemolyticus* in oysters to understand the range of climate change impacts on food and water safety (Smith et al., 2015).

7.5.3 DISEASE SURVEILLANCE

Human and animal epidemiological surveillance is an important component of public health. It is not only of worth in identifying emerging diseases but helps in better resource planning and measuring the effects of diseases control strategies. Epidemiological surveillance is important at both national and international level and hence needs collaboration between human, animal, and environmental health professionals. In Arctic, in 1999, the International Circumpolar Surveillance system was established (Parkinson et al., 1999) which monitors there all the infectious diseases of concern. The IC Surveillance system helped in identifying an invasive pneumococcal disease outbreak in young adults and this in turn helped to begin subsequent drive of pneumococcal vaccine implementation in three regions of northern Canada (Public Health Agency of Canada, 2002). If this system is coordinated with suitable climate data, it could be used to monitor and detect the early emergence of climate-sensitive infections (Parkinson and Butler, 2005).

7.5.4   RISK ASSESSMENT

To combat the effects of climate change on food safety, provision of scientific risk assessments provides the basis for the development and adoption of food safety standards and for guidance on other food safety measures (Food safety, 2019). A number of joint FAO/WHO expert bodies came into existence so as to execute the risk assessments on food contaminants, food additives, pesticide residues, veterinary drug residues and microbiological hazards (FAO/WHO, 2018).

7.5.5   APPLICATION OF GOOD PRACTICES

To address the challenges of food safety caused by changing climate, implementation of good practices involving hygiene, agricultural, animal husbandry, veterinary, and aquaculture have always remained as the basis of national food safety management.

7.5.6   EARLY WARNING AND EMERGENCY RESPONSE SYSTEMS

To minimize the threat of lives and earnings of people exposed to natural disasters and emergencies due to climate change, enhanced early warning systems are necessary to ensure adequate consideration of food safety management issues in those situations. When natural disasters strike, food safety is a crucial public health concern. The following issues require immediate attention under the situations of natural disasters and emergencies (WHO, 2005): (1) food safety assurance following preventive measures. (2) inspection and salvaging of food. (3) making provisions for safe water and food. (4) immediate recognition and response to food-borne disease outbreaks. (5) provision of food safety information and education to affected people.

7.5.7   BETTER COMMUNICATION WITH THE PUBLIC

Food safety is guaranteed when all the levels from farm to fork are aware of the control measures. People should be well communicated regarding the ill effects of climate change on food safety. To ensure the implementation of food safety management strategies and other control measures among consumers, it is worth important that they are having the knowledge of the risks with foods. Thus, their education should be given prior importance for which governments have a key role to play in the education of general populations along with the implementation of food safety measures.

7.6 IMPLEMENTATION OF NEWER TECHNOLOGIES

The implementation of newer technologies and scientific innovations such as molecular biological methods for the characterization of complex communities of microorganisms, the various applications of nanotechnology in food and agriculture sectors, etc. are believed to widen our understanding and deal with the challenges of food safety posed by climate change. The excessive use of fertilizers, pesticides, and other chemicals may be reduced and prior detection of various pathogens to crops in agricultural fields is possible by the application of precision farming techniques of nanotechnology so as to ensure not only food production and quality but also food safety (War et al., 2020).

7.7 CONCLUSION

Climate change is the major issue in the present and if untreated, will prove to be highly devastating for the whole life on this biosphere in near future. Climate change is a matter of concern that affects all natural systems and has its adverse consequences for the developmental agenda at global level. Global warming has been the hot topic in the present century as the temperature of planet earth is increasing at an unprecedented rate because of human-induced activities. It has created a number of challenges related to health and communicable disease control and prevention. All the climatic variables directly or indirectly affect human health. Elevation in temperature and the dynamism in precipitation have a great impact on the pattern and persistence of microbial world, such as bacteria, viruses, parasites, etc. as well as on the patterns of food-borne diseases caused by these organisms. The changes in climatic variables also influence microbial ecology, their growth and host susceptibility that may result in the redistribution, emergence, incidence and intensity of the diseases they cause. The burden of food-borne diseases is associated with a huge economic cost that the countries are facing. Thus, the food safety challenges created due to climate change highlight the need for continued emphasis on food safety capacity building particularly to developing countries. The key aspects of ensuring rapid mitigative responses to food-borne disease outbreaks include integrated and cross-sectoral surveillance, monitoring and transparent data sharing, risk assessment and predictive modeling, etc. The International Food Safety Authorities Network (INFOSAN) is among the suitable platforms that directs the swift exchange of information related to the threats of food-borne diseases, thus enabling countries to set up procedures for risk management on time so that the occurrence of food-borne disease outbreaks can be prevented. The application of scientific innovations, production of stress-resistant crops, improved food traceability, development of early warning systems, and prediction of extreme events are some other ways of mitigating the negative effects of climate change on food safety and food production systems. The coordination among international organizations and donor agencies for funds and technical assistance has always remained as backbone in this area.

KEYWORDS

- **microbial foodborne diseases**
- **climate change**
- **transmission factors**
- **causal organisms**
- **vectors**

REFERENCES

Abanto, M.; Gavilan, R. G.; Baker-Austin, C.; Gonzalez-Escalona, N.; Martinez-Urtaza, J. Global Expansion of Pacific Northwest Vibrio Parahaemolyticus Sequence Type 36. *Emerg. Infect. Dis.* **2020,** *26,* 323–326.

Agunos, A.; Waddell, L.; Léger, D.; Taboada, E. A Systematic Review Characterizing On-Farm Sources of *Campylobacter* spp. for Broiler Chickens. *PLoS One.* **2014,** *9,* e104905.

Ahmed, S. M.; Lopman, B. A.; Levy, K. A Systematic Review and Meta-Analysis of the Global Seasonality of Norovirus. *PLOS ONE.* **2013,** *8,* e75922.

Akil, L.; Ahmad, H. A.; Reddy, R. S. Effects of Climate Change on Salmonella Infections. *Foodborne Pathog. Dis.* **2014,** *11*, 974–980.

Allard, R.; Plante, C.; Garnier, C.; Kosatsky, T. The Reported Incidence of Campylobacteriosis Modelled as a Function of Earlier Temperatures and Numbers of Cases, Montreal, Canada, 1990–2006. *Int. J. Biometeorol.* **2011,** *55*, 353–60.

Aminharati, F.; Dallal, M. M. S.; Ehrampoush, M. H.; Dehghani-Tafti, A.; Yaseri, M.; Memariani, M.; Rajabi, Z. The Effect of Environmental Parameters on Incidence of Shigella Outbreaks in Yazd Province, Iran. Water Sci. Tech. Water Supply. 2018, 18, 1388–1395.

Ansari, S. A.; Springthorpe, V. S.; Sattar, S. A. Survival and Vehicular Spread of Human Rotaviruses: Possible Relation to Seasonality of Outbreaks. *Rev. Infect. Dis.* **1991,** *13*, 448–461.

Atchison, C. J.; Tam, C. C.; Hajat, S.; van Pelt, W.; Cowden, J. M.; Lopman, B. A. Temperature-Dependent Transmission of Rotavirus in Great Britain and the Netherlands. *Proc. Biol. Sci.* **2010,** *277*, 933–942.

Baker-Austin, C.; Trinanes, J. A.; Taylor, N. G. H.; Hartnell, R.; Siitonen, A.; Martinez-Urtaza, J. Emerging Vibrio Risk at High Latitudes in Response to Ocean Warming. *Nat. Clim. Change.* **2013,** *3*, 73–77.

Bardgett, R. D.; Freeman, C.; Ostle, N. J. Microbial Contributions to Climate Change through Carbon Cycle Feedbacks. *ISME. J.* **2008,** *2*, 805–814.

Bardgett, R. D.; Manning, P.; Morrie¨n, E.; de Vries, F. T. Hierarchical Responses of Plant Soil Interactions to Climate Change: Consequences for the Global Carbon Cycle. *J. Ecol.* **2013,** *101*, 334–343.

Bartoszewicz, M.; Hansen, B. M.; Swiecicka. I. The Members of the *Bacillus cereus* Group Are Commonly Present Contaminants of Fresh and Heat-Treated Milk. *Food Microbiol.* **2008,** *25*, 588–596.

Battilani, P.; Toscano, P.; Van der Fels-Klerx, H. J.; Moretti, A.; Camardo Leggieri, M.; Brera, C.; Rortais, A.; Goumperis, T.; Robinson, T. Aflatoxin B1 Contamination in Maize in Europe Increases Due to Climate Change. *Sci. Rep.* **2016,** *6*, 24328.

Bencze, S.; Puskas, K.; Vida, G.; Karsai, I.; Balla, K.; Komaromi, J.; Veisz, O. Rising Atmospheric CO2 Concentration May Imply Higher Risk of Fusarium Mycotoxin Contamination of Wheat Grains. *Mycotoxin Res.* **2017,** *33*, 229–236.

Bi, P.; Hiller, J. E.; Cameron, A. S.; Zhang, Y.; Givney, R. Climate Variability and Ross River Virus Infections in Riverland, South Australia, 1992–2004. *Epidemiol. Infect.* **2009,** *137*, 1486–1493.

Britton, E.; Hales, S.; Venugopal, K.; Baker, M. G. Positive Association between Ambient Temperature and Salmonellosis Notifications in New Zealand, 1965–2006. *Aust. NZ. J. Public Health.* **2010,** *34*, 126–129.

Caburlotto, G.; Bianchi, F.; Gennari, M.; Ghidini, V.; Socal, G.; Aubry, F. B.; Bastianini, M.; Tafi, M.; Lleo, M. M. Integrated Evaluation of Environmental Parameters Influencing Vibrio Occurrence in the Coastal Northern Adriatic Sea (Italy) Facing the Venetian Lagoon. *Microb. Ecol.* **2012,** *63*, 20–31.

Carlton, E. J.; Woster, A. P.; DeWitt, P.; Goldstein, R. S.; Levy, K. A Systematic Review and Meta-Analysis of Ambient Temperature and Diarrhoeal Diseases. *Int. J. Epidemiol.* **2016,** *45*, 117–130.

Chan, N. Y.; Ebi, K. L.; Smith, F.; Wilson, T. F.; Smith, A. E. An Integrated Assessment Framework for Climate Change and Infectious Diseases. *Environ. Health Perspect.* **1999,** *107*, 329–337.

Chandler, D. S.; Craven, J. A. Relationship of Soil Moisture to Survival of *Escherichia coli* and *Salmonella typhimurium* in Soils Bacteria. *Aust. J. Agric. Res.* **1980,** *31*, 547–555.

Chersich, M. F.; Scorgie, F.; Rees, H.; Wright, C. Y., How Climate Change Can Fuel Listeriosis Outbreaks in South Africa. *SA Med. J.* **2018,** *108*, 453–454.

Cody, E. M.; Dixon, B. P. Hemolytic Uremic Syndrome. *Pediatr. Clin. North Am.* **2019,** *66*, 235–246.

Cook, S.; Glass, R.; Le Baron, C.; Ho, M. Global Seasonality of Rotavirus Infections. *Bull. World Health Organ.* **1990,** *68*, 1281–1289.

Cooley, M.; Carychao, D.; Crawford-Miksza, L.; Jay, M. T.; Myers, C.; Rose, C.; Keys, C.; Farrar, J.; Mandrell, R. E. Incidence and Tracking of *Escherichia coli* O157:H7 in a Major Produce Production Region in California. *PLoS One.* **2007,** *2*, e1159.

Crump, J. A.; Sjölund-Karlsson, M.; Gordon, M. A.; Parry, C. M. Epidemiology, Clinical Presentation, Laboratory Diagnosis, Antimicrobial Resistance, and Antimicrobial Management of Invasive Salmonella Infections. *Clin. Microbiol. Rev.* **2015,** *28*, 901–937.

Cullen, E. The Impact of Climate Change on the Future Incidence of Specified Foodborne Diseases in Ireland. *Epidemiology.* **2009,** *20*, S227–S228.

D'Souza, D. H.; Moe, C. L.; Jaykus, L. A. Foodborne Viral Pathogens. In *Food Microbiology, Fundamentals and Frontiers*; Washington, DC., ASM Press, 2007.

D'Souza, R.; Becker, N.; Hall, G.; Moodie, K. Does Ambient Temperature Affect Foodborne Disease? *Epidemiol. Infect.* **2004,** *15*, 86–92.

Data Sheet on Foodborne Biological Hazards/Rotavirus, French Agency for Food, Environmental and Occupational Health and Safety, April 2012. https://www. anses. fr/ en/content/data-sheet-foodborne-biological-hazards-rotavirus.

Derolez, V.; Soudant, D.; Fiandrino, A.; Cesmat, L.; Serais, O. Impact of Weather Conditions on *Escherichia coli* Accumulation in Oysters of the Thau Lagoon (the Mediterranean, France). *J. Appl. Microbiol.* **2013,** *114*, 516–525.

Dobolyi, C.; Sebők, F.; Varga, J.; Kocsubé, S.; Szigeti, G.; Baranyi, N.; Szécsi, Á.; Tóth, B.; Varga, M.; Kriszt, B.; Szoboszlay, S.; Krifaton, C.; Kukolya, J. Occurrence of Aflatoxin Producing Aspergillus Flavus Isolates in Maize Kernel in Hungary. *Acta Alimentaria.* **2013,** *42*, 451–459.

Drexler, J. F.; Corman, V. M.; Lukashev, A. N.; van den Brand, J. M. A.; Gmyl, A. P.; Brunink, S.; Rasche, A.; Seggewi, N.; Feng, H.; Leijten, L. M.; Vallo, P.; Kuiken, T.; Dotzauer, A.; Ulrich, R. G.; Lemon, S. M.; Drosten, C.; Hepatovirus Ecology Consortium. Evolutionary Origins of Hepatitis A Virus in Small Mammals. *Proc. Natl. Acad. Sci.* **2015,** *112*, 15190–15195.

Ebi, K. Climate Change and Health Risks: Assessing and Responding to Them through 'Adaptive Management'. *Health Aff. (Millwood).* **2011,** *30*, 924–930.

Ebi, K. L.; Mills, D. M.; Smith, J. B.; Grambsch, A. Climate Change and Human Health Impacts in the United States: An Update on the Results of the US. National Assessment. *Environ. Health Persp.* **2006,** *114*, 1318–1324.

Ehrnst, A.; Eriksson, M. Epidemiological Features of Type 22 Echovirus Infection. *Scand. J. infect. Dis.* **1993,** *25*, 275–281.

FAO. Climate Change and Food Security, a Framework Document, 2008a. http://www. fao. org/forestry/15538-079b31d45081fe9c3dbc6 ff34de4807e4.

FAO. Climate Change: Implications for Food Safety, 2008b. http://www. fao. org/food/ food-safety-quality/a-z-index/ climate-change1/en/.

FAO. Food Safety and Climate Change. FAO Conference on Food Security and the Challenges of Climate Change and Bioenergy, 2008c. http://www. fao. org/ag/agn/agns/ files/ HLC1_Climate_Change_and_Food_Safety.

FAO/WHO Framework for the Provision of Scientific Advice on Food Safety and Nutrition. 2018. https://www. who. int/foodsafety/publications/nutrition-advice/en/.

Farmer, J. J.; Janda, J. M.; Brenner, F. W.; Cameron, D. N.; Birkhead, K. M. Genus I. Vibrio Pacini 1854. In *Bergey's Manual of Systematic Bacteriology*, Vol 2, 2nd ed.; Brenner, D. J., Krieg, N. R., Staley, J. R., Eds.; Springer Science Business Media, Inc.: New York, 2005; pp 494–546.

Fathalla, S. E.; Al-Jama, A. A.; Al-Sheikh, I. H.; Islam, S. I. Seroprevalence of Hepatitis Virus Markers in Eastern Saudi Arabia. *Saudi Med. J.* **2000,** *21*, 945–949.

Ferreira, V.; Wiedmann, M.; Teixeira, P.; Stasiewicz, M. J. *Listeria Monocytogenes* Persistence in Food-Associated Environments: Epidemiology, Strain Characteristics, and Implications for Public Health. *J. Food Prot.* **2014,** *77*, 150–170.

Fleury, M.; Charron, D. F.; Holt, J. D.; Allen, O. B.; Maarouf, A. R. A Time Series Analysis of the Relationship of Ambient Temperature and Common Bacterial Enteric Infections in Two Canadian Provinces. *Int. J. Biometeorol.* **2006,** *50*, 385–391.

Food Safety, Climate Change and Role of WHO. Department of Food Safety and Zoonoses, 2019. https://www. who. int/publications/i/item/food-safety-climate-change-and-the-role-of-who.

Foodborne Diseases Active Surveillance Network (Food Net): Food Net Surveillance Report for 2011 (Final Report). Atlanta (GA): US Department of Health and Human Services, Centres for Disease Control and Prevention, CDC, 2012.

Gould, E. A.; Higgs, S. Impact of Climate Change and Other Factors Emerging Arbovirus Diseases. *Trans. R. Soc. Tropical Med. Hyg.,* **2009,** *103*, 109–121.

Goulet, V.; Jacquet, C.; Martin, P.; Vaillant, V.; Laurent, E.; de Valk, H. Surveillance of Human Listeriosis in France, 2001–2003. *Euro. Surveill.* **2006,** *11*, 79–81.

Greening, G. E. Human and Animal Viruses in Food (Including Taxonomy of Enteric Viruses). *Viruses Foods.* **2006,** 5–42.

Gubler, D. J.; Reiter, P.; Ebi, K. L.; Yap, W.; Nasci, R.; Patz, J. A. Climate Variability and Change in the United States: Potential Impacts on Vector and Rodent Borne Diseases. *Environ. Health Perspect.* **2001,** *109*, 223–233.

Gutierrez-Rodriguez, E.; Gundersen, A.; Sbodio, A. O.; Suslow, T. V. Variable Agronomic Practices, Cultivar, Strain Source and Initial Contamination Dose Differentially Affect Survival of *Escherichia coli* on Spinach. *J. Appl. Microbiol.* **2012,** *112*, 109–118.

Haley, B. J.; Cole, D. J.; Lipp, E. K. Distribution, Diversity, and Seasonality of Waterborne Salmonellae in a Rural Watershed. *Appl. Environ. Microbiol.* **2009,** *75*, 1248–1255.

Harvell, C. D.; Mitchell, C. E.; Ward, J. R.; Altizer, S.; Dobson, A. P.; Ostfeld, R. S.; Samuel, M. D. Climate Warming and Disease Risks for Terrestrial and Marine Biota. *Science.* **2002,** *296*, 2158–2162.

Hashizume, M.; Armstrong, B.; Wagatsuma, Y.; Faruque, A. S.; Hayashi, T.; Sack, D. A. Rotavirus Infections and Climate Variability in Dhaka, Bangladesh: A Time-Series Analysis. *Epidemiol. Infect.* **2008**, *136*, 1281–1289.

Heiman, K. E.; Mody, R. K.; Johnson, S. D.; Griffin, P. M.; Gould, L. H. *Escherichia coli* O157 Outbreaks in the United States, 2003–2012. *Emerg. Infect. Dis.* **2015**, *21*, 1293–1301.

Hellberg, R. S.; Chu, E. Effects of Climate Change on the Persistence and Dispersal of Foodborne Bacterial Pathogens in the Outdoor Environment: A Review. *Crit. Rev. Microbiol.* **2016**, *42*, 548–572.

Hernandez, F.; Monge, R.; Jimenez, C.; Taylor, L. Rotavirus and Hepatitis A Virus in Market Lettuce (Lactuca sativa) in Costa Rica. *Int. J. Food Microbiol.* **1997**, *37*, 221–223.

Heymann, D. L. ed.; *Control of Communicable Diseases Manual*, 18th ed.; American Public Health Association: Washington, DC, 2004.

Hocking, A.; Arnalod, G.; Jenson, I.; et al., eds. *Foodborne Microorganisms of Public Health Significance.* Sydney: Australian Institute of Food Science and Technology, Food Microbiology Group: NSW Branch, 1997.

Hoelzer, K.; Isabel, A.; Switt, M.; Wiedmann, M. Animal Contact as a Source of Human Non-Typhoidal Salmonellosis. *Vet. Res.* **2011**, *42*, 34.

Holley, R. A.; Arrus, K. M.; Ominski, K. H. Salmonella Survival in Manure-Treated Soils during Simulated Seasonal Temperature Exposure. *J. Environ. Qual.* **2006**, *35*, 1170–1180.

Horseman, M. A.; Surani, S. A. Comprehensive Review of Vibrio Vulnificus: An Important Cause of Severe Sepsis and Skin and Soft-Tissue Infection. *Int. J. Infect. Dis.* **2011**, *15*, e157–e166.

Hossain, A.; Juani, R.; Shams, S.; Rokonujjaman, M.; Shafiuddin, K. The Challenges and Alternatives of Water Supply and Sanitation in flood Prone Area: A Case Study for Bhuapur in Bangladesh. 5th Brunei International Conference on Engineering and Technology (BICET), **2014**. https://www. researchgate. net/publication/272041704.

Hsiao, H. I.; Jan, M. S.; Chi, H. J. Impacts of Climatic Variability on Vibrio Parahaemolyticus Outbreaks in Taiwan. *Int. J. Environ. Res. Public Health.* **2016**, *13*, 188.

Hsieh, J. L.; Fries, J. S.; Noble, R. T. Dynamics and Predictive Modelling of Vibrio spp. in the Neuse River Estuary, North Carolina, USA. *Environ. Microbiol.* **2008**, 10, 57–64.

Huang, D. S.; Peng, G.; Guo, J. Q.; Ping, W.; Zhou, B. S. Investigating the Effects of Climate Variations on Bacillary Dysentery Incidence in Northeast China Using Ridge Regression and Hierarchical Cluster Analysis. *BMC Infect. Dis.* **2008**, *8*, 130.

Humphrey, T.; O'Brien, S.; Madsen, M. Campylobacters as Zoonotic Pathogens: A Food Production Perspective. *Int. J. Food Microbiol.* **2007**, *117*, 237–257.

Hunter, P. R. Climate Change and Waterborne and VectorBborne Disease. *J. App Microbiol.* **2003**, *94*, 37S–46S.

International Agency for Research on Cancer (IARC). Monographs on the Evaluation of Carcinogenic Risks of Chemicals to Humans. *IARC Press, Lyon, France,* **1993**, *56*, 489–521.

International Agency for Research on Cancer (IARC). Monographs on the Evaluation of Carcinogenic Risks of Chemicals to Humans: Chemical Agents and Related Occupations: A Review of Human Carcinogens. *IARC Press, Lyon, France,* **2012**, *100.*

International Agency for Research on Cancer (IARC). Monographs on the Evaluation of Carcinogenic Risks of Chemicals to Humans: Some Traditional Herbal Medicines, Some Mycotoxins, Naphthalene and Styrene, *IARC Press, Lyon, France,* **2002**, *82.*

IPCC. Summary for Policymakers. In.;*Climate Change 2013*: *The Physical Science Basis Contribution of Working Group I to the Fifth Assessment Report of the Intergovernmental Panel on Climate Change Cambridge*; Stocker, T. F.; Qin, D.; Plattner, G-K., et al., Eds.; Cambridge University Press: New York, NY, 2018.

Jajić, I.; Krstović, S.; Jakšić, S.; Vuković, G.; Bursić, V.; Guljaš, D. Deoxynivalenol Occurrence in Serbian Maize under Different Weather Conditions. *Zbornik Matice srpske za prirodne nauke*. **2017**, *133*, 37–46.

Jaykus, L. A.; D 'Souza, D. H.; Moe, C. L. Foodborne Viral Pathogens. In *Food Microbiology*: *Fundamentals and Frontiers*, 4th ed.; Doyle, M. P.; Buchanan, R. L., Eds.; ASM Press: Washington, DC, **2013**; pp 619–649.

Johnson, C. N.; Bowers, J. C.; Griffitt, K. J.; Molina, V.; Clostio, R. W.; Pei, S.; Laws, E.; Paranjpye, R. N.; Strom, M. S.; Chen, A.; Hasan,N. A.;Huq, A.; Noriea, N. F.; Grimez, D. J.; Colwell, R. R. Ecology of Vibrio Parahaemolyticus and Vibrio Vulnificus in the Coastal and Estuarine Waters of Louisiana, Maryland, Mississippi, and Washington (United States). *Appl. Environ. Microbiol*. **2012**, *78*, 7249–7257.

Jokinen, C. C.; Schreier, H.; Mauro, W.; Taboada, E.; Issac-Renton, J. L.; Topp, E.; Edge, T.; Thomas, J. E.; Gannon, V. P, J. The Occurrence and Sources of Campylobacter spp., *Salmonella enterica* and *Escherichia coli* O157:H7 in the Salmon River, British Columbia, Canada. *J. Water Health*. **2010**, *8*, 374–386.

Jothikumar, N.; Kang, G.; Hill, V. R. Broadly Reactive TaqMan Assay for Real Time RT-PCR Detection of Rotavirus in Clinical and Environmental Samples. *J. Virol. Methods*. **2009**, *155*, 126–131.

Kapperud, G.; Aasen, S. Descriptive Epidemiology of Infections Due to Thermotolerant Campylobacter spp. in Norway, 1979–1988. *APMIS*. **1992**, *100*, 883–890.

Kelly-Hope, L. A.; Alonso, W. J.; Thiem, V. D.; Canh, D. G.; Anh, D. D.; Lee, H.; Miller, M. A. Temporal Trends and Climatic Factors Associated with Bacterial Enteric Diseases inVietnam, 1991–2001. *Environ. Health Perspect*. **2008**, *116*, 7.

Kendrovski, V.; Gjorgjev, D. Climate Change: Implication for Food-Borne Diseases (*Salmonella* and Food Poisoning among Humans in R. Macedonia). *Structure and Function of Food Engineering*.; Intech: Rijeka, Croatia, 2012; pp. 151–170.

Kim, J. Y.; Lee, H.; Lee, J. E.; Chung, M. S.; Ko, G. P. Identification of Human and Animal Fecal Contamination after Rainfall in the Han River, Korea. Microbes and Environments. *JSME*. **2013**, *28*, 187–194.

Kim, Y. S.; Park, K. H.; Chun, H. S.; Choi, C.; Bahk, G. J. Correlations between Climatic Conditions and Foodborne Disease. *Food Res. Int.,* **2015**, *68*, 24–30.

Kolstad, E. H. W.; Johansson, K. A. Uncertainties Associated with Quantifying Climate Change Impacts on Human Health: A Case Study for Diarrhea. *Environ. Health Perspect*. **2011**, *119*, 299–305.

Koluman, A. Detection of *Campylobacter jejuni* Contamination in Poultry Houses and Slaughter Houses. *Turk. Hij. Den. Biyol. Derg*. **2010**, *67*, 57–64.

Koluman, A.; Celik, G.; Unlu, T. Salmonella Identification from Foods in Eight Hours: A Prototype Study with *Salmonella Typhimurium*. *Iran J. Microbiol*. **2012**, *4*, 15–24.

Koluman, A.; Dikici, A.; Kahraman, T.; Incili, G. K. Food Safety and Climate Change: Seasonality and Emerging Food Borne Pathogens. *J. Gastroenterol*. *Res*. **2017**, *1*, 24–29.

Kos, J.; Janic Hajnal, E.; Mandic, A.; Duragić, O.; Jovanov, P.; Milovanovic, L. J. Mycotoxins in Maize: Annual Variations and the Impact of Climate Change. *Zbornik Matice srpske za prirodne nauke*. **2017**, *133*, 63–70.

Kotloff, K. L.; Winickoff, J. P.; Ivanoff, B.; Clemens, J. D.; Swerdlow, D. L.; Sansonetti, P. J.; Adak, G. K.; Levine, M. M. Global Burden of Shigella Infections: Implications for Vaccine Development and Implementation of Control Strategies. *Bull. World Health Organ.* **1999**, *77*, 651–666.

Kovats, R. S.; Edwards, S. J.; Charron, D.; Cowden, J.; D'Souza, R. M.; Ebi, K. L.; Gauci, C.; Gerner-Smidt, P.; Hajat, S.; Hales, S.; Pezzi, G. H.; Kriz, B.; Kutsar, K.; Mckeown, P.; Mellou, K.; Menne, B.; O`Brien, S.; Van Pelt, W.; Schmid, H. Climate Variability and Campylobacter Infection: An International Study. *Int. J. Biometeorol.* **2005**, *49*, 207–214.

Kovats, R. S.; Edwards, S. J.; Hajat, S.; Armstrong, B. G.; Ebi, K. L.; Menne, B. The Effect of Temperature on Food Poisoning: A Time Series Analysis of Salmonellosis in Ten European Countries. *Epidemiol. Infect.* **2004**, *132*, 443–453.

Kumar, V.; Kumar, S.; Kumar, S.; Singh, H.; Singh, R. K. Climate Change and Its Impact on Environment. *International Conference on Climate Change and Its Implications on Crop Production and Food Security*, 2016.; pp 62.

Lanata, C. F.; Fischer-Walker, C. L.; Olascoaga, A. C.; Torres, C. X.; Aryee, M. J.; Black, R. E.; Child Health Epidemiology Reference Group of the WHO and UNICEF. Global Cause of Diarrheal Disease Mortality in Children < 5 Years of Age: A Systematic Review. *PloS One*. **2013**, *8*, e72788.

Lauber, C.; Gorbalenya, A. E. Partitioning the Genetic Diversity of a Virus Family: Approach and Evaluation through a Case Study of Picornavirues. *J Virol*. **2012**, *86*, 3890–3904.

Lee, H. J.; Ryu, D. Worldwide Occurrence of Mycotoxins in Cereals and Cereal-Derived Food Products: Public Health Perspectives of Their Co-Occurrence. *J. Agric. Food Chem*. **2017**, *65*, 7034–7051.

Lee, H. S.; Hoang, T. T. H.; Pham-Duc, P.; Lee, M.; Grace, D.; Phung- Dac, C.; Thuc, V. M.; Nguyen-Viet, H. Seasonal and Geographical Distribution of Bacillary Dysentery (Shigellosis) and Associated Climate Risk Factors in Kon Tum Province in Vietnam from 1999 to 2013. *Infect. Dis. Poverty*. **2017**, *6*, 113.

Lees, D. Viruses and Bivalve Shellfish. *Int. J. Food Microbiol*. **2000**, *59*, 81–116.

Levantesi, C.; Bonadonna, L.; Briancesco, R.; Grohmann, E.; Toze, S.; Tandoi, V. Salmonella in Surface and Drinking Water: Occurrence and Water-Mediated Transmission. *Food Res. Int*. **2012**, *45*, 587–602.

Levic, J.; Gošic-Dondo, S.; Ivanovic, D.; Stankovic, S.; Krnjaja, V.; Bocarov-Stancic, A.; Stepanic, A. An Outbreak of Aspergillus Species in Response to Environmental Conditions in Serbia. *Pest. Phytomed*. **2013**, *28*, 167–179.

Levy, K.; Hubbard, A. E.; Eisenberg, J. N. S. Seasonality of Rotavirus in the Tropics: A Systematic Review and Meta-Analysis. *Int. J. Epidemiol*. **2009**, *38*, 1487–1496.

Levy, K.; Woster, A. P.; Goldstein, R. S.; Carlton, E. J. Untangling the Impacts of Climate Change on Waterborne Diseases: A Systematic Review of Relationships between Diarrheal Diseases and Temperature, Rainfall, Fooding, and Drought. *Environ. Sci. Technol*. **2016**, *50*, 4905–4922.

Lipp, E. K.; Kurz, R.; Vincent, R.; Rodriguez-Palacios, C.; Farrah, S. R.; Rose, J. B. The Effects of Seasonal Variability and Weather on Microbial Fecal Pollution and Enteric Pathogens in a Subtropical Estuary. *Estuaries*. **2001**, *24*, 266–276.

Liu, J. N.; Wu, X.; Li, C.; Xu, B.; Hu, L.; Chen, J.; Dai, S. Identification of Weather Variables Sensitive to Dysentery in Disease-Affected County of China. *Sci. Total Environ.* **2017a,** *575*, 956–962.

Liu, X.; Liu, Z.; Zhang, Y.; Jiang, B. The Effects of Foods on the Incidence of Bacillary Dysentery in Baise (Guangxi Province, China) from 2004 to 2012. *Int. J. Environ. Res. Public Health.* **2017b,** *14*, 179.

Logan, C.; O'Leary, J. J.; O'Sullivan, N. Real-Time Reverse Transcription-PCR for Detection of Rotavirus and Adenovirus as Causative Agents of Acute Viral Gastroenteritis in Children. *J. Clin. Microbiol.* **2006,** *44*, 3189–3195.

Lopman, B.; Armstrong, B.; Atchison, C.; Gray, J. J. Host, Weather and Virological Factors Drive Norovirus Epidemiology: Time-Series Analysis of Laboratory Surveillance Data in England and Wales. *PLoS ONE.* **2009,** *4*, e6671.

Louise, K.; Watkins, F.; Appiah, G. D. Travel-Related Infectious Diseases: In Yellow Book, Centers for Disease Control and Prevention (CDC), **2020.** https://wwwnc.cdc.gov/travel/yellowbook/2016/infectious-diseases-related-to-travel/ shigellosis.

Ma, J.; Ibekwe, A. M.; Crowley, D. E.; Ching-Hong, Y. Persistence of Escherichia coli O157:H7 in major leafy green producing soils. *Environ. Sci. Technol.* **2012,** *46*, 12154–12161.

Marasas, W. S. O.; Riley, R. T.; Hendricks, K. A.; Stevens, V. L.; Sadler, T. W.; Waes, J. G.; Missmer, S. A.; Cabrera, J.; Torres, O.; Gelderblom, W. C. A.; Allegood, J.; et al. Fumonisins Disrupt Sphingolipid Metabolism, Folate Transport, and Neural Tube Development in Embryo Culture and In Vivo: A Potential Risk Factor for Human Neural Tube Defects among Populations Consuming Fumonisin-Contaminated Maize. *J. Nutr.* **2002,** *134*, 711–716.

Markland, S. M.; Ingram, D.; Kniel, K. E.; Sharma, M. Water for Agriculture: The Convergence of Sustainability and Safety. *Microbiol. Spectr.* **2017,** *5*, 1–11.

Marshall, J. A.; Bruggink, L. D. The Dynamics of Norovirus Outbreak Epidemics: Recent Insights. *Int. J. Environ. Res. Public Health.* **2011,** *8*, 1141–1149.

Martinez-Urtaza, J.; Saco, M.; de Novoa, J.; Perez-Pineiro, P.; Peiteado, J.; Lozano-Leon, A.; Garcia-Martin, O. Influence of Environmental Factors and Human Activity on the Presence of *Salmonella serovars* in a Marine Environment. *Appl. Environ. Microbiol.* **2004,** *70*, 2089–2097.

McLaughlin, H. P.; Casey, P. G.; Cotter, J.; Gahan, C. G. M.; Hill, C. Factors Affecting Survival of Listeria Monocytogenes and *Listeria innocua* in Soil Samples. *Arch. Microbiol.* **2011,** *11*, 775–785.

McMeekin, T.; Baranyi, J.; Bowman, J.; Dalgaard, P.; Kirk, M.; Ross, T.; Schmid, S.; Zwietering, M. Information Systems in Food Safety Management. *Int. J. Food Microbiol.* **2006,** *112*, 181–194.

McMichael, J. A.; Haines, A.; Slooff, R.; Kovats, S. eds.; *Climate Change and Human Health: An Assessment Provided by a Task Group on Behalf of the WHO, the World Meteorological Association and the UN Environment Programme.*; Geneva: WHO; 1996.

Medeiros, L.; Hillers, V.; Kendall, P.; Mason, A. Evaluation of Food Safety Education for Consumers. *J. Nutr. Educ. Behav.* **2001,** *33*, S27–S34.

Melby, K.; Dahl, O. P.; Penner, J. L. Clinical and Serological Manifestations in Patients during a Waterborne Epidemic Due to *Campylobacter jejuni. J. Infect.* **1990,** *21*, 309–316.

Miettinen, H.; Wirtanen, G. Ecology of Listeria spp. in a Fish Farm and Molecular Typing of Listeria Monocytogenes from Fish Farming and Processing Companies. *Int. J. Food. Microbiol.* **2006,** 112, 138–46.

Miettinen, M. K.; Siitonen, A.; Heiskanen, P.; Haajanen, H.; Bjorkroth, K. J.; Korkeala, H. J. Molecular Epidemiology of an Outbreak of Febrile Gastroenteritis Caused by *Listeria Monocytogenes* in Cold-Smoked Rainbow Trout. *J. Clin. Microbiol.* **1999,** *37*, 2358–2360.

Milazzo, A.; Giles, L. C.; Zhang, Y.; Koehler, A. P.; Hiller, J. E.; Bi, P. Factors Influencing Knowledge, Food Safety Practices and Food Preferences during Warm Weather of Salmonella and Campylobacter Cases in South Australia. *Foodborne Pathog. Dis.* **2017,** *14*, 125–131.

Milićević, D.; Jurić, V.; Stefanović, S.; Jovanović, M.; Janković, S. Survey of Slaughtered Pigs for Occurrence of Ochratoxin A and Porcine Nephropathy in Serbia. *Int. J. Mol. Sci.* **2008,** *9*, 2169–2183.

Milicevic, D.; Lakicevic, B.; Petronijevic, R.; Petrovic, Z.; Jovanovic, J.; Stefanovic, S.; Jankovic, S. Climate Change: Impact on Mycotoxins Incidence and Food Safety. *Theor. Pract. Meat Proc.* **2019,** *4*, 9-16.

Milicevic, D.; Nedeljkovic-Trailovic; Jelena.; Mašic, Z. Mycotoxins in Food Chain—Risk Assessment and Importance for Public Health. *Tehnologija mesa.* **2014,** *55*, 22–38.

Milicevic, D.; Nikšic, M.; Baltic, T.; Stefanovic, S.; Jankovic, S. Presence of Moulds and Mycotoxins in Pigs' Feed—Significance in Risk Assessment. *Tehnologija mesa.* **2009,** *50*, 261–270.

Miossec, L.; Le Guyader, F.; Haugarreau, L.; Comps, M. A.; Pommepuy, M. Possible Relationship between a Winter Epidemic of Acute Gastroenteritis in France and Viral Contamination of Shellfish. *J. Shellfish Res.* **1998,** *17*, 1661–1664.

Miraglia,M.;Marvin,H. J. P.;Kleter,G. A.;Battilani,P.;Brera,C.;Coni,E.; Cubadda, F.; Croci, L.; De Santis, B.; Dekkers, S.; Fillippi, L.; Hutjes, R. W. A.; Noordam, M. Y.; Pisante, M.; Piva, G.; Prandini, A.; Toti, L.; Van den Born, G. J.; Vespermann, A. Climate Change and Food Safety: An Emerging Issue with Special Focus on Europe. *Food Chem. Toxicol.* **2009,** *47*, 1009–1021.

Moe, K.; Shirley, J. A. The Effects of Relative Humidity and Temperature on the Survival of Human Rotavirus in Faeces. *Arch. Virol.* **1982,** *72*, 179–186.

Moniruzzaman, S. Crop Choice as Climate Change Adaptation: Evidence from Bangladesh. *Ecol. Econ.* **2015,** *118*, 90–98.

Montville, T. J.; Matthews, K. R.; Kniel, K. E. *Food Microbiology: An Introduction.*; ASM Press: Washington, DC, 2012.

Moretti, A.; Pascale, M.; Logrieco, A. Mycotoxin Risks under a Climate Change Scenario in Europe. *Trends Food Sci. Technol.* **2018,** *84*, 38–40.

Morley, J. W.; Selden, R. L.; Latour, R. J.; Frolicher, T. L.; Seagraves, R. J.; Pinsky, M. L. Projecting Shifts in Thermal Habitat for 686 Species on the North American Continental Shelf. *PLoS One.* **2018,** *13*, e0196127.

Motsoaledi, A. Media Statement by the Minister of Health Dr. Aaron Motsoaledi Regarding the Update on the Listeriosis Outbreak in South Africa., 2017. http://www. kznhealth. gov. za/Listeriosis/Mediastatement-NDOH-04032018. (accessed May 15, 2018).

Musengimana, G.; Mukinda, F. K.; Machekano, R.; Mahomed, H. Temperature Variability and Occurrence of Diarrhoea in Children under Five-Years-Old in Cape Town Metropolitan Sub-Districts. *Int. J. Environ. Res. Public Health.* **2016,** *13*, 859.

Naicker, P. R. The Impact of Climate Change and Other Factors on Zoonotic Diseases. *Arch. Clin. Microbiol.* **2011,** *2,* 1–8.

NASA. http://climate. nasa. gov/solutions/adaptation-mitigation/. (accessed Dec 15, 2015).

Neimann, J.; Engberg, J.; Molbak, K.; Wegener, H. C. A Case Control Study of Risk Factors for Sporadic Campylobacter Infections in Denmark. *Epidemiol. Infect.* **2003,** *130,* 353–366.

Neukom, R.; Steiger, N.; Gómez-Navarro, J. J.; Wang, J.; Werner, J. P. No Evidence for Globally Coherent Warm and Cold Periods over the Preindustrial Common Era. *Nature.* **2019,** *571,* 550–554.

Newton, A.; Kendall, M.; Vugia, D. J.; Henao, O. L.; Mahon, B. E. Increasing Rates of Vibriosis in the United States, 1996–2010: Review of Surveillance Data from 2 Systems. *Clin. Infect. Dis.* **2012,** *54,* S391–S395.

Nylen, G.; Dunstan, F.; Palmer, S.; Andersson, Y.; Bager, F.; Cowden, J.; Feierl, G.; Galloway, Y.; Kapperud, G.; Megraud, F.; Molbak, K.; Petersen, L.; Ruutu, P. The Seasonal Distribution of Campylobacter Infection in Nine European Countries and New Zealand. *Epidemiol. Infect.* **2002,**. *128,* 383–390.

Oberbeckmann, S.; Fuchs, B. M.; Meiners, M.; Wichels, A.; Wiltshire, K.; Gerdts, G. Seasonal Dynamics and Modelling of a Vibrio Community in Coastal Waters of the North Sea. *Microb. Ecol.* **2012,** *63,* 543–551.

Odjadjare, E. E.; Obi, L. C.; Okoh, A. I. Municipal Wastewater Effluents as a Source of Listerial Pathogens in the Aquatic Milieu of the Eastern Cape Province of South Africa: A Concern of Public Health Importance. *Int. J. Environ. Res. Public Health.* **2010,** *7,* 2376–2394.

Olaniran, A. O.; Nzimande, S. B.; Mkize, N. G. Antimicrobial Resistance and Virulence Signatures of Listeria and Aeromonas Species Recovered from Treated Wastewater Effluent and Receiving Surface Water in Durban, South Africa. *BMC Microbiol.* **2015,** *15,* 234.

Pangloli, P.; Dje, Y.; Ahmed, O.; Doane, C. A.; Oliver, S. P.; Draughon, F. A. Seasonal Incidence and Molecular Characterization of Salmonella from Dairy Cows, Calves, and Farm Environment. *Foodborne Pathog. Dis.* **2008,** *5,* 87–96.

Papaevangelou, G.; Roumeliotou-Karayannis, A.; Contoyannis, P. Changing Epidemiological Characteristics of Acute Viral Hepatitis in Greece. *Infection.* **1982,** *10,* 1–4.

Park, M. S.; Park, K. H.; Bahk, G. J. Interrelationships between Multiple Climatic Factors and Incidence of Foodborne Diseases. *Int. J. Environ. Res. Public Health.* **2018,** *15,* 2482.

Heaney, A. K.; Shaman, J.; Alexander, K. A. El Niño-Southern Oscillation and under-5 Diarrhea in Botswana. *Nat. Commun.* **2019,** *10,* 5798.

Parkinson, A. J.; Butler, J. C. Potential Impacts of Climate Change on Infectious Diseases in the Arctic. *Int. J. Circumpolar Health.* **2005,** *64,* 478–486.

Parkinson, A. J.; Bell, A.; Butler, J. C. International Circumpolar Surveillance of Infectious Diseases: Monitoring Community Health in the Arctic. *Int. J. Circumpolar Health.* **1999,** *58,* 222–225.

Patel, M. M.; Hall, A. J.; Vinje, J.; Parashar, U. D. Noroviruses: A Comprehensive Review. *J. Clin. Virol.* **2009,** *44,* 1–8.

Patel, M. M.; Widdowson, M. A.; Glass, R. I.; Akazawa, K.; Vinje, J.; Parashar, U. D. Systematic Literature Review of Role of Noroviruses in Sporadic Gastroenteritis. *Emerg. Infect. Dis.* **2008,** *14,* 1224–1231.

Patriarca, A.; Fernández Pinto, V. Prevalence of Mycotoxins in Foods and Decontamination. *Curr. Opin. Food Sci.* **2017**, *14*, 50–60.

Patrick, M.; Christiansen, L.; Wain, M.; Ethelberg, S.; Madsen, H.; Wegener, H. Effects of Climate on Incidence of Campylobacter spp. in Humans and Prevalence in Broiler Flocks in Denmark. *Appl. Environ. Microbiol.* **2004**, *70*, 7474–7480.

Paz, S.; Bisharat, N.; Paz, E.; Kidar, O.; Cohen, D. Climate Change and the Emergence of *Vibrio vulnificus* Disease in Israel. *Environ. Res.* **2007**, *103*, 390–396.

Perracia, M.; Radic, B.; Lucic, A.; Pavlovic, M. 1999. Toxic Effects of Mycotoxins in Humans. *Bull. World Health Organ.* **1999**, *77*, 754–766.

Phillips, C. A.; Aronson, M. D.; Tomkow, J.; Philips, M. E. Enterovirus in Vermont, 1969–1978: An Important Cause of Illness throughout the Year. *J. infect. Dis.* **1980**, *141*, 162–164.

Public Health Agency of Canada. *Disease and Conditions.*; PHAC: Ottawa, ON, 2019. www. canada. ca/en/public-health/ services/diseases. html.

Public Health Agency of Canada. Pneumonia Epidemic Caused by a Virulent Strain of *Streptococcus pneumoniae* Serotype 1 in Nunavik, Quebec 2000. *Commun. Dis. Rep.,* **2002**, 28–16.

Ravel, A.; Smolina, E.; Sargeant, J. M.; Cook, A.; Marshall, B.; Fleury, M. D.; Pollari, F. Seasonality in Human Salmonellosis: Assessment of Human Activities and Chicken Contamination as Driving Factors. *Foodborne Pathog. Dis.* **2010**, *7*, 785–794.

Reddy, K. R. N.; Abbas, H. K.; Abel, C. A.; Shier, W. T.; Oliveira, C. A. F.; Raghavender, C. R. Mycotoxin Contamination of Commercially Important Agricultural Commodities. *Toxin Rev.* **2009**, *28*, 154–168.

Reid, P. C.; Gorick, G.; Edwards, M. Climate Change and European Marine Ecosystem Research. Sir Alister Hardy Foundation for Ocean Science, Plymouth, UK, 2011.

Reidl, J.; Klose, K. E. Vibrio Cholerae and Cholera: Out of the Water and into the Host. *FEMS Microbiol. Rev.* **2002**, *26*, 125–139.

Robertson, B. H.; Jansen, R. W.; Khanna, B.; Totsuka, A.; Nainan, O. V.; Siegl, G.; Widell, A.; Margolis, H. S.; Isomura, S.; Ito, K.; Ishizu, T.; Moritsugu, Y.; Lemon, M. Genetic Relatedness of Hepatitis A Virus Strains Recovered from Different Geographical Regions. *J. Gen. Virol.* **1992**, *73*, 1365–1377.

Robine, J. M.; Cheung, S. L. K.; Le Roy, S.; Van Oyen, H.; Griffiths, C.; Michel, J. P.; Herrmann, F. R. Death Toll Exceeded 70,000 in Europe during the Summer of 2003. *C. R. Biol.* **2008**, *331*, 171–178.

Rodrigues, L. C.; Cowden, J. M.; Wheeler, J. G.; Sethi, D.; Wall, P. G.; Cumberland, P.; Tompkins, D. S., Hudson, M. J., Roberts, J. A., Roderick, P. J. The Study of Infectious Intestinal Disease in England: Risk Factors for Cases of Infectious Intestinal Disease with *Campylobacter jejuni* Infection. *Epidemiol. Infect.* **2001**, *127*, 185–193.

Rodriguez-Castro, A.; Ansede-Bermejo, J.; Blanco-Abad, V.; Pet, J. V.; Garcia-Martin, O.; Martinez-Urtaza, J. Prevalence and Genetic Diversity of Pathogenic Populations of Vibrio Parahaemolyticus in Coastal Waters of Galicia, Spain. *Environ. Microbiol. Rep.* **2010**, *2*, 58–66.

Rohayem, J. Norovirus Seasonality and the Potential Impact of Climate Change. *Clin. Microbiol. Infect.* **2009**, *15*, 524–527.

Rose, J. B.; Epstein, P. R.; Lipp, E. K.; Sherman, B. H.; Bernard, S. M.; Patz, J. A. Climate Variability and Change in the United States: Potential Impacts on Water and Foodborne

Diseases Caused by Microbiologic Agents. *Environ. Health Perspect.* **2001,** *109,* 211–221.

Rosenberg, A.; Weinberger, M.; Paz, S.; Valinsky, L.; Agmon, V.; Peretz, C. Ambient temperature and age-related notified Campylobacter infection in Israel: A 12-Year Time Series Study. *Environ. Res.* **2018,** *164,* 539–545.

Rui, J.; Li, J.; Wang, S.; An, J.; Liu, W.; Lin, Q.; Yang, Y.; He, Z.; Li, X. Responses of Bacterial Communities to Simulated Climate Changes in Alpine Meadow Soil of the Qinghai-Tibet Plateau. *Appl. Environ. Microbiol.* **2015,** *81,* 6070–6077.

Salvadori, M.; Bertoni, E. Update on Hemolytic Uremic Syndrome: Diagnostic and Therapeutic Recommendations. *World J. Nephrol.,* **2013,** *2,* 56–76.

Sanchıs, V.; Magan, N. Environmental Profiles for Growth and Mycotoxin Production. In…; *Mycotoxins in Food*: *Detection and Control*; Magan, N.; Olsen, M. Eds.; Woodhead Publishing Ltd., 2014.

Savelli, C. J.; Bradshaw, A.; Embarek, P. B.; Mateus, C. The FAO/WHO International Food Safety Authorities Network in Review, 2004–2018: Learning from the Past and Looking to the Future. *Foodborne Pathog. Dis.* **2019,** *16,* 480–488.

Scallan, E.; Hoekstra, R. M.; Angulo, F. J.; Tauxe, R. V.; Widdowson, M. A.; Roy, S. L.; Jones, J. L.; Griffin. P. M. Foodborne Illness Acquired in the United States–Major Pathogens. *Emerg. Infect. Dis.* **2011,** *17,* 7–15.

Sinton, L. W.; Braithwaite, R. R.; Hall, C. H.; Mackenzie, M. L. Survival of Indicator and Pathogenic Bacteria in Bovine Feces on Pasture. *Appl. Environ. Microbiol.* **2007,** *73,* 7917–7925.

Smith, B. A.; Fazil, A. How Will Climate Change Impact Microbial Foodborne Disease in Canada? *Can. Commun. Dis. Rep.* **2019,** *45,* 108–113.

Smith, B. A.; Ruthman, T.; Sparling, E.; Auld, H.; Comer, N.; Young, I.; Lammerding, A. M.; Fazil, A. A Risk Modeling Framework to Evaluate the Impacts of Climate Change and Adaptation on Food and Water Safety. *Food Res. Int.* **2015,** *68,* 78–85.

Snel, S. J.; Baker, M. G.; Venugopal, K.; French, N.; Learmonth, J. A Tale of Two Parasites: The Comparative Epidemiology of Cryptosporidiosis & Giardiasis in New Zealand 1997–2006. *Epidemiol. Infect.* **2009,** *137,* 1641–1650.

Snelling, W. J.; Matsuda, M.; Moore, J. E.; Dooley, J. S. G. *Campylobacter jejuni. Lett. Appl. Microbiol.* **2005,** *41,* 297–302.

Solo-Gabriele, H. M.; Wolfert, M. A.; Desmarais, T. R.; Palmer, C. J. Sources of *Escherichia coli* in a Coastal Subtropical Environment. *Appl. Environ. Microbiol.* **2000,** *66,* 230–237.

Solomon, S.; Qin, D.; Manning, M.; Chen, Z.; Marquis, M.; Averyt, K. B.; Tignor, M.; Miller, H. L., Eds. *Climate Change*: *The Physical Science Basis. Contribution of Working Group I to the Fourth Assessment Report of the Intergovernmental Panel in Climate Change.*; Cambridge University Press: Cambridge and New York, 2007.

Soneja, S.; Jiang, C.; Upperman, C. R.; Murtugudde, R.; Mitchell, C. S.; Blythe, D.; Sapkota, A. R.; Sapkota, A. Extreme Precipitation Events and Increased Risk of Campylobacteriosis in Maryland, U. S. A. *Environ. Res.* **2016,** *14,* 216–221.

Song, Y-J.; Cheong, H-K.; Ki, M.; Shin, J-Y.; Hwang, S-S.; Park, M.; Ki, M.; Lim, J. The Epidemiological Influence of Climatic Factors on Shigellosis Incidence Rates in Korea. *Int. J. Environ. Res. Public Health.* **2018,** *15,* 2209.

Spies, D. WHO: South Africa's Listeriosis Outbreak 'Largest Ever'. *News 24*, 13 January 2018. https://www.news24.com/SouthAfrica/News/who-south-africas-listeriosis-outbreak-largest-ever-20180113 (accessed May 10, 2018).

Springmann, M.; Mason- D'Croz, D.; Robinson, S.; Garnett, T.; Godfray, H. C.; Gollin, D.; Rayner, M.; Ballon, P.; Scarborough, P. Global and Regional Health Effects of Future Food Production under Climate Change: A Modelling Study. *Lancet.* **2016,** *387,* 1937–1946.

Stefanović, S. Comparative Investigation of Aflatoxin B1 Presence in Fed and Aflatoxin M1 Presence in Milk. PhD Dissertation, Faculty of Veterinary Medicine, Belgrade, 2014.

Strawn, L. K.; Fortes, E. D.; Bihn, E. A.; Nightingale, K. K.; Worobo, R. W.; Wiedmann, M.; Bergholz, P. W. Landscape and Meteorological Factors Affecting Prevalence of Three Food-Borne Pathogens in Fruit and Vegetable Farms. *Appl. Environ. Microbiol.* **2013,** *79,* 588–600.

Studahl, A.; Andersson, Y. Risk Factors for Indigenous Campylobacter Infection: A Swedish Case-Control Study. *Epidemiol. Infect.* **2000,** *125,* 269– 275.

Su, Y. C.; Liu, C. Vibrio Parahaemolyticus: A Concern of Seafood Safety. *Food Microbiol.* **2007,** *24,* 549–558.

Taiwan Food and Drug Administration. Food-Borne Disease Data by Years. http://www. fda. gov. tw/TC/includes/GetFile. ashx?mID=133&id=39089&chk=8fdff827-a191-4e4b-8b8111b876c0b7ce (accessed October 25, 2015).

Tam, C. C.; Rodrigues, L. C.; O'Brien, S. J.; Hajat, S. Temperature Dependence of Reported Campylobacter Infection in England, 1989–1999. *Epidemiol. Infect.* **2006,** *134,* 119–125.

Tarek, F.; Hassou, N.; Benchekroun, M. N.; Boughribil, S.; Hafid, J.; Ennaji, M. M. Impact of Rotavirus and Hepatitis A Virus by Worldwide Climatic Changes during the Period between 2000 and 2013. *Bioinformation.* **2019,** *15,* 194–200.

Thomas, M. K.; Murray, R.; Flockhart, L.; Pintar, K.; Pollari, F.; Fazil, A.; Nesbitt, A.; Marshall, B. Estimates of the Burden of Foodborne Illness in Canada for 30 Specified Pathogens and Unspecified Agents, Circa 2006. *Foodborne Pathog. Dis.* **2013,** *10,* 639–648.

Tirado, M. C.; Clarke, R.; Jaykus, L. A.; Mc Quatters-Gollop, A.; Frank, J. M. 2010. Climate Change and Food Safety: A Review. *Food Res. Int.* **2010,** *43,* 1745–1765.

Uçar, A.; Yilmaz, M. V.; Çakıroğlu, F. P. Food Safety–Problems and Solutions. In *Significance, Prevention and Control of Food Related Diseases.*; Makun, H. A., Ed.; IntechOpen, 2016. http://dx. doi. org/10. 5772/63176.

US EPA. Climate Change: Basic nformation: http://www3. epa. gov/climate change/basics/. (Accessed Dec 15, 2015).

Valcour, J. E.; Charron, D. F.; Berke, O.; Wilson, J. B.; Edge, T.; Waltner-Toews, D. A Descriptive Analysis of the Spatiotemporal Distribution of Enteric Diseases in New Brunswick, Canada. *BMC Public Health.* **2016,** *16,* 204.

Vasickova, p.; Pavlik, I.; Verani, M.; Carducci, A. Issues Concerning Survival of Viruses on Surfaces. *Food Environ. Virol.* **2010,** *2,* 24–34.

Venegas-Vargas, C.; Henderson, S.; Khare, A.; Mosci, R. E.; Lehnert, J. D.; Singh, P.; Ouellette, L. M.; Norby, B.; Funk, J. A.; Rust, S.; Bartlett, P. C.; Grooms, D.; Manning, S. D. Factors Associated with Shiga Toxin-Producing *Escherichia coli* Shedding by Dairy and Beef Cattle. *Appl. Environ. Microbiol.* **2016,** *82,* 5049–5056.

Vereen Jr, E.; Lowrance, R. R.; Cole, D. J.; Lipp, E. K. Distribution and Ecology of Campylobacters in Coastal Plain Streams (Georgia, United States of America). *Appl. Environ. Microbiol.* **2007,** *73,* 1395–1403.

Vereen Jr, E.; Lowrance, R. R.; Jenkins, M. B.; Adams, P.; Rajeev, S.; Lipp, E. K. Landscape and Seasonal Factors Influence Salmonella and Campylobacter Prevalence in a Rural Mixed Use Watershed. *Water Res.* **2013**, *47*, 6075–6085.

Vermeulen, L.; Hofstra, N. Influence of Climate Variables on the Concentration of *Escherichia coli* in the Rhine, Meuse, and Drentse Aa during 1985–2010. *Reg. Environ. Change.* **2013**, *14*, 1–13.

Vezzulli, L.; Colwell, R. R.; Pruzzo, C. Ocean Warming and Spread of Pathogenic Vibrios in the Aquatic Environment. *Microb. Ecol.* **2013**, *65*, 817–825.

Villar, L. M.; De Paula, V. S.; Gaspar, A. M. C. Seasonal Variation of Hepatitis A Virus Infection in The City of Rio De Janeiro, Brazil. *Rev. Inst. Med. Trop. S. Paulo.* **2002**, *44*, 289–292.

Walia, M.; Gaind, R.; Mehta, R.; Paul, P.; Aggarwal, P.; Kalaivani, M. Current Perspectives of Enteric Fever: A Hospital-Based Study from India. *Ann. Trop. Paediatr.* **2005**, *25*, 161–174.

Walker, C. L. F.; Rudan, I.; Liu, L.; Nair, H.; Theodoratou, E.; Bhutta, Z. A.; O'Brien, K. L.; Campbell, H.; Black, R. E. Global Burden of Childhood Pneumonia and Diarrhoea. *Lancet.* **2013,**. *381*, 1405–1416.

Wallace, J. S.; Stanley, K. N.; Currie, J. E.; Diggle, P. J.; Jones, K. Seasonality of Thermophilic Campylobacter Populations in Chickens. *J. Appl. Microbiol.* **1997**, *82*, 219–224.

Wang, X. Y.; Du, L.; Seidlein, L. V.; Xu, Z. Y.; Zhang, Y. L.; Hao, Z. Y.; Han, O. P.; Ma, J. C.; Lee, H. J.; Ali, M.; Han, C. Q.; Xing, Z. C.; Chen, J. C.; Clemens, J. Occurrence of Shigellosis in the Young and Elderly in Rural China: Results of a12-Month Population-Based Surveillance Study. *Ann. Trop. Med. Parasitol.* **2005**, *73*, 416–422.

War, J. M.; Fazili, M. A.; Mushtaq, W.; Wani, A. H.; Bhat, M. Y. Role of Nanotechnology in Crop Improvement. In *Nanobiotechnology in Agriculture, Nanotechnology in the Life Sciences*;…Hakeem, K. R.;…Pirzadah, T. B., Eds.; Springer Nature: Switzerland, 2020; pp 63–97.

Watts, N.; Amann, M.; Arnell, N.; et al. The 2018 report of the Lancet Countdown on Health and Climate Change: Shaping the Health of Nations for Centuries to Come. *Lancet.* **2018**, *392*, 2479-2514.

Wei, Y.; Kouse, A. B.; Murphy, E. R. Transcriptional and Post Transcriptional Regulation of Shigella shu T in Response to Host-Associated Iron Availability and Temperature. *Microbiologyopen.* **2017**, *6*, e00442.

Weisent, J.; Seaver, W.; Odoi, A.; Rohrbach, B. The Importance of Climatic Factors and Outliers in Predicting Regional Monthly Campylobacteriosis Risk in Georgia, USA. *Int. J. Biometeorol.* **2014**, *58*, 1865–1878.

WHO. Food Safety in Natural Disasters: Ensuring Food Safety in the Aftermath of Natural Disasters. World Health Organization, Geneva, 2005. http://www. who. int/foodsafety/ boodborne_disease/emergency/en/.

Wilkes, G.; Edge, T.; Gannon, V.; Jokinen, C. C.; Lyautey, E.; Medeiros, D. T.; Neumann, N.; Ruecker, N.; Topp, E.; Lapen, D. R. Seasonal Relationships among Indicator Bacteria, Pathogenic Bacteria, Cryptosporidium Oocysts, Giardia Cysts, and Hydrological Indices for Surface Waters within an Agricultural Landscape. *Water Res.* **2009**, *43*, 2209–2223.

World Health Organization. Food Safety: Fact sheet No 399. World Health Organization, Geneva.2017. http://www. who. int/media centre/factsheets/fs399/en/.

World Health Organization. Rotavirus Vaccines: WHO Position Paper-January 2013. *Week. Epidemiol. Rec.* **2013,** *88,* 49–64. https://apps. who. int/iris/handle/10665/ 242024.

World Health Organization. *WHO Estimates of the Global Burden of Foodborne Diseases: Foodborne Disease Burden Epidemiology Reference Group 2007–2015.*; WHO: Geneva, CH, 2015. https//www. who. int/foodsafety/publications/foodbornedisease/fergreport/en/.

Wu, J.; Rees, P.; Storrer, S.; Alderisio, K.; Dorner, S. Fate and Transport of Potential Pathogens: The Contribution from Sediments. *J. Am. Water Resour. Assoc.* **2009,** *45,* 35–44.

Yang, F.; Yang, J.; Zhang, X.; Chen, L.; Jiang, Y.; Yan, Y.; Tang, X.; Wang, J.; Xiong, Z.; Dong, J.; Xue, Y.; Zhu, Y.; Xu, X.; Sun, L.; Chen, S.; Nie, H.; Peng, J.; Xu, J.; Wang, Y.; Yuan, Z.; Wen, Y.; Yao, Z.; Shen, Y.; Qiang, B.; Hou,Y.; Yu, J.; Jin, Q. Genome Dynamics and Diversity of Shigella Species, the Etiologic Agents of Bacillary Dysentery. *Nucl. Acids Res.* **2005,** *33,* 6445–6458.

Zhang, Y.; Bi, P.; Hiller, J. E. Climate Variations and Salmonellosis Transmission in Adelaide, South Australia: A Comparison between Regression Models. *Int. J. Biometeorol.* **2007,** *52,* 179–187.

Zhang, Y.; Bi, P.; Hiller, J. E. Projected Burden of Disease for Salmonella Infection Due to Increased Temperature in Australian Temperate and Subtropical Regions. *Environ. Int.* **2012,** *44,* 26–30.

Zhang, Y.; Bi, P.; Hiller, J. E.; Sun, Y.; Ryan, P. Climate Variations and Bacillary Dysentery in Northern and Southern Cities of China. *J. Inf. Secur.* **2007,** *55,* 194–200.

Zhao, C.; Liu, B.; Piao, S.; Wang, X.; Lobell, D. B.; Huang, Y., Huang, M.; Yao, Y.; Bassu, S.; Ciais, P.; Durand, J.; Elliott, J.; Ewert. F.; Janssens, I. A.; Li, T.; Lin, E.; Liu, Q.; Martre, P.; Müller, C.; Penga, S.; Peñuelas, J.; Ruane, A. A.; Wallach, D.; Wang, T.; Wu, D.; Liu, Z.; Zhu, Y.; Zhu, Z.; Asseng, S. Temperature Increase Reduces Global Yields of Major Crops in Four Independent Estimates. *PNAS.,* **2017,** *114,* 9326–9331.

Zhou, J.; Xue, K.; Xie, J.; Deng, Y.; Wu, L.; Cheng, X.; Fei, S.; Deng, S.; He, Z.; Van Nostrand, J. D.; Luo, Y. Microbial Mediation of Carbon-Cycle Feedbacks to Climate Warming. *Nat. Clim. Change.* **2012,** *2,* 106–110.

Zimmerman, M.; De Paola, A.; Bowers, J. C.; Krantz, J. A.; Nordstrom, J. L.; Johnson, C. N.; Grimes, D. J. Variability of Total and Pathogenic Vibrio Parahaemolyticus Densities in Northern Gulf of Mexico Water and Oysters. *Appl. Environ. Microbiol.* **2007,** *73,* 7589–7596.

CHAPTER 8

Algae as Indicators of Climate Change

CHRISTY B. K. SANGMA[1] and SABIRA SULTANA[2]

[1]*ICAR Research Complex for NEH Region, Nagaland Centre, Jharnapani, Medziphema, Nagaland, India*

[2]*Gauhati University, Guwahati, Assam, India*

Corresponding author. E-mail: christysangma@gmail.com

ABSTRACT

Over the last few decades, the studies on the factors of climate change viz. temperature, carbon dioxide etc. and their impact on different ecosystems have been gaining importance, as they are threatening the ecological balance. The most immediate effect of change in climate is expected to be seen in water bodies with change in temperature affecting their pH, salinity, solubility, diffusion rate and viscosity. Water acidification due to rise in CO_2 concentration, and nutrient load (eutrophication) as a result of anthropogenic activities are the major forces which are expected to have negative impact on water quality. They modify the structure, function and algal diversity (toxic or nontoxic- harmful algal blooms), having detrimental effects on the aquatic ecosystems. Algae (Latin- alga, seaweed), the diverse group of "https://en.wikipedia.org/wiki/Photosynthesis"photosynthetic "https://en.wikipedia.org/wiki/Eukaryotic"eukaryotic, unicellular or multicellular "https://en.wikipedia.org/wiki/Organism" organisms (kingdom "https://www.britannica.com/science/protist"Protista), can be found in vast habitats including rivers, streams, lakes, oceans and other exclusive habitats like ice or snow, thermal vents, and also in terrestrial ecosystems. Besides being used as a food and a substrate for biofuel production, they possess characteristics like rapid reproduction rates, short life cycle, sensitivity and ability to accumulate pollutants, and responsiveness to natural or environmental disturbances at Spatio-temporal scales, which qualify them as one of the biological indicators for changing climate. There are several species of algae which respond to the change in environmental factors, and some of them are used as indicators of change. Besides being the indicators, the large-scale production of algal foods and biofuels also have a significant impact on global energy requirements and greenhouse gas emissions. Algae as indicators of climate change have been studied only in few locations and started very recently, so without the past knowledge, the present and the future effects are very difficult to quantify.

8.1   CLIMATE CHANGE AND ITS IMPACT ON WATER BODIES

As the age of the earth increases, the variables of climate including temperature, humidity, and precipitation, etc. change globally with incidences of changing patterns, frequencies, and magnitudes (Cooney, 2010; Medvigy and Beaulieu, 2012). These changes are influenced by natural (e.g., volcanoes, forest fire, continental drift, ocean currents, comets, and meteorites, etc.) as well as the anthropogenic factors (e.g., fossil fuel emissions, industrial emissions etc.), because these factors are responsible for the production of a huge quantity of polluting gases including the major greenhouse gases. These gases (Table 8.1) are responsible for the ever changing climatic conditions.

The rise in concentrations of these atmospheric greenhouse gases are expected to raise the global surface temperature by 2°C–4.5°C above preindustrial levels by 2100 (IPCC, 2007; Noyes et al., 2009), and are predicted to have a serious impact on the structure, function, and health of different ecosystems, affecting the physiological functioning of plants, their evolutionary processes and interactions with other organisms, therefore threatening the whole ecological balance which sustains the life on our planet (Miraglia et al., 2009; Lau and Lennon, 2012). This rising global surface temperature was also observed to raise the normal air and water temperature, a resulting factor for the large-scale melting of glaciers and subsequent rising of the worldwide sea level. It is expected that the global mean sea level can increase @1.8 mm/yr due to this widespread melting of ice caps and glaciers (IPCC, 2007), which is liable to erode and recede the existing coastlines. But, the degree of severity and the occurrence of such events will depend on many natural (e.g., elevation, coastal materials or sediments, ocean currents, and storm waves, etc.) and anthropogenic factors (e.g., faulty land management systems

TABLE 8.1 Major Greenhouse Gases.

Sl. No.	Greenhouse gases	Lifespan in atmosphere	Present level	Increase from preindustrial levels (1750)	Global warming potential (in 100 yrs)
1.	Carbon dioxide (CO_2)	~100 yrs	405.5 ppm	41%	1
2.	Methane (CH_4)	12 yrs	1748–1870 ppb	144%–162%	25
3.	Nitrous Oxide (N_2O)	114 yrs	329.9 ppb	20%	298
4.	Chloroflouro Carbons (CFCs)	12–100 yrs	75–534 ppt	*	5000–10,900
5.	Ozone (O_3)	hours to days	34 ppb	42%	n.a.

Notes: *ppm*, Parts per million; *ppb*, parts per billion; *ppt*, parts per trillion.
*Pre-industrial concentration is zero as they are industrially manufactured compounds.
N.a. Short lifetime, so cannot be calculated.
Source: IPCC (2007), NOAA (2011), WMO (2018), and Blasing (2014).

or agricultural practices causing enhance erosions) which accelerate or decelerate the process. Nearing et al. (2004) stated that because of climate change, the rainfall patterns are expected to change, with variation in magnitude, and intensity of rains, leading to enhanced rate of runoffs with huge amount of soil loss. This soil loss or erosion will be approximately 1.7% higher for each 1% variation in annual rainfall, and this will have a serious impact on water bodies, as, greater the erosion level higher will be the sediment deposition and nutrient load which will directly affect the aquatic organisms. Other than the climatic factors, erosion contributing to water pollution will occur primarily due to other human interferences like excess use of fertilizers in agricultural system, exhaust from vehicles, and urban runoff, etc.

Globally, 97% of earth's water (hydrosphere) is present in the ocean, which is saline (including mangroves, coral reefs and coastal estuaries, to the open ocean) and the remaining 3% is fresh water (75% of 3% is glaciers and ice and rest 25% consists of rivers, groundwater, lakes, streams, etc.) enveloping 71% of global area. An estimated 10% of the surface is covered by cryosphere referring to the frozen components, that is, glaciers, seasonal snowfalls, permafrost, and ice sheets. So, directly or indirectly the life on our planet earth is affected by the water bodies. Water bodies on the earth support unique habitats that are intertwined or interlinked to other elements of a climatic system through the universal cyclic process of water, energy, and carbon. The upper layer of oceans is a sink of CO_2 and other greenhouse gases and the deeper layers are a storehouse of carbon in the world (Sarmiento and Gruber, 2002). Field et al. (1998) and Singh et al. (2010) stated that 60 gigatons of carbon in the ocean is processed through microbial activities, and ocean acts as a sink itself for 3 billion tons of carbon and 40% of CO_2 emission enters by the combustion of fossil fuel and can be absorbed by the ocean alone.

8.1.1 DIRECT EFFECT OF CLIMATE CHANGE ON WATER BODIES

The concentration of elevated amount of greenhouse gases will have a serious effect on various stabilizing factors of water bodies, namely, nutrient inputs, water acidification, oceanic temperature, stratification, mixing, and thermohaline circulation, etc. These factors will in turn have a major impact on microbiota of marine ecosystem, changing the productivity, food chain and food webs and the other nutrient elements that are present in the oceanic environment. The rise in temperature of the ocean water is predicted to be 2°C–6°C due to climate change (Sarmento et al., 2010), with fluctuation of 1.18°C (at low CO_2 concentration) and 6.48°C (at high CO_2 concentration) (Meehl et al., 2007). This rising global temperatures directly affect the water chemistry and reduce the water density, thereby affecting the stratification and circulation and in turn organismal dispersal and nutrient transport system (Glöckner et al., 2012). Such fluctuations will have a probable impact on oceanic food webs and ultimately affecting the biogeochemical cycles (Walsh, 2015). Besides this, sea water rise in temperature reduces the oxygen level creating the death zones, and affect the water viscosity, in turn affecting the carrying capacity and growth rate of living organisms in water (Beveridge et al., 2010). Intensity of rainfall and wind, and electrical conductivity of water (salinity) also affect the oceanic mixing and circulation as well as stratification. Besides warming, increased concentration of CO_2 in water bodies or oceans have increased the acidification of the water by ~0.1 pH units from preindustrial era, with pH of the ocean's surface waters reduced from 8.21–8.10 pH. The additional decline of 0.3–0.4 units has been projected by Intergovernmental Panel on Climate Change (IPCC) if the atmospheric CO_2 reaches 800 ppm by the end of this century. Even though 90% of marine biomass is known to be microbial, there is lack of study and research on how the marine life, the microbial community compositions and functional groups will respond to this changing pH and other physical factors of water bodies.

8.1.2 INDIRECT EFFECT OF CLIMATE CHANGE ON WATER BODIES

Structure and functioning of ecosystem is complex in nature, and the extent of impact on ecosystem by climate change is of critical importance. The factors like acidification and nutrient load are projected to be the two major forces that will have a negative effect on the coastal marine species and the ecosystem as a whole. Both these factors are expected to stimulate growth and productivity of opportunistic, fast-growing, ephemeral macro algae at the expense of foundational species, such as corals, sea grasses, and long-lived perennial macro algae (i.e., kelps). Temperature to some extent, can select the survivability of certain aquatic organisms by influencing the process of respiration and production (Glöckner et al., 2012). CO_2 also has a potential to control the biodiversity in the aquatic ecosystem (Radford,

2013). High level of CO_2 in the sea water might further lower the pH and thus increases the acidification which in turn might stimulate the primary producers (Glöckner et al., 2012). Higher level CO_2 is capable of directly affecting the microorganisms by speeding up their reproduction system, with subsequent additional utilization of nutrients especially the elements like iron (Fe) and phosphorus (P) that are not easily accessible in the marine ecosystem. The deficiency of these elements has the probability of causing the species to adapt, migrate, be replaced or go extinct including the fishes. In this way, the whole food chain will be disrupted and might lead to the evolution of harmful microbes and the biodiversity will be harmed (Howard, 2015).

8.2   CLIMATE CHANGE AND ALGAE AS THE INDICATOR OF CHANGING CLIMATE

Several factors, namely, ocean warming, acidification, eutrophication, and excess of fishing and tourism will have a severe impact on the basic qualities of water which are vital for aquatic environment. The fluctuations in these factors can trigger the dramatic changes in life-forms, their competitiveness and their existence and the extinction (NASA, 2015). The changing climatic conditions might cause the disappearances of the coral reefs at the same time, might favor the macro algae and benthic cyanobacterial mats. The factors like increased temperature, acidification, and reduced nutrient inputs will accelerate the release of extracellular dissolved organic matter from phytoplanktons, in turn increasing the microbial production. Warming is projected to affect the iron requirement of nitrogen-fixing cyanobacteria, enhanced the thermal stratification and minimizes the mixing, hence creating a favorable environment for toxic cyanobacteria. Due to warming, 20% more proliferation of algal blooms in lakes and 5% higher toxic blooms are projected, which will be poisonous to fish and other aquatic animals (NASA, 2015). This was evident from the studies of Mudie et al. (2002) and Johnk et al. (2008), where they observed a rise in the toxic algal blooms due to warming during the last few decades on the Pacific and Atlantic sides of North America and observed increased occurrences of toxic cyanobacterium *Microcystis* in eutrophic lakes during hot summers. Climate change is also projected to promote smaller plankton types over larger ones which will indirectly affect the biogeochemical cycles of aquatic ecosystems (Huertas et al., 2011). The rates of multiplication, cell density as well as the decay process of marine phytoplankton will also be higher at the higher temperature (CMFRI, 2011). So survivability, at elevated temperature will depend on the natural selection, mutation, and acclimation.

Algae (*Latin*-alga, seaweed), the diverse group of photosynthetic eukaryotic, unicellular or multicellular organisms (kingdom Protista) qualify, as the indicator species to climate change as they are ubiquitous in nature,, that is, it flourished in almost every habitats ranging from freshwater to marine and hot springs to ice caps. They are the key primary producers in all kinds of aquatic ecosystems, and can act both as the pollutants as well as the self-purifier of water depending on the species. As a pollutant, under enrichment of nutrients in water, algae can produce massive surface growths called "blooms," and can advance the mass mortality of other organisms by reducing the water quality. As self-purifier, algae flourished in water polluted with organic waste, and purify the water bodies and generate oxygen, which is a life support system for other aquatic animals and also contribute to the economic well-being in the form of food, medicine, and other products.

8.2.1   CRITERIA OF ALGAE AS THE BIOLOGICAL INDICATORS

Biological indicators are those species which are used to monitor the quantifiable environmental health, or habitat restoration or to see the effect of pollution and contamination. These indicators provide the early warning signals of disturbances in the ecosystem and allow decision for cost-efficient operations to be taken well in advance (Spellerberg, 2005). The indicator species can be dynamics of single population (e.g., abundance, density, growth rate etc.) or group of population of one or more taxa. Biological indicators can be known in different terms, namely, ecological indicators, ecological index, environmental index, biomonitor, indicator species, etc.

The criteria for the qualification of algae as the indicator species (Stevenson and Smol, 2003) are: (1) they are easily monitorable, (2) present all year round, (3) diverse group with large quantities, (4) fast rate of dispersal, short life cycles, and sensitivity to pollutants and changing climatic conditions, (5) cost-effective, (6) well-defined taxa to limnological variables and previously measured, (7) high tolerance level, and (8) well correlation with environmental factors. Other than these criteria, algal assemblages and species display broad allocation in different ecological and geographical regions. The sampling of algal assemblages can be carried out easily, with minimal disturbances to the habitat, and their culturing in the laboratory is inexpensive. But their community is directly influenced by the changing physical and chemical factors of the growing environment (McCormick and Cairns, 1994; Stevenson and Pan, 1999).

8.2.2   ALGAL INDICATOR PARAMETERS

Selection of indicator parameters depends on the assessment goals and conceptual models used. In algal indicators, parameters like algal biomass, species composition and functions, toxins, etc. can be used during assessment (APHA, 2012; Kelly et al., 2009).

8.2.2.1   ALGAL BIOMASS

Algal biomass is the first signal for the quality of drinking water, productivity of fisheries, recreation, aesthetics, and level of dissolved oxygen and pH, which will directly affect the other biodiversity in water bodies. The algal biomass increases or decreases in the water bodies according to the presence of resources and stressors in particular location (Dodds et al., 1998). Algal biomass can be measured in the field, as well as in the laboratory by collecting the samples. In field, visual observations are taken for taxonomic studies by employing Sechi depth in lakes, rapid periphyton surveys in streams, and metaphyton and epiphyton surveys in wetlands, are carried out (Wetze and Likens, 2000). These rapid surveys provide estimates of algal biomass and identification up to genus level (Stevenson et al., 2012). Under laboratory condition, assays on chlorophyll content are analyzed on collected samples. Chlorophyll a (chl a), dry mass, ash-free dry mass, algal cell density, biovolume, or chemical mass of samples are considered to be the common indicator for planktonic and benthic algal biomass. Algal biomass study is considered to be the good indicator for human disturbances and a response to the toxic stressors. But this parameter has a spatial and temporal variation which can be resolved by large sampling sizes (Stevenson et al., 2012). Another way to quantify the algal biomass studies are through the use of remote sensing with fitted imaging system in satellites (SeaWifs, MERIS, LandSat, and MODIS), aircrafts and drones which can be operated in seas, large lakes, rivers, streams, etc.

8.2.2.2   ALGAL DIVERSITY STUDIES

Algal diversity studies are one of the good indicators for human disturbances or change in the environmental factors (Blanco et al., 2012). Algal diversity studies are carried out by considering species richness and evenness of taxa abundance by employing the Shannon diversity or Hurlbert's evenness index (Shannon, 1948; Hurlbert, 1971). Richness is considered to be decrease by the disturbances, at the same time evenness found to be favored by the disturbances in the environmental factors which retard the growth of the dominant taxa in the area (Patrick, 1973). But this indicator is many a time is not consistent with the human disturbances with failed correlations and some researchers discourage the use of this index for evaluation (Blanco et al., 2012).

8.2.2.3   TOXINS AND NUTRIENT CHEMISTRY STUDIES

The measure of algal toxins provides the potential risk of the palatability of water for consumption, fishes and other aquatic living beings. Some of the toxins produced by the algae are microcystin, cyanotoxin, etc. Toxins are one of the best indicators for changing water quality (Botana et al., 2009). In nutrient chemistry studies, elemental ratios of carbon, nitrogen and phosphorus (C:N:P ratio @106:16:1) concentration in algae and water are determined to know the abundance and limiting nutrient in the water bodies. The relative supply of these nutrients determined the deficiency and the excess concentration when it comes to the growth of algae (Downing et al., 2001). With this nutrient assessment studies heavy metal contamination in water bodies and their bioaccumulation by some of the species of algae can be monitored.

8.2.2.4   TAXONOMIC STUDIES

Records of species on habitat and ecological preferences, their detailed characteristics on taxonomic traits provide information on the presence of the species on specific locations according to the environmental conditions (van Dam et al., 1994). Taxonomic composition of algal assemblages acts as the precise and accurate indicator for environmental degradations. Taxonomic studies are carried out by microscopic identification of samples, their counting, and gathering the basic information like morphology (filamentous, heterocystous, motile, stalked, monoraphid, biraphid) and physiological traits, growth forms (colonial, unicellular, planktonic, benthic), genus, family, order, and class levels of taxonomy, tolerances, habitat preferences, etc. (Stevenson et al., 2010). Presently, the knowledge on these taxonomic studies has greatly advanced, with the recent developments in electron microscopy and DNA sequencing technologies (Cavalier-Smith, 2007). Periodical studies on taxonomy and functional group of algae will provide the shift or any change in the species in particular habitat. Record of species and their ecological preferences or tolerances to pollution, salinity, temperature, and nutrients can also be presented in relative ranking systems (e.g., 1, 2 ... 5) across regions.

8.2.2.5   METRICS AND MULTIMETRIC INDICES

Metrics either with parametric or nonparametric statistical methods are tested for separate attribute at numerous locations for varying level of anthropogenic disturbances. The disturbances are characterized and simplified with the use of ordination techniques and axis scores for ranking scale. These disturbances are characterized by the use of reference sites, the source which is not affected by human activities and preferably upstream to the point source. In this kind of studies "*Paleolimnological*" techniques offer direct assessment of long-term environmental trends at a particular location, and it gives the estimation of the speed and degree of degradation of the targeted system. This type of studies require the long-term data for estimating the critical load of stressors, reconstruction of past conditions, and provides a target for restoration. Multimetric indices, pioneered by James Karr (1981) is usually applied for bioassessment of algae (Hill et al., 2000). Multimetric index is analyzed by averaging the values of ranges of metrics. There are various multimetric indices developed by different researchers, namely, diatom metrics, Saprobic index, cyanobacteria index, etc. (van Dam et al., 1994; Stevenson et al., 2013; Wu et al., 2012). Multimetric indices are widely used for the assessment of aquatic resources in different counties.

8.2.2.6   OTHER ALGAL INDICATORS

Other traits in algae which can be used as the indicators are estimation of the rate of photosynthesis, respiration, nutrient uptake and cycling, phosphatase activity, and net primary productivity (Hill et al., 1997;

Whitton et al., 1998). Bioassays studies in microcosm and mesocosm studies like *Selenastrum* bottle assay can also be carried out to check the effect of environment on the growth of algae (McCormick et al., 1996; Pan et al., 2000).

8.2.3 ALGAL INDICATOR SPECIES

Some of the algal indicator species which are commonly used in the environmental studies are as follows:

8.2.3.1 DIATOMS

Diatoms possess various characteristics which make them ideal indicator species for environmental change, as they are common in various habitats like benthic, shallow waters, etc. Diatoms are diverse group of algae with 10,000–1,00,000 or more taxa (Mann et al., 1996) and various species have been used for monitoring water variables like pH, salinity, nutrient loads, and other water quality and climate-related parameters (Stoermer and Smol, 1999). Water quality assessments using diatoms are considered to be the most precise indicator compared with other chemical and zoological assessments (Leclercq, 1988). Examples of diatom species are *Nitzschia, Navicul,* and *Cyclotella*, etc.

8.2.3.2 BENTHIC OR ATTACHED ALGAE

Benthic algae consisting of diatoms and non-diatoms are also a useful indicator of ecological conditions or water qualities assessment in streams, rivers, lakes, and saline waters. They are usually attached to some surfaces like bedrocks, sediments, sand, etc. (Porter, 2008; Smucker et al., 2013) and are sampled by scraping the hard or firm substrata. Depending on the attachment on substrata they can be classified as adnate, mucilage pad, and mucilage stalk. Examples of some of the species are *Cocconeis*, *Ulnaria*, and *Gomphonema*, etc.

8.2.3.3 CHRYSOPHYTES

Chrysophytes are diverse group with 1000 described species and mostly referred to as the "golden brown algae." They mostly exist as single, flagellated cells, motile, and spherical colonies. They are found in most aquatic habitats ranging from fresh water to saline waters. Chrysophytes are used less commonly than diatoms, but it holds significant potential in paleolimnological studies. In high latitude water bodies, Chrysophytes often dominate the planktons, so it is considered to be one of the indicator species for environmental change studies (Wilkinson et al., 1997). Examples of *Chrysophytes* are *Mallomonas*, *Synura*, etc.

8.2.3.4   PERIPHYTON

Periphytons have also been widely used for monitoring the aquatic ecosystems (Cosgrovea et al., 2004). They exhibit diverse group, and are observed to be susceptible to the environmental conditions, and portray responses with changes in species composition, cell density, ash-free dry mass, chlorophyll and enzyme activities. Periphytons have many advantages as the indicator species, which include quick re-colonization ability upon disturbance, easy for sampling, and they cannot avoid pollution as they have a fixed habitat. For example, *Plectonema spp.*

8.2.3.5   BLUE GREEN ALGAE (CYANOBACTERIA)

Blue-green algae are aquatic organisms which have characteristic of both bacteria and algae. They are not true algae. The water bodies which are warm, poorly mixed, with low dissolved oxygen and rich in nutrients have high density of cyanobacteria (Teta et al., 2017). When cyanobacteria disintegrate, they usually release toxins like cyanotoxin or microcystin which are harmful to the aquatic life as well as to the human beings. Examples of cyanobacteria are *Oscillatoria*, *Phormidium*, etc.

8.2.3.6   MACROALGAE

Macroalgae or seaweed occupies a habitat ranging from coastal to marine wetlands. They are the functional group in coral reef communities and comprise of major phyla like Chlorophyta (green algae), Heterokontophyta (brown algae), and Rhodophyta (red algae). Macroalgae belonging to the genus Cladophora is considered to be one of the best bio-indicators of water quality (Chmielewská and Medved, 2001). Examples of macroalgae is *Ulva* spp., *Enteromorpha*, *Padina, etc.*

8.3 SPECIES OF ALGAE FOUND BASED ON THE ENVIRONMENTAL FACTORS

8.3.1 WATER DEGRADATION

Algae (benthic and pelagic) serve as the reliable bioindicators of aquatic environment globally (Lange et al., 2016; Wu et al., 2012) because of their strong sensitivity to environmental perturbations (Dong et al., 2016; Stevenson, 2014). Algae are extensively used as indicators to monitor the biotic conditions of water due to the following reasons: (1) they are highly sensitive to ecosystem situations, (2) easier to detect and sample, and (3) algal communities are sessile, species-rich and cosmopolitan in distribution (Gökçe, 2016; Wu et al., 2017). Algae are the unicellular as well as multicellular photosyntnetic organisms which comprises of two groups: Macroalgae and microalgae. Macroscopic filamentous algae are being used as bioindicators for assessing the environmental condition and water pollution caused by nutrients and heavy metals (Lange et al., 2016). The filamentous macroalgae belonging to the genus *Cladophora* was documented to be one of the best bioindicators of water pollution (Lange et al., 2016). Pikosz et al. (2017) assessed the quality of freshwater ecosystem by utilizing filamentous algae as bioindicators. The macroalgal species of the genus *Ulva* are reported to be good indicators of waters polluted with organic content and trace elements (Ustunada et al., 2010). Various *Ulva* species have been reported to be utilized as indicators of trace elements by several researchers (Conti and Cecchetti, 2003; Shams El-Din and Mohamedein, 2014). In Italy, the alga *Ulva lactuca* has been reported to be an indicator of water contaminated with trace element (Bonanno et al., 2020). Microalgae are the freshwater borne unicellular eukaryotic protist and blue-green algae (cyanobacteria) which account for half of the world's photosynthetic activity (Day et al., 1999). The microalgal diversity in the Western Himalayan region was estimated to be efficient indicators of freshwater ecosystem (Kadam et al., 2020). The investigators recorded a total of 33 algal taxa belonging to 27 microalgal genera. The species observed in the lentic and lotic zones belonged to the Bacillariophyta, Chlorophyta, Dinophyta, Euglenophyta, and Myxophyta divisions. This data of microalgal diversity was utilized as specific bioindicators to assess the water degradation status of the sampled sites. Diatoms are one of the dominant groups of microalgae inhabiting the fresh and marine waters (Mann and

Vanormelingen, 2013). They are successfully used as water quality bioindicators of freshwater ecosystem. In England, five algal species, namely,
Stigeoclonium tenu, *Nitzschia palea*, *Gomphonema parvulum*, *Cocconeis*,
and *Chamaesiphon* were utilized as indicators of polluted rivers (Butcher,
1949). The algae *Navicula accomoda* and *Gomphonema* have been
reported to be sewage pollution tolerant species [as reviewed by Sen et
al. (2013)]. The diatoms *Amphora ovalis* and *Gyrosigma attenuatum* were
documented to be good indicators of enriched waters [as reviewed by Sen
et al. (2013)]. In Equador, five diatom species were reported to be suitable
ecological indicators of freshwater environment. The diatom *Sellaphora
seminulum* (strain JA01b, c), *Nitzschia fonticola* (strain SP02a) and *N.
palea* (strain CA01a) were found to be tolerant to high organic content
of waters (Ballesteros et al., 2020). The term blue-green algae generally refers to cyanobacteria, which are group of prokaryotic autotrophic
bacteria containing photosynthetic pigments. Several studies have reported
the use of cyanobacteria as indicators of aquatic health (Mateo et al., 2015;
Teta et al., 2017). The unicellular cyanobacteria (*Synechococcus* spp.
and *Prochlorococcus* spp.) were recorded from subtropical and tropical
oligotrophic ecosystems (Teta et al., 2017). In Iran, the cyanobacterial
species *Chroococcus minor, Oscillatoria chlorine*, *Phormidium tenue*,
Lyngbya kuetzingii were evaluated as water quality indicators for monitoring eutrophication (Soltani et al., 2012). Also the cyanobacterial species
Lyngbya mesotrichia, Lyngbya infixa were observed to be bioindicators of
unpolluted water. Many species of blue-green algae were reported to be
present in nutrient poor waters and some have been found to be tolerant
to organically polluted waters Sen et al. (2013). Several species such
as *Oscillatoria chlorina*, *O. putrida*, *O. splendida*, and *O. ornate* were
described to be suitable indicators of organically polluted waters Sládecek
(1973), Rott et al. (1997). By monitoring cyanobacterial blooms that occur
due to eutrophication in aquatic bodies, the water quality can be assessed
(Bag et al., 2019). Cyanobacteria have also been recently used in treating
wastewater. Many algal species serve to be good bioindicators of water
pollution. Phytoplanktons can act as the reflective mirror of the quality
of aquatic ecosystem (Ali and Khairy, 2016). The species of freshwater
phytoplankton are rapidly affected to environmental changes and so
provide good indication to the assessment of water quality in river and
stream ecosystems (Bellinger and Sigee, 2015). Many algal species have
been reported to grow well in organically polluted water. The green algae

belonging to genera (*Chlamydomonas, Euglena)*, diatoms (*Navicula, Synedra),* and blue-green algae (*Oscillatoria, Phormidium)* were found to be tolerant to organic pollution (Palmer, 1969). Also *Euglena viridis, Oscillatoria limosa, O.tenuis, O.princeps* and *Phormidium uncinatum, Nitzschia palea* were reported to be used as bioindicators for monitoring polluted waters (Palmer, 1960). Algae are also employed in detecting clean waters as many species have been found to be present predominantly in clean waters. Approximately, 46 algal taxa comprising certain green and blue-green algae, diatoms and flagellates have been reported to be the indicators of clean water. However, as compared with larger algae, the flagellates were found to be better indicators of unpolluted water (Palmer, 1969). The green algae (*Ulothrix zonata* and *Microspora amoena)*, the flagellates (*Chromulina rosanoffi, Mallomonas caudate)* were also reported as organisms of unpolluted zones (Liebmann et al., 1951). Also, the algal groups Cryptophyta and Chrysophyta abundantly occurred in unpolluted waters Sen et al, (2013).

8.3.2 CHANGE IN TEMPERATURE

Climate change is the major cause for the intense heat and rising temperatures of the oceans. Temperature plays a significant role in the growth of algae (Yu et al., 2018). Algae are organisms ranging from microscopic to macroscopic organisms belonging to kingdom Plantae. Algae are very reactive to environmental changes such as increasing ocean temperatures or run off from glacial melt (Dong et al., 2016), nutrient loadings (Miranda and Krishnakumar, 2015). Because of the strong response to environmental changes algae serve as good and reliable biological indicators of aquatic environment (Kadam et al., 2020). Water temperature is very essential in controlling the geographical distribution and abundance of macroalgae as it affects the survival, growth, and reproduction of macroalgae (Lüning, 1990; Wernberg et al., 2012). The variations in the survival and growth of macroalgae under changing environmental conditions serve as sensitive indicators to temperature change (Merzouk and Johnson, 2011). In Nova Scotia, the influence of elevated surface temperature of water on the growth of rockweeds (*Ascophyllum nodosum, Fucus vesiculosus*), Irish moss (*Chondrus crispus*), kelp (*Laminaria digitata*), and the invasive *Codium fragile* ssp. *tomentosoides* was investigated by Wilson et al. (2015). All the species were exposed for 9 weeks to increasing water temperatures

(12°C, 16°C, 20°C, 23°C, 26°C, and 29°C). The authors reported that the highest growth occurred at 12°C (*C. crispus* and *L. digitata)*, at 16°C (*F. vesiculosus* and *Codium)* and at 20°C (*A. nodosum)*. Species-specific response to increasing water temperatures was observed. Coralline algae *Corralina officinalis* L. is a macroalgae occurring in the intertidal habitats and are highly exposed to temperature fluctuations. The physiological response of *Corralina officinalis* to elevating temperatures was investigated by Rendina et al. (2019). The authors observed the significant effect of temperature on the *C. oficinalis* physiology and thus suggested that the continuous increase in the temperature due to global warming might result in lowering the survival rate and growth of *C. officinalis.* Phytoplanktons are the primary producers that can affect the higher trophic levels in the ecosystem. Temperature is a crucial factor influencing the growth, nutrient status, and the spatial and temporal distribution (Yvon-Durocher, et al., 2017) of phytoplanktons. The response of different phytoplanktons varied under diverse temperature regimes (Schabhuettl et al., 2012). The response of phytoplankton to temperature has been studied experimentally in the laboratories and was found to be species–specific since long times Reynolds (1984). In European lakes, the significant effects of heat waves in the composition and diversity of phytoplankton and zooplankton had been reported by de Senerpont Domis et al. (2013). Weisse et al. (2016) investigated the effect of the variations in temperature and nutrients in phytoplanktons. The authors observed that two species, that is, the cryptophyte *Cryptomonas* sp. and the dinoflagellate *Peridinium* sp., indicated more sensitivity to increasing surface water temperatures and proposed that the community composition of phytoplanktons are highly influenced by higher temperatures. One of the most common types of filamentous algae found in all types of aquatic ecosystems is the *Cladophora* species, which has been considered to be one of the best bioindicators of contamination in aquatic ecosystem (Chmielewská and Medved, 2001). In Poland, the growth rate of the common freshwater filamentous algae *Cladophora glomerata* was studied (Pikosz et al., 2017). The investigators observed that growth curve of *C. glomerata* decreased under elevating temperatures. Nalley et al. (2018) surveyed the effects of six different temperatures (ranging between 9°C and 32°C) on 26 algal species from five different functional groups. The algal taxa respond differently to varying temperatures but among the functional group cyanobacteria were found to show the highest temperature optima at 30.6°C± 2.3°C, followed by chlorophytes (25.7°C±0.1°C), and diatoms

(24.0°C±0.4°C). The combined effects of temperature and diversity on the rate of growth of phytoplanktons were studied by Schabhuttl et al. (2013). A total of 15 phytoplankton taxa (cyanobacteria, diatoms and green algae) and 25 mixed communities comprising 2–12 species were grown under constant temperatures (12°C, 18°C and 24°C). Better performance of green algae and diatoms was observed under lower temperatures, while with an increase in temperature in mixed communities of cyanobacteria showed stronger response. The foundation species of most of the temperate rocky reefs are formed by the canopy forming macroalgae where the abundant group comprises of kelps and fucoids (Schiel and Foster, 1986). The biological aspects (reproduction, growth, and development, etc.) of these macroalgae are negatively affected by the increasing water temperatures (Wernberg et al., 2010). Significant variation in the habitat structure of macroalgae (kelp, fucoids, mixed canopies, etc.) along a latitudinal gradient in ocean temperature was investigated by Wernberg et al. (2011). *Ostreopsis ovata* Fukuyo, is a toxic, benthic dinoflagellate found in the temperate waters of the world (Parsons et al., 2012). The influence of elevated sea surface temperatures on the growth and toxicity of *O. ovata* was evaluated (Granéli et al., 2011). The cells of *O. ovata* were grown at varying temperature ranging from 16°C to 30°C and the growth rate, cell densities, and toxicities were studied. The authors reported that the cell growth of *O. ovata* increased under increasing temperature of 26°C–30°C. This study suggested that increase in water temperatures by several degrees have paved the way for the increased accumulation in the algal blooms of *O. ovata*. Higher temperatures stimulate the growth of cyanobacteria in temperate lakes (Lürling et al., 2013). *Cylindrospermopsis raciborskii* is an invasive, nitrogen-fixing, filamentous cyanobacterium found in many tropical and subtropical regions of the world (Isvánovics et al., 2000). The growth responses of *C. raciborskii* to temperature changes were studied by Thomas and Litchman (2016). The authors observed that increase in the lake temperatures would likely favor the growth of *C. raciborskii* because of the inability of *C. raciborskii* to survive at lower water temperatures. The influence of temperature and UV-B radiation on *Microcystis aeruginosa* CS558 and *Anabaena circinalis* CS537 was studied by Islam and Beardall (2020). The cultures of these algae were exposed to UV-B irradiance (1.4 W/m^2) and grown at temperatures (25°C or 30°C). There was difference in the growth rates of both species under the two different temperatures. Higher growth rates (0.41 ± 0.02 day^{-1}) of *M. aeruginosa* and *A. circinalis* (0.38 ± 0.01 day^{-1}) were recorded at 25°C and 30°C, respectively. Singh and Singh (2015) compiled literature data on the effect of temperature and light on some species of green algae (viz. *Chlorella, Enteromorpha, Spirogyra, Chlamydomonas, Botryococcus, Scenedesmus, Neochloris, Haematococcus, Ulva*, etc.) blue-green algal species (viz. *Microcystis, Synechocystis*), red algal species (*Tichocarpus, Skeletonema, Heterosigma, Chondrus, Porphyra, Chattonella, Porphyridium*, etc.), brown algal species *(Sargassum, Undaria, Nitzschia)*, and phytoplanktons (*Euglena gracilis, Scrippsiella trochoidea, Protoceratium reticulatum*, etc.).

8.3.3 HIGH NUTRIENT LOAD OR ORGANIC CONTENT

Nutrient loading has been a major concern since it promotes eutrophication and other problems related to the aquatic ecosystem (Winter et al., 2007). Nutrient enrichment, specifically the over enrichment of nitrogen and phosphorus has led to the degradation of the coastal water ecosystem (Burkholder et al., 2007). Many algal species and groups can serve as biological indicators to monitor the aquatic ecosystems with high nutrient loads. Such algal indicators can be used to find evidence of eutrophication in waters and to designate the trophic state in aquatic ecosystems. In China, the change in the physiological response of sea grass species, *Thalassia hemprichii* (Ehrenb.) Ascher was suggested to be suitable bioindicators of eutrophic coastal waters (Zhang et al., 2014). The cyanobacterial bloom caused by cyanobacterial genera *Microcystis* has been recognized to be an indicator of high nutrient load (O'Neil et al., 2012). Studies on the quality of water in the stream of Norway suggested the utility of cyanobacteria *Stigonema mamillosum* as bioindicator for recording high nutrient load (Lindstrøm, 1999). From Iran, the use of cyanobacterial species, namely, *Chroococcus minor, Oscillatoria chlorine, Phormidium tenue*, and *Lyngbya kuetzingii* was confirmed to be suitable bioindicators of eutrophic waters (Soltani et al., 2012). Some cyanobacterial taxa reported to serve as enriched water bioindicators include: *Oscillatoria limosa, O. princeps*, and *O. tenuis* (Palmer, 1969; Gutowski et al., 2004; Sierra and Go´mez, 2007), *Leptolyngbya foveolarum* (Gutowski et al., 2004; Stancheva et al., 2012), *Phormidium terebriforme* (Loza et al., 2013; Rott et al., 1999; Sládecek, 1973), and *P. ambiguum* (Rott et al., 1999). Diatoms have been frequently used in the assessment of eutrophication (Bennion et al., 2014; Hall and Smol,

1999). In tropical streams, the benthic diatoms were reported to be useful as indicators of high nutrient load (Bellinger et al., 2006). The authors reported the identification of diatom taxa belonging to 20 genera and 54 species. The diatom species *Encyonema minutum, Nitzschia palea, Amphora copulata, Cocconeis placentula* and *Gomphonema parvulum* were recorded as occurring in nutrient-enriched waters (Bellinger et al., 2006; Soininen, 2002; Potapova et al., 2004). The diatom species (*Pseudostaurosira brevistriata* (Grun.) Williams and Round, *Achnanthidium minutissimum* (Kütz.) Czarnecki, and *Cocconeis placentula* Ehrenb) were reported to be indicators of meso eutrophic lakes (Toporowska et al., 2008). The cyanobacterial species of genera *Aphanizomenon, Anabaena, Microcystis* and the diatoms *Fragilaria, Aulacoseira, Asterionella* and *Stephanodiscus* were found to be present in nutrient-rich waters (Reynolds, 1998). Phytoplankton communities have been reported to be suitable indicators in the assessment of nutrient-enriched waters (O'Farrell et al., 2002; Lemley et al., 2016). Phtoplankton taxa belonging to 30 genera and 24 species were recognized from organically enriched Museum Lake of Thiruvananthapuram Botanical Garden, Kerela, India. Some of the algal species found in the high nutrient waters of the lake included *Nitzschia palea, Pandorina morum, Scenedesmus quadricauda, Synedra acus*, and *Trachelomonas volvocina* (Ajayan and Kumar, 2017).

8.3.4 CO_2 CONCENTRATION

Algae play an important role in the moderation of the atmospheric CO_2 concentration (Fourqurean et al., 2012). Algal species dynamically respond to the changes in the CO_2 concentration. Several algal species such as *Chlorella* sp., *Nannochloropsis* sp., *Scenedesmus* sp., *Chaetoceros* sp., *Ulva rigida, Gracilariopsis chorda, Spirulina* sp., etc. were described as being tolerant to different CO_2 concentration [as reviewed by Asadian et al. (2018)]. Seagrass ecosystems are the sites of high rates of carbon sequestration (Jiang et al., 2018). The ability of the miocroalgae (*Nannochloropsis salina* and *Isochrysis galbana)* and the macroalgae *(Gracilaria corticata, Sargassum polycystum* and *Ulva lactuca)* to sequester carbon was estimated by Kaladharan et al. (2009). The authors observed that the seaweed *Ulva lactuca* was able to utilize 100% of dissolved CO_2 at 15 mg/L whereas the microalgae registered only 27.7% of dissolved CO_2 at

15 mg/L. The bark inhabiting green algae *Scenedesmus komarekii* was found to tolerate an atmosphere up to 30% CO_2 (Hanagata et al., 1992). In Korea, the influence of elevated CO_2 levels and NH_{4+} on the pH, oxygen evolution, and rates of nutrient uptake, chlorophyll fluorescence, growth, and C/N ratio of *Ulva pertusa* Kjellman (Chlorophyta) was analyzed. The investigators found that the physiological reactions of this macroalgae increased under the exposure of elevated combination of CO_2/NH_{4+} (Kang and Chang, 2017). Some microalgae have been reported to grow easily at high CO_2 concentration. The impact of CO_2 concentration on the growth of microlagal species (*Chlorella* sp., *Zygnema* sp., *Scenedesmus* sp., *Hizikia fusiforme, Chaetoceros* sp., *Microcystis aeruginosa, Botryococcus braunii, Nannochloropsis* sp., *Ulvarigida, Chlorococcum* sp., *Spirulina* sp., *Prorocentrum minimum, Mytilusedulis, Botryococcus braunii, Nannochloropsis* sp., *Ulvarigida, Chlorococcum* sp.) Singh and Singh (2014). In India, two green algal strains identified as *Asterarcys quadricellulare* and *Chlorella sorokiniana* have been isolated from the water bodies near a steel plant. These strains were found to be tolerant to high levels of CO_2 and nitric oxide (Varshney et al., 2018).

8.4   HARMFUL ALGAL BLOOMS AND THEIR EFFECT

Harmful algal blooms (HABs) are the algal occurrences that cause toxicity or deleterious physiological and environmental effects (Moore et al., 2009; Wells et al., 2015). Approximately, 300 harmful algal species have been documented form the phytoplankton groups (Anderson et al., 2015; Berdalet et al., 2016). These phytoplanktons include diatoms, dinoflagellates, and cyanobacteria (Pal et al., 2020) which are responsible for the formation of blooms in both marine and freshwater ecosystem. These HABs produce potent toxins causing toxicity to higher trophic level species, including fish, shellfish, marine mammals, and humans (Wells et al., 2015). These toxins accumulate in the food chain and cause illness and death in animals and humans (Davidson et al., 2014). Some of the cyanobacterial species of genera *Anabaena, Aphanizomenon, Cylindrospermopsis, Microcystis, Planktothrix*, etc. synthesize potent hepatotoxins and neurotoxins that have the potential to kill animals and human beings (Chia et al., 2018). Certain species of diatoms produce domoic acid that can kill marine mammals and sea birds by causing

Amnesic Shellfish Poisoning (Tatters et al., 2012). Dinoflagellate species of the genera *Gambierdiscus* and *Fukuyoa* have been reported to produce neurotoxins (ciguatoxins and maitotoxins) that concentrate into the marine food chain causing death of the aquatic animals (Tester et al., 2020). Blooms formed by *Alexandrium fundyense* and *Dinophysis acuminate* have been documented to synthesize okadaic acid, that causes human health syndromes Paralytic and Diarrhetic Shellfish Poisoning, respectively (Gobler et al., 2017). HABs formed by cyanobacteria pose serious threat to human health by degrading the quality of water supplies and causing difficulty in the treatment of potable water (Chapra et al., 2017). Twenty-three classes of cyanotoxins comprising diverse range of metabolites have been identified which negatively affect the aquatic plants along with invertebrates and fish habitats (Manning and Nobles, 2017). Increasing cases of human hepatocellular carcinoma caused by cyanotoxins have been reported from China (Ren et al., 2017). HABs cause severe depletion of oxygen levels in the aquatic ecosystem leading to the death of several marine organisms (Anderson et al., 2008). The hindrance in the penetration of light caused due to HABs results into the reduced density of submerged aquatic vegetation (Anderson, 2009). Some toxins of HABs become aerosolized, causing respiratory harm to humans (Cheng et al., 2007). The high biomass of HABs causes depletion of oxygen and results in fish kills to most of the aquaculture fish and shellfish that can have dire impacts on coastal economies (Proenca et al. 2017; Griffith and Gobler, 2019). Also the increase in oxygen consumption that occurs from the death and decay of algal blooms affects and harms the aquatic organisms (Brosnahan et al., 2020). HABs have also been reported to cause air contamination by releasing toxin (Pal et al., 2020). Tester et al. (2020) reported the human health risks associated with the harmful dinoflagellate genera like *Gambierdiscus, Fukuyoa* and *Ostreopsis* species. Recently, Brosnahan et al. (2020) reviewed evidence of temperature-mediated control of dormancy duration from the cyst-forming dinoflagellates like *Alexandrium catenella* and *Pyrodinium bahamense*, causing illness and death to humans by the production of Paralytic Shellfish Poisoning. Thus, the global expansion of harmful algal blooms in recent decades have posed a threat to ecological integrity, ecosystem services, sustainability of coastal waters (Anderson et al., 2015; Paerl et al., 2016) along with global risk to human health and economies (Townhill et al., 2018).

KEYWORDS

- ecological balance
- harmful algal blooms
- aquatic ecosystems
- algal diversity
- indicators of change

REFERENCES

Ajayan, A. P.; Kumar, K. G. A. Phytoplankton as Biomonitors: A Study of Museum Lake in Government Botanical Garden and Museum, Thiruvananthapuram, Kerala India. *Lakes Reserv. Res. Manag.* **2017**, *22*, 1–13.

Ali, E.; Khairy, H. Environmental Assessment of Drainage Water Impacts on Water Quality and Eutrophication Level of Lake Idku, Egypt. *Environ. Pollut.* **2016**, 216.

Anderson, C. R.; Moore, S. K.; Tomlinson, M. C.; Silke, J.; Cusack, C. K. Living with Harmful Algal Blooms in a Changing World: Strategies for Modelling and Mitigating Their Effects in Coastal Marine Ecosystems. In *Coastal and Marine Hazards, Risks, and Disasters*; Shroder, J. F. Ellis, J. T.; Sherman, D. J., Eds.; Elsevier: The Netherlands, 2015.

Anderson, D. M. Approaches to Monitoring, Control and Management of Harmful Algal Blooms (HABs). *Ocean Coast Manag.* **2009**, *52*, 342–347.

Anderson, D. M.; Burkholder, J. M.; Cochlan, W. P.; Glibert, P. M.; Gobler, C. J.; Heil, C. A.; Kudela, R. M., Parsons, M. L.; Jack Rensel, J. E.; Townsend, D. W.; Trainer V. l.; Vargo, G. A. Harmful Algal Blooms and Eutrophication: Examining Linkages from Selected Coastal Regions of the United States. *Harmful Algae* **2008**, *8*, 39–53.

APHA. Standard Methods for the Evaluation of Water and Wastewater, 22nd ed.; American Public Health Association: Washington, DC, 2012.

Asadian, M.; Fakheri, B. A.; Mahdinezhad, N.; Gharanjik, S.; Beardal, J.; Talebi, A. F. Algal Communities: An Answer to Global Climate Change. *Clean Soir Air Water* **2018**, *46*, 1800032.

Bag, P.; Ansolia, P.; Mandotra, S. K.; Bajhaiya, A. K. Potential of Blue-Green Algae in Wastewater Treatment. In *Application of Microalgae in Wastewater Treatment*; Gupta, S. K.; Bux, F., Eds.; Springer Nature: Switzerland AG, 2019; p 363.

Ballesteros, I.; Castillejo, P.; Haro, A. P.; Montes, C. C.; Heinrich, C.; Lobo, E. A. Genetic Barcoding of Ecuadorian Epilithic Diatom Species Suitable as Water Quality Bioindicators. *Comptes. Rendus. Biol.* **2020**, *343*, 41–52.

Bellinger, B. J.; Cocquyt, C.; O'Reilly, C. M. Benthic Diatoms as Indicators of Eutrophication in Tropical Streams. *Hydrobiologia* **2006**, *573*, 75–87.

Bellinger, E. G.; Sigee, D. C. *Freshwater Algae Identification, Enumeration and Use as Bioindicators*, 2nd ed.; John Wiley and Sons Ltd.: West Sussex, 2015, ISBN 9780470058145.

Bennion, H.; Kelly, M. G.; Juggins, S.; Yallop, M. L.; Burgess, A.; Jamieson, B. J.; Krokowski, J. Assessment of Ecological Status in UK Lakes Using Benthic Diatoms. *Freshw. Sci.* **2014**.

Berdalet, E.; Fleming, L. E.; Gowen, R.; Davidson, K.; Hess, P.; Backer, L. C.; Moore, S. K.; Hoagland, P.; Enevoldsen, H. Marine Harmful Algal Blooms, Human Health and Wellbeing: Challenges and Opportunities in the 21st Century. *J. Mar. Biol. Assoc. U. K.* **2016**, *96*, 61–91.

Beveridge, O. S.; Petchey, O. L.; Humphrie, S. Direct and Indirect Effects of Temperature on Population Dynamics and Ecosystem Functioning of Aquatic Microbial Ecosystem. *J. Animal Ecol.* **2010**, *76* (6), 1324–1331.

Blanco, S.; Cejudo-Figueiras, C.; Tudesque, L.; Bécares, E.; Hoffmann, L.; Ector, L. Are Diatom Diversity Indices Reliable Monitoring Metrics? *Hydrobiologia* **2012**. doi:10.1007/s10750-012-1113-1.

Blasing, T. J. Recent Greenhouse Gas Concentrations. Carbon Dioxide Information and Analysis Center; Oak Ridges, TN, 2014. http://dx. doi. org/10. 3334/CDIAC/atg. 032.

Bonanno, G.; Veneziano, V.; Piccione, V. The Alga *Ulva lactuca* (Ulvaceae, Chlorophyta) as a Bioindicator of Trace Element Contamination along the Coast of Sicily, Italy. *Sci. Total Environ.* **2020**, *699*, 134329.

Botana, L.; Louzao, M. C.; Alfonso, A.; Botana, A.; Vieytes, M.; Viñariño, N.; Vale, C. Measurement of Algal Toxins in the Environment, 2009. doi: 10.1002/9780470027318. a9058.

Brosnahan, M. A.; Fischer, A. D.; Lopez, C. B.; Moore, S. K.; Anderson, D. M. Cyst-Forming Dinoflagellates in a Warming Climate. *Harmful Algae* **2020**, *91*, 101728.

Burkholder, J. M.; Tomasko, D. A.; Touchette, B. W.; Seagrasses and Eutrophication. *J. Exp. Mar. Biol. Ecol.* **2007**, *350*, 46–72.

Butcher, R. W. Pollution and Re-purification as Indicated by the Algae. *Fourth Int. Cong. Microbiol.* 1949, 149–150.

Cavalier-Smith, T. Evolution and Relationships of Algae: Major Branches of the Tree of Life. In *Unravelling the Algae*: *The Past, Present and Future of Algal Systematics*; CRC Press: New York, 2007; pp 21–55.

Chapra, S. C.; Boehlert, B.; Fant, C.; Bierman, V. J.; Henderson, J.; Mills, D.; Mas, D. M. L.; Rennels, L.; Jantarasami, L.; Martinich, J.; Strzepek, K. M.; Paerl, H. W. Climate Change Impacts on Harmful Algal Blooms in U. S. Freshwaters: A Screening-Level Assessment. *Environ. Sci. Technol.* **2017**, *51*, 8933–8943.

Cheng, Y.; Yue, Z.; Irvin, C.; Kirkpatrick, B.; Backer, L. Characterization of Aerosols Containing Microcystin. *Mar. Drugs* **2007**, *5*, 136–150.

Chia, M. A.; Jankowiak, J. G.; Kramer, B. J.; Goleski, J. A.; Huang, I. -S.; Zimba, P. V.; do Carmo Bittencourt-Oliveira, M.; Gobler, C. J. Succession and Toxicity of *Microcystis* and *Anabaena* (*Dolichospermum*) Blooms Are Controlled by Nutrient-Dependent Allelopathic Interactions. *Harmful Algae* **2018**, *74*, 67–77.

Chmielewská, E.; Medved, J. Bioaccumulation of Heavy Metals by Green Algae Cladophora Gramerata in a Refinery Sewage Lagoon. *Croat. Chem.* **2001**, *74* (1), 135–145.

CMFRI. *Annual Report 2010–11*; Central Marine Fisheries Research Institute, Cochin, 2011; p. 122.

Conti, M. E.; Cecchetti, G. A Biomonitoring Study: Trace Metals in Algae and Molluscs from Tyrrhenian Coastal Areas. *Environ. Reveres.* **2003**, *93*, 99–112.

Cooney, C. M. The Perception Factor; Climate Change Gets Personal. *Environ. Health Persp.* **2010**, *118* (11), 484–489.

Cosgrovea, J. D. W.; Morrison, P.; Hillman, K. Periphyton Indicate Effects of Wastewater Discharge in the Near-Coastal Zone, Perth, Western Australia. *Estuarine, Coastal Shelf Sci.* **2004**, *61* (2), 331–338.

Davidson, K.; Gowen, R. J.; Harrison, P. J.; Fleming, L. E.; Hoagland, P.; Moschonas, G. Anthropogenic Nutrients and Harmful Algae in Coastal Waters. *J. Environ. Manage.* **2014**, *146*, 206–216.

Day, J. G.; Benson, E. E.; Fleck, R. A. In Vitro Culture and Conservation of Microalgae: Applications for Aquaculture, Biotechnology and Environmental Research. *In Vitro Cell. Develop. Biol. Plant* **1999**, *35*, 127–136.

de Senerpont Domis, L. N.; Elser, J. J.; Gsell, A. S.; Huszar, V. L. M.; Ibelings, B. W.; Jeppesen, E.; Kosten, S.; Mooij, W. M.; Roland, F.; Sommer, U.; Van Donk, E.; Winder, M.; Lürling, M. Plankton Dynamics under Different Climatic Conditions in Space and Time. *Freshwater Biol.* **2013**, *58*, 463–482.

Dodds, W. K.; Jones, J. R.; Welch, E. B. Suggested Classification of Stream Trophic State: Distributions of Temperate Stream Types by Chlorophyll, Total Nitrogen, and Phosphorus. *Water Res.* **1998**, *32*, 1455–1462.

Dong, X. Y.; Li, B.; He, F. Z.; Gu, Y.; Sun, M. Q.; Zhang, H. M.; Tan, L.; Xiao, W.; Liu, S. R.; Cai, Q. H. Flow Directionality, Mountain Barriers and Functional Traits Determine Ddiatom Metacommunity Structuring of High Mountain Streams. *Sci. Rep.* **2016,** *6*, 24711.

Dong, X. Y.; Li, B.; He, F. Z.; Gu, Y.; Sun, M. Q.; Zhang, H. M.; Tan, L.; Xiao, W.; Liu, S. R.; Cai, Q. H. Flow Directionality, Mountain Barriers and Functional Traits Determine Diatom Metacommunity Structuring of High Mountain Streams. *Sci. Rep.* **2016,** *6*, 24711.

Downing, J.; Watson, S.; McCauley, E. Predicting Cyanobacterial Dominance in Lakes. *Can. J. Fisheries Aquat. Sci.* **2001**, *58*, 1905–1908. doi: 10.1139/cjfas-58-10-1905.

Field, C. B.; Behrenfeld, M. J.; Randerson, J. T.; Falkowski, P. Primary Production of the Biosphere: Integrating Terrestrial and Oceanic Components. *Science* **1998**, *281*, 237–240.

Fourqurean, J. W.; Duarte, C. M.; Kennedy, H.; Marba, N.; Holmer, M.; Mateo, M. A.; Serrano, O. Seagrass Ecosystems as a Globally Significant Carbon Stock. *Nat. Geosci.* **2012**, *5*, 505–509.

Glöckner, F. O.; Stal, L. J.; Sandaa, R. A.; Gasol, J. M.; O'Gara, F.; Hernandez, F.; Labrenz, M.; Stoica, E.; Varela, M. M.; Bordalo, A.; Pitta, P. Marine Microbial Diversity and Its Role in Ecosystem Functioning and Environmental Change. In *Marine Board Position Paper 17*; Calewaert, J. B.; McDonough, N., Eds.; Marine Board-ESF: Ostend, Belgium, 2012; pp 13–25.

Gobler, C. J.; Doherty, O. M.; Hattenrath-Lehmann, T. K.; Griffith, A. W.; Kang, Y.; Litaker, R. W. Ocean Warming ince 1982 Has Expanded the Niche of Toxic Algal Blooms in the North Atlantic and North Pacific Oceans. *Proc. Natl. Acad. Sci.* 2017, *114*, 201619575.

Gökçe, D. Algae as an Indicator of Water Quality. In *Algae-Organisms for Imminent Biotechnology*; Dhanasekaran, D., Ed.; InTech, 2016.

Granéli, E.; Vidyarathna, N. K.; Funari, E.; Cumaranatunga, P. R. T.; Scenati, R. Can Increases in Temperature Stimulate Blooms of the Toxic Benthic Dinoflagellate *Ostreopsis ovata*? *Harmful Algae.* **2011**, *10*, 165–172.

Griffith, A. W.; Gobler, C. J. Harmful Algal Blooms: A Climate Change Co-Stressor in Marine and Freshwater Wcosystems. *Harmful Algae* 2019, *91*, 101590.

Gutowski, A.; Foerster, J.; Schaumburg, J. Use of Benthic Algae Excluding Diatoms and Charales for the Assessment of the Ecological Status of Running Freshwaters: A Case History from Germany. *Oceanol. Hydrobiol. Stud.* **2004**, *33*, 3–15.

Hall, R. I.; Smol, J. P. Diatoms as Indicators of Lake Eutrophication. In *The Diatoms: Applications for the Environmental and Earth Sciences*; Stoermer, E. F.; Smol, J. P., Eds.; Cambridge University Press: Cambridge, 1999; pp 128–168.

Hanagata, N.; Takeuchi, T.; Fukuju, Y.; Barnes, D. J.; Karube, I. Tolerance of Microalgae to High CO_2 and High Temperature. *Phytochemistry* **1992**, *31*, 3345–3348.

Hill, B. H.; Herlihy, A. T.; Kaufmann, P. R.; Stevenson, R. J.; McCormick, F. H.; Burch, J. C. Use of Periphyton Assemblage Data as an Index of Biotic Integrity. *J. N. Am. Benthol. Soc.* **2000**, *19*, 50–67.

Hill, B. H.; Lazorchak, J. M.; McCormick, F. H.; Willingham, W. T. The Effects of Elevated Metals on Benthic Community Metabolism in a Rocky Mountain Stream. *Environ. Pollut.* **1997**, *95*, 183–190.

Howard, E. http://www.theguardian.com/environment/2015/sep/02/climatechangewillalteroceanmicroorganismscrucialtofoodchainsayscientists, 2015.

Huertas, E.; Rouco, M.; Lopez-Rodas, V., Costas, E. Warming Will Affect Phytoplankton Differently: Evidence through a Mechanistic Approach. *Proc. R. Soc. B.* **2011**, *278*, 3534–3543.

Hurlbert, S. H. The Concept of Species Diversity: A Critique and Alternate Parameters. *Ecology* **1971**, *52*, 577–586.

IPCC. Climate Change 2007: Synthesis Report. Contribution of Working Groups I, II and III to the Fourth Assessment Report of the Intergovernmental Panel on Climate Change; Reisinger, A., Ed.; IPCC: Geneva, Switzerland, 2007; p 104.

Islam, M. A.; Beardall, J. Effects of Temperature on The UV-B Sensitivity of Toxic Cyanobacteria *Microcystis aeruginosa* CS558 and *Anabaena circinalis* CS53. *Photochem. Photobiol.* **2020**, *10*.

Isvánovics, V.; Shafik, H. M.; Pre´sing, M.; Juhos, S. Growth and Phosphate Uptake Kinetics of the Cyanobacterium, *Cylindrospermopsis raciborskii* (*Cyanophyceae*) in through Flow Cultures. *Freshw. Biol.* **2000**, *43*, 257–275.

Jiang, Z.; Liu, S.; Zhang, J.; Wu, Y.; Zhao, C.; Lian, Z.; Huang, X. Eutrophication Indirectly Reduced Carbon Sequestration in a Tropical Seagrass Bed. *Plant Soil* **2018**, *426*, 135–152.

Johnk, D.; Huisman, J.; Sharples, J.; Sommeijer, B.; Visser, P. M.; Stroom, J. M. Summer Heatwaves Promote Blooms of Harmful Cyanobacteria. *Global Change Biol.* **2008**, *14* (3), 495–512.

Kadam, A. D.; Kishore, G.; Mishra, D. K.; Arunachalam, K. Microalgal Diversity as an Indicator of the State of the Environment of Water Bodies of Doon Valley in Western Himalaya, India. *Ecol. Indic.* **2020**, *112*, 106077.

Kadam, A. D.; Kishore, G.; Mishra, D. K.; Arunachalam, K. Microalgal Diversity as an Indicator of the State of the Environment of Water Bodies of Doon Valley in Western Himalaya, India. *Ecol. Indic.* **2020**, *112*, 106077.

Kaladharan, P.; Veena, S.; Vivekanandan, E. Carbon Sequestration by a Few Marine Algae: Observation and Projection. *J. Mar. Biol. Ass. India* **2009**, *51*, 107–110.

Kang, J. W., Chung, I. K. The Effects of Eutrophication and Acidification on the Ecophysiology of *Ulva pertusa* Kjellma. *J. Appl. Phycol.* **2017**, *29*, 2675–2683.

Karr, J. R. Assessment of Biotic Integrity Using Fish Communities. *Fisheries* **1981**, *6* (6), 21–27.

Kelly, M.; Bennion, H.; Burgess, A., et al. Uncertainty in Ecological Status Assessments of Lakes and Rivers Using Diatoms. *Hydrobiologia* **2009**, *633*, 5–15.

Lange, K.; Townsend, C. R.; Matthaei, C. D. A Trait-Based Framework for Stream Algal Communities. *Ecol. Evol.* **2016**, *6*, 23–36.

Lau, J. A.; Lennon, J. T. Rapid Responses of Soil Microorganisms Improve Plant Fitness in Novel Environments. *Proceed. Nat. Acad. Sci. USA* **2012**, *109*, 14058–14062.

Leclercq, L. Utilization de trios indices, chimique, diatomique et biocénotique, pour l'évaluation de la qualité de l'eau de la Joncquiere, riviére calcaire polluée par le village de Doische (Belgique, Prov. Namur). *Mém Soc Roy Bot Belg.* **1988**, *10*, 26–34.

Lemley, D. A.; Adams, J. B.; Bate, G. C. A Review of Microalgae as Indicators in South African Estuaries. *S. Afr. J. Bot.* **2016**, *107*, 12–20.

Liebmann, H. Handbuch der Frischwasser-, und Abwasserbiologie. R. Oldenburg, München, Germany, 1951; p 539.

Lindstrøm, E.-A. Attempts to Assess Biodiversity of Epilithic Algae in Running Water in Norway. In *Use of Algae for Monitoring Rivers III*; Prygiel, J.; Whitton, B.; Bukowska, J., Eds.; Douai: France, 1999; pp 253–260.

Loza, V.; Perona, E.; Carmona, J.; Mateo, P. Phenotypic and Genotypic Characteristics of Phormidiumlike Cyanobacteria Inhabiting Microbial Mats Are Correlated with the Trophic Status of Running Waters. *Eur. J. Phycol.* **2013**, *48*, 235–252.

Lu¨rling, M.; F. Eshetu, E.; Faassen, J.; Kosten, S.; Huszar, V. L. M. Comparison of Cyanobacterial and Green Algal Growth Rates at Different Temperatures. *Freshw. Biol.* **2013**, *58*, 552–559.

Lüning, K. *Seaweeds Their Environment, Biogeography, and Ecophysiology* (trans: Yarish C, Kirkan H); Wiley: New York (Original work published 1985), 1990.

Mann, D. G.; Droop, S. J. M. Biodiversity, Biogeography and Conservation of Diatoms. *Hydrobiologia* **1996**, *336*, 19–32.

Mann, D.; Vanormelingen, P. An Inordinate Fondness? The Number, Distributions, and Origins of Diatom Species. *J. Eukaryot. Microbiol.* **2013**, *60*, 414–420.

Manning, S. R; Nobles, D. R. Impact of Global Warming on Water Toxicity: Cyanotoxins. *Curr. Opin. Food Sci.* **2017**. https://doi.org/10.1016/j.cofs.2017.09.013

Mateo, P.; Leganés, F.; Perona, E.; Loza, V.; Fernández-Piñas, F. Cyanobacteria as Bioindicators and Bioreporters of Environmental Analysis in Aquatic Ecosystems. *Biodivers. Conserv.* **2015**, *24*, 909–948.

McCormick, P. V.; Cairns, J. Jr. Algae as Indicators of Environmental Change. *J. Appl. Phycol.* **1994**, *6*, 509–526.

McCormick, P. V.; Rawlik, P. S.; Lurding, K.; Smith, E. P.; Sklar, F. H. Periphyton-Water Quality Relationships along a Nutrient Gradient in the Northern Florida Everglades. *J. North Am. Benthol. Soc.* **1996**, *15*, 433–449.

Medvigy, D.; Beaulieu, C. Trends in Daily Solar Radiation and Precipitation Coefficients of Variation since 1984. *J. Clim.* **2012**, *25*, 1330–1339.

Meehl, G., et al. The WCRP CMIP3 Multimodel Dataset: A New Era in Climate Change Research. *Bull. Am. Meteorol. Soc.* **2007**, *88*, 1383–1394.

Merzouk, A.; Johnson, L. E. Kelp Distribution in the Northwest Atlantic Ocean under a Changing Climate. *J. Exp. Mar. Biol. Ecol.* **2011,** *400,* 90–98.

Miraglia, M.; Marvin, H. J. P.; Kleter, G. A.; Battilani, P.; Brera, C.; Coni, E., et al. Climate Change and Food Safety: An Emerging Issue with Special Focus on Europe. *Food Chem. Toxicol.* **2009,** *47,* 1009–1021.

Miranda, J.; Krishnakumar, G. Microalgal Diversity in Relation to the Physico hemical Parameters of Some Industrial Sites in Mangalore, South India. *Environ. Monit. Assess.* **2015,** *187,* 664–688.

Moore, S. K.; Mantua, N. J.; Hickey, B. M.; Trainer, V. L. Recent Trends in Paralytic Shellfish Toxins in Puget Sound, Relationships to Climate, and Capacity for Prediction of Toxic Events. *Harmful Algae* 2009, 8, 463–477.

Mudie, P. J.; Rochon, A.; Levac, E. Palynological Records of Red Tide Producing Species in Canada: Past Trends and Implications for the Future. *Palaeogeogr. Palaeoclimatol. Palaeoecol.* **2002,** *180,* 159–186.

Nalley, J. O.; O'Donnell, D. R.; Litchman, E. Temperature Effects on Growth Rates and Fatty Acid Content in Freshwater Algae and Cyanobacteria. *Algal Res.* **2018,** *35,* 500–507.

NASA. http://climate.nasa.gov/, 2015.

Nearing, M. A.; Pruski, F. F.; O'Neal, M. R. Expected Climate Change Impacts on Soil Erosion Rates: A Review. *J. Soil Water Conserv.* **2004,** *59,* 43–50.

NOAA. Natural Variability Main Culprit of Deathly Russian Heat Wave That Killed Thousands. NOAA News Online, 2011. www. noaanews. noaaa. gov/stories2011.

Noyes, P. D.; McElwee, M. K.; Miller, H. D.; Clark, B. W., van Tiem, L. A.; Walcott, K. C.; Erwin, K. N., Levin, E. D. The Toxicology of Climate Change: Environmental Contaminants in a Warming World. Environ. Intern. **2009,** *35* (6), 971–986.

O'Farrell, I.; Lombardo, R. J.; Pinto, P. T.; Loez, C. The Assessment of Water Quality in the Lower Luján River (Buenos Aires, Argentina): Phytoplankton and Algal Bioassays. *Environ. Pollut.* **2002,** *120,* 207–218.

O'Neil, J. M.; Davis, T. W.; Burford, M. A.; Gobler, C. J. The Rise of Harmful Cyanobacteria Blooms: The Potential Roles of Eutrophication and Climate Change. *Harmful Algae* **2012,** *14,* 313–334.

Paerl, H. W.; Gardner, W. S.; Havens, K. E.; Joyner, A. R.; McCarthy, M. J.; Newell, S. E.; Qin, B.; Scott, J. T. Mitigating Cyanobacterial Harmful Algal Blooms in Aquatic Ecosystems Impacted by Climate Change and Anthropogenic Nutrients. *Harmful Algae* **2016,** *54,* 213–222.

Pal, M.; Yesankar, P. J.; Dwivedi, A.; Qureshi, A. Biotic Control of Harmful Algal Blooms (HABs): A Brief Review. *J. Environ. Manage.* **2020,** *268,* 110687.

Palmer, C. M. A Composite Rating of Algae Tolerating Organic Pollution. *J. Phycol.* **1969,** *5,* 78–82.

Palmer, C. M. *Algae and Water Pollution*; Castle House Publications Ltd., 1980; p 110.

Pan, Y.; Stevenson, R. J.; Hill, B. H.; Herlihy, A. T. Ecoregions and Benthic Diatom Assemblages in Mid-Atlantic Highland Streams, USA. *J. North Am. Benthol. Soc.* **2000,** *19,* 518–540.

Parsons, M. L.; Aligizaki, K.; Bottein, M. Y. D.; Fraga, S.; Morton, S. L. *Gambierdiscus* and *Ostreopsis*: Reassessment of the State of Knowledge of Their Taxonomy, Geography, Ecophysiology, and Toxicology. *Harmful Algae* **2012,** *14,* 107–129.

Patrick, R. Use of Algae, Especially Diatoms, in the Assessment of Water Quality. In *Biological methods for the assessment of water quality, Special Technical Publication*; Cairns, J. Jr.; Dickson, K. L., Eds., Vol. 528; American Society for Testing and Materials: Philadelphia, PA, 1973, pp 76–95.

Pikosz, M.; Messyasz, B.; Gąbka, M. Functional Structure of Algal Mat (*Cladophora glomerata*) in a Freshwater in Western Poland. *Ecol. Indic.* **2017,** *74*, 1–9.

Porter, S. D. Algal Attributes: An Autecological Classification of Algal Taxa Collected by the National Water-Quality Assessment Program. *US Geol. Survey Data Ser.* **2008,** *329*, 18.

Potapova, M.; Charles, D. F.; Ponander, K. C.; Winter, D. M. Quantifying Species Indicator Values for Trophic Diatom Indices: A Comparison of Approaches. *Hydrobiologia* **2004,** *517*, 25–41.

Proenca, L. A. O.; Hallegraeff, G. M. Marine and Fresh-Water Harmful Algae. *Proceedings of the 17th International Conference on Harfmul Algae*, 2017.

Radford, I. J. Fluctuating Resources, Disturbance and Plant Strategies: Diverse Mechanisms Underlying Plant Invasions. *J. Arid Land* **2013,** *5* (3), 284–297.

Ren, L. X.; Wang, P. F.; Wang, C.; Chen, J.; Hou, J.; Qian, J. Algal Growth and Utilization of Phosphorus Studied by Combined Mono-Culture and Co-Culture Experiments. *Environ. Pollut.* **2017,** *220*, 274–285.

Rendina, F.; Bouchet, P. J.; Appolloni, L.; Russo, G. F.; Sandulli, R.; Kolzenburg, R.; Putra, A. Physiological Response of the Coralline Alga *Corallina officinalis* L. to Both Predicted Long-Term Increases in Temperature and Short-Term Heatwave Events. *Mar. Environ. Res.* **2019,** *150*, 104764.

Reynolds, C. S. *The Ecology of Freshwater Phytoplankton*; University Press: Cambridge, 1984.

Reynolds, C. S. What Factors Influence the Species Composition of Phytoplankton in lakes of Different Trophic Status? *Hydrobiologia* **1998,** *369*, 11–26.

Rott, E.; Hofmann, G.; Pall, K.; Pfister, P.; Pipp, E. *Indikationslisten für Aufwuchsalgen in Österreichischen Fliessgewässern. Teil 1: Saprobielle Indikation Wasserwirtschaftskataster, Bundeministerium f. Landu;* Forstwirtschaft: Wien, 1997.

Rott, E.; Pipp, E.; Pfister, P.; Van Dam, H.; Ortler, K.; Binder, N.; Pall, K. *Indikationslisten für Aufwuchsalgen in österreichischen Fliessgewässern. Teil 2*: Trophieindikation (*sowie geochemische Präferenzen, taxonomische und toxikologische Anmerkungen*); Wasser-wirtschaftskataster, Bundesministerium f. Land- u. Forstwirtschaft: Wien, 1999; p 248.

Sarmento, H.; Montoya, J. M.; Vazquez-Domınguez, E.; Vaque, D.; Gasol, J. M. Warming Effects on Marine Microbial Food Web Processes: How Far Can We Go When It Comes to Predictions? *Phil. Trans. R. Soc. B* **2010,** *365*, 2137–2149. doi: 10.1098/rstb.2010.0045.

Sarmiento, J. L.; Gruber, N. Anthropogenic Carbon Sinks. *Phys. Today* **2002,** *55*, 30–36.

Schabhuettl, S.; Hingsamer, P.; Weigelhofer, G, Hein T.; Weigert, A.; Striebel, M. Temperature and Species Richness Effects in Phytoplankton Communities. *Oecologia* **2013,** *171*, 527–536.

Schabhuettl, S.; Hingsamer, P.; Weigelhofer, G.; Hein, T.; Weigert, A.; Striebel, M. Tempera-ture and Species Richness Effects in Phytoplankton Sommunities. *Oecologia* **2012,** *171*, 527–536.

Schiel, D. R.; Foster, M. S. The Structure of Subtidal Algal Stands in Temperate Waters. *Oceanogr. Mar. Biol. Ann. Rev.* **1986,** *24*, 265–307.

Sen, B., Sonmez, F.; Kocer, M. A. T.; Alp, M. T.; Canpolat, O. *Relationship of Algae to Water Pollution and Waste Water Treatment;* InTech, 2013.

Shams El-Din, N. G.; Mohamedein, L. I. Seaweeds as Bioindicators of Heavy Metals Off a Hot Spot Area on the Egyptian Mediterranean Coast during 2008–2010. *Environ. Monit. Assess.* 2014, *186*, 5865–5881.

Shannon, C. E. A Mathematical Theory of Communication. *Bell System Technical J.* **1948**, *27*, 379–423.

Sierra, M. V.; Go´mez, N.; Structural Characteristics and Oxygen Consumption of the Epipelic Biofilm in Three Lowland Streams Exposed to Different Land Uses. *Water Air Soil Pollut.* **2007**, *186*, 115–127.

Singh, B. K.; Bardgett, R. D.; Smith, P.; Reay, D. S. Microorganisms and Climate Change: Terrestrial Feedbacks and Mitigation Options. *Nat. Rev. Microbiol.* **2010**, *8*, 779–790.

Singh, S. P., Singh, P. Effect of CO_2 Concentration on Algal Growth: A Review. *Renew. Sust. Energ. Rev.* **2014**, *38*, 172–179.

Singh, S. P.; Singh, P. Effect of Temperature and Light on the Growth of Algae Species: A Review. *Renew. Sustain. Energy Rev.* **2015**, *50*, 431–444.

Sládecek, V. System of Water Quality from the Biological Point of View. *Ergeb Limnol.* **1973**, *7*, 1–18.

Sládecek, V. System of Water Quality from the Biological Point of View. *Ergeb Limnol.* **1973**, *7*, 1–18.

Smucker, N. J.; Becker, M.; Detenbeck, N. E.; Morrison, A. C. Using Algal Metrics and Biomass to Evaluate Multiple Ways of Defining Concentrationbased Nutrient Criteria in Streams and Their Ecological Relevance. *Ecol. Indic.* **2013**, *32*, 51–61. doi:10.1016/j.ecolind.2013.03.018.

Soininen, J. Responses of Epilithic Diatom Communities to Environmental Gradients in Some Finnish Rivers. *Int. Rev. Hydrobiol.* **2002**, *87*, 11–24.

Soltani, N.; Khodaei, K.; Alnajar, N.; Shahsavari, A.; Ashja Ardalan, A. Cyanobacterial Community Patterns as Water Quality Bioindicators. *Iran J. Fish. Sci.* **2012**, *11*, 876–891.

Soltani, N.; Khodaei, K.; Alnajar, N.; Shahsavari, A.; Ashja Ardalan, A. Cyanobacterial Community Patterns as Water Quality Bioindicators. *Iran J. Fish. Sci.* **2012**, *11*, 876–891.

Spellerberg, I. F. *Monitoring Ecological Change*; Cambridge University Press: Cambridge, 2005.

Stancheva, R.; Fetscher, A. E.; Sheath, R. G. A Novel Quantification Method for Stream-Inhabiting, Nondiatom Benthic Algae, and Its Application in Bioassessment. *Hydrobiologia* **2012**, *684*, 225–239.

Stevenson, J. Ecological Assessments with Algae: A Review and Synthesis. *J. Phycol.* **2014**, *50*, 437–461.

Stevenson, R. J.; Bennette, B. J.; Jordan, D. N.; French, R. D. Phosphorus Regulate Stream Injury by Filamentous Algae, DO and pH with Threshold in Responses. *Hydrobiologia* **2012**, *695*, 25–42.

Stevenson, R. J.; Pan, Y. Assessing Ecological Conditions in Rivers and Streams with Diatoms. In *The Diatom: Applications to the Environmental and Earth Science*; Stoemer, E. F.; Smol, J. P., Eds.; Cambridge University Press: Cambridge, 1999; pp 11–40.

Stevenson, R. J.; Pan, Y.; van Dam, H. Assessing Environmental Conditions in Rivers and Streams with Diatoms. In *The Diatoms: Applications for the Environmental and*

Earth Sciences; Smol, J. P.; Stoermer, E. F., Eds. 2nd ed.; Cambridge University Press: Cambridge, MA, 2010.

Stevenson, R. J.; Smol, J. P. Use of Algae in Environmental Assessments; Academic Press: San Diego, 2003.

Stevenson, R. J.; Zalack, J. T.; Wolin, J. A Multimetric Index of Lake Diatom Condition Based on Surface-Sediment Assemblages. *Freshwater Sci.* **2013**, *32*, 1005–1025.

Stoermer, E. F.; Smol, J. P., Eds. *The Diatoms*: *Applications for the Environmental and Earth Sciences*; Cambridge University Press: Cambridge, 1999; p 484.

Tatters, A. O.; Fu, F. -X.; Hutchins, D. High CO_2 and Silicate Limitation Synergistically Increase the Toxicity of *Pseudo-nitschia fradulenta*. *PLoS One* **2012**, *7*, e32116.

Tester, P. A.; Litakerb, R. W.; Berdalet, E. Climate Change and Harmful Benthic Microalgae. *Harmful Algae* **2020**, *91*, 101655.

Teta, R.; Romano, V.; Della Sala, G.; Picchio, S.; De Sterlich, C.; Mangoni, A.; Di Tullio, G.; Costantino, V.; Lega, M. Cyanobacteria as Indicators of Water Quality in Campania Coasts, Italy: A Monitoring Strategy Combining Remote/Proximal Sensing and in Situ Data. *Environ. Res. Lett.* **2017**, *12*, 024001.

Thomas, M. K.; Litchman, E. Effects of Temperature and Nitrogen Availability on the Growth of Invasive and Native Cyanobacteria. *Hydrobiologia* **2016**, *763*, 357–369.

Toporowska, M.; Pawlik-Skowronska, B.; Wojtal, A. Z. Epiphytic algae on *Stratiotes aloides* L., *Potamogeton lucens* L., *Ceratophyllum demersum* L. and *Chara* spp. in a Macrophyte-Dominated Lake. *Oceanol. Hydrobiol. St.* **2008**, *37*, 51–63.

Townhill, B. L.; Tinker, J.; Jones, M.; Potois, S.; Creach, V.; Simpson, S. D.; Dye, S.; Bear, E.; Pennegar, J. K. Harmful Algal Blooms and Climate Change: Exploring Future Distribution Changes. *ICES J. Mar. Sci.* 2018, *75*, 1882–1893.

Ustunada, M.; Erdugan, H.; Yilmaz, S.; Akgul, R.; Aysel, V. Seasonal Concentrations of Some Heavy Metals (Cd, Pb, Zn, and Cu) in *Ulva rigida* J. Agardh (Chlorophyta) from Dardanelles (Canakkale, Turkey). *Environ. Monit. Assess.* **2010**, *177*, 337–342.

van Dam, H.; Mertens, A.; Sinkeldam, J. A Coded Check List and Ecological Indicator Values of Freshwater Diatoms from the Netherlands. *Netherlands J. Aquat. Ecol.* **1994**, *28*, 117–133. doi: 10.1007/BF02334251.

Varshney, P.; Beardall, J.; Bhattacharya, S.; Wangikar, P. P. Isolation and biochemical characterisation of two thermophilic green algal species-*Asterarcys quadricellulare* and *Chlorella sorokiniana*, which are tolerant to high levels of carbon dioxide and nitric oxide. Algal Res. 2018, 30, 28–37.

Walsh, D. A. Consequences of Climate Change on Microbial Life in the Ocean. *Microbiol Today* **2015**, Microbiology Society, England.

Weisse, T.; Gröschl, B.; Bergkemper, V. Phytoplankton Response to Short-Term Temperature and Nutrient Changes. *Limnologica* **2016**, *59*, 78–89.

Wells, M. L.; Trainer, V. L.; Smayda, T. J.; Karlson, B. S. O.; Trick, C. G.; Kudela, R. M.; Ishikawa, A.; Bernard, S.; Wulff, A.; Anderson, D. M.; Cochlan, W. P. Harmful Algal Blooms and Climate Change: Learning from the Past and Present to Forecast the Future. *Harmful Algae* **2015**, *49*, 68–93.

Wernberg, T.; Smale, D. A.; Thomsen, M. S. A Decade of Climate Change Experiments on Marine Organisms: Procedures, Patterns and Problems. *Global Change Biol.* **2012**, *18*, 1491–1498.

Wernberg, T.; Thomsen, M. S.; Tuya, F.; Kendrick, G. A. Biogenic Habitat Structure of Seaweeds Change along a Latitudinal Gradient in Ocean Temperature. *J. Exp. Mar. Biol. Ecol.* **2011,** *400,* 264–271.

Wernberg, T.; Thomsen, M. S.; Tuya, F.; Kendrick, G. A.; Staehr, P. A.; Toohey, B. D. Decreasing Resilience of Kelp Beds along a Latitudinal Temperature Gradient: Potential Implications for a Warmer Future. *Ecol. Lett.* **2010,** *13,* 685–694.

Wetzel, R. G.; Likens, G. E. *Limnological Analyses,* 3rd ed.; Springer: New York, 2000.

Whitton, B. A.; Yelloly, J. M.; Christmas, M.; Hernandez, I. Surface Phosphatase Activity of Benthic Algae in a Stream with Highly Variable Ambient Phosphate Concentrations. In *International Association of Theoretical and Applied Limnology—Proceedings*; Williams, W. D.; Sladeckova, A., Eds., vol 26, Pt 3; E Schweizerbart'sche: Stuttgart, 1998; pp 967–972.

Wilkinson, A. N.; Zeeb, B. A.; Smol, J. P.; Douglas, M. S. V. Chrysophyte Stomatocyst Assemblages Associated with Periphytic, High Arctic Pond Environments. *Nord. J. Bot.* **1997,** *17,* 95–112.

Wilson, K. L.; Kay, L. M.; Schmidt, A. L.; Lotze, H. K. Effects of Increasing Water Temperatures on Survival and Growth of Ecologically and Economically Important Seaweeds in Atlantic Canada: Implications for Climate Change. *Mar. Biol.* **2015,** *162,* 2431–2444.

Winter, J.; Eimers, C.; Dillon, P.; Scott, L.; Scheider, W.; Willox, C. Phosphorus Inputs to Lake Simcoe from 1990 to 2003: Declines in Tributary Loads and Observations on Lake Water Quality. *J. Great Lakes Res.* **2007,** *33,* 381–396.

World Meteorological Organization (WMO). Greenhouse Gas Levels in Atmosphere Reach New Record, 2018. https://public. wmo. int/.

Wu, N. C.; Schmalz, B.; Fohrer, N. Development and Testing of a Phytoplankton Index of Biotic Integrity (P-IBI) for a German Lowland River. *Ecol. Indic.* **2012,** *13,* 158–167.

Wu, N.; Cai, Q.; Fohrer, N. Development and Evaluation of a Diatom-Based Index of Biotic Integrity (D-IBI) for Rivers Impacted by Run-of-River Dams. *Ecol. Indic.* **2012,** *18,* 108–117.

Wu, N.; Dong, X.; Liu, Y.; Wang, C.; Baattrup-Pedersen, A.; Riis, T. Using River Microalgae as Indicators for Freshwater Biomonitoring: Review of Published Research and Future Directions. *Ecol Indic.* **2017,** *81,* 124–131.

Yu, J.; Wang, P.; Wang, Y.; Chang, J.; Deng, S.; Wei, W. Thermal Constraints on Growth, Stoichiometry and Lipid Content of Different Groups of Microalgae with Bioenergy Potential. *J. Appl. Phycol.* **2018,** *30,* 1503–1512.

Yvon-Durocher, G.; Schaum, C.-E.; Trimmer, M. The Temperature Dependence of Phytoplankton Stoichiometry: Investigating the Roles of Species Sorting and Local Adaptation. *Front. Microbiol.* **2017,** *8,* 2003.

Zhang, J.; Huang, X.; Jiang, Z. Physiological Responses of the Seagrass Thalassia Hemprichii (Ehrenb.) Aschers as Indicators of Nutrient Loading. *Mar. Pollut. Bull.* **2014,** *83,* 508–515.

CHAPTER 9

Regulation of Ethylene Levels with 1-Aminocyclopropane-1-Carboxylate Deaminase Produced by Plant Growth Promoting Rhizobacteria

APARNA B. GUNJAL

Department of Microbiology, Dr. D.Y. Patil, Arts,
Commerce and Science College, Pimpri, Pune, Maharashtra, India

*Corresponding author. E-mail: aparnavsi@yahoo.com

ABSTRACT

The plant growth promoting rhizobacteria promote the plant growth by direct and indirect mechanisms. These include production of siderophores; solubilization of minerals, namely, phosphorus and potassium; chitinase, cellulase, 1-aminocyclopropane-1-carboxylate deaminase enzyme production; production of plant hormones, namely, indole acetic acid, gibberellins and cytokinins, etc. The review here focuses on the important aspect of regulation of ethylene levels by 1-amino-1-cyclopropane carboxylate deaminase enzyme produced by plant growth promoting rhizobacteria. The ethylene levels can be toxic if the concentration is more than the required level in the plants, and therefore, the regulation of ethylene levels becomes very essential. This is significant as it is a biological, economical, easy and eco-friendly approach of regulation of ethylene levels in plants.

9.1 INTRODUCTION

The plants require macro (carbon, nitrogen, phosphorus, and potassium) and micro nutrients (iron, magnesium, manganese, sodium, calcium, and vanadium, etc.) for their growth (Etienne et al., 2018). There are many plant growth hormones, namely., indole acetic acid (IAA), gibberellins, cytokinins, ethylene, etc. Ethylene is important in the growth mechanism of many crops. The ethylene levels are necessary to maintain the plants; if exceeded becomes toxic to the plants. The ethylene level can be maintained by the enzyme 1-aminocyclopropane-1-carboxylate (ACC) deaminase. This enzyme is produced by beneficial bacteria which improve the growth of the plants and are known as plant growth promoting rhizobacteria (PGPR) (Egamberdieva and lugtenberg, 2014). The ACC deaminase enzyme protects the plants and other crops from the enhanced levels of ethylene which is harmful.

There are very less reports on the exact role of ACC deaminase enzyme in monitoring ethylene levels in plants, and hence, the present review focuses on this aspect.

9.2   PLANT GROWTH PROMOTING RHIZOBACTERIA

Plant growth promoting rhizobacteria are present near the root region of plants where there is maximum population of these beneficial bacteria (Grobelak et al., 2018; Nagargade et al., 2018). The PGPR improve the plant growth by two ways, namely, direct and indirect. These are nitrogen fixation, production of hormones, solubilization of macro elements,, that is, phosphorus and potassium, and production of siderophores (Egamberdieva and lugtenberg, 2014; Shameer and Prasad, 2018). One of the ways where PGPR minimize the stress is by maintaining minimum level of ethylene in the plants, and the ACC deaminase enzyme plays here an essential role. PGPR with the help of ACC deaminase enzyme brings the conversion of ACC to ammonia and α-ketobutyrate, thereby reducing the ethylene level in the plants and other crops (Gamalero and Glick, 2015; Raghuwanshi and Prasad, 2018).

The PGPR producing ACC deaminase enzyme are *Pseudomonas putida*, *P. marginalis*, *P. syringae*, *Acetobacter* sp., *Bacillus* sp., *Enterobacter aerogenes*, *Rhodococcus* sp., *Penicillium citrinum*, *Trichoderma asperellum*, *Acidovorax facilis*, *Bukholderia* sp., etc. (Jia et al., 2006; Saraf et al., 2010).

The fungi producing ACC deaminase enzyme are *Penicillium citrinum* and *Trichoderma asperellum* (Jia et al., 2006; Singh and Kashyap, 2012).

9.3 PIVOTAL ROLE OF ETHYLENE IN THE PLANT GROWTH AND DEVELOPMENT

Ethylene production by microorganisms was reported for the first time in the pathogen *Ralstonia solanacearum* (Freebairn and Buddenhagen, 1964). Ethylene helps both in plant growth and development. Ethylene helps to regulate cell size and cell division in plants. In plants, ethylene also enhances the processes, namely, ripening, senescence, and abscission (Schaller, 2012). It is termed "regulatory hormone" and performs an essential role in the developmental and regulatory processes of various plants. Another role of ethylene is of root initiation in plants, and promote flowering. Ethylene also inhibits the formation of storage organs, namely, tubers and bulbs (Schaller, 2012). Ethylene is also reported to balance the equilibrium between the macro elements (iron and potassium) and it improves root growth under different stress conditions in *Arabidopsis* plant (Li et al., 2015).

9.3.1 ETHYLENE SIGNALING PATHWAY IN THE PLANTS (CHANG, 2016)

Ethylene when detected, a signal is given and ETR1 cleaves ETYHLENE INSENSITIVE2 (EIN2) which is an insoluble protein. EIN2 stops the translation of F-box proteins EBF1/2. EIN3/EIL1 becomes firm; the transcriptional cascade follows including ETR1. The ethylene response genes help in ethylene responses (Fig. 9.1a). In the ethylene signaling pathway, when ethylene is not detected, ETR1 activates the protein CONSECUTIVE RESPONSE1 (CTR1) kinase which is an insoluble protein and degrades EIN3/EIL1. The ethylene response genes do not allow ethylene responses (Fig. 9.1b).

9.3.2 ETHYLENE TOXICITY TO PLANTS

The activity of ethylene is seen even at least concentrations of 0.01–1.0 parts per million (ppm) (Chang, 2016). There exist a study where 0.7 ppm of ethylene near the Beltway (the highway near Washington DC and University of Maryland) had toxic impact on the vegetation (Abeles and Heggestad, 1973).

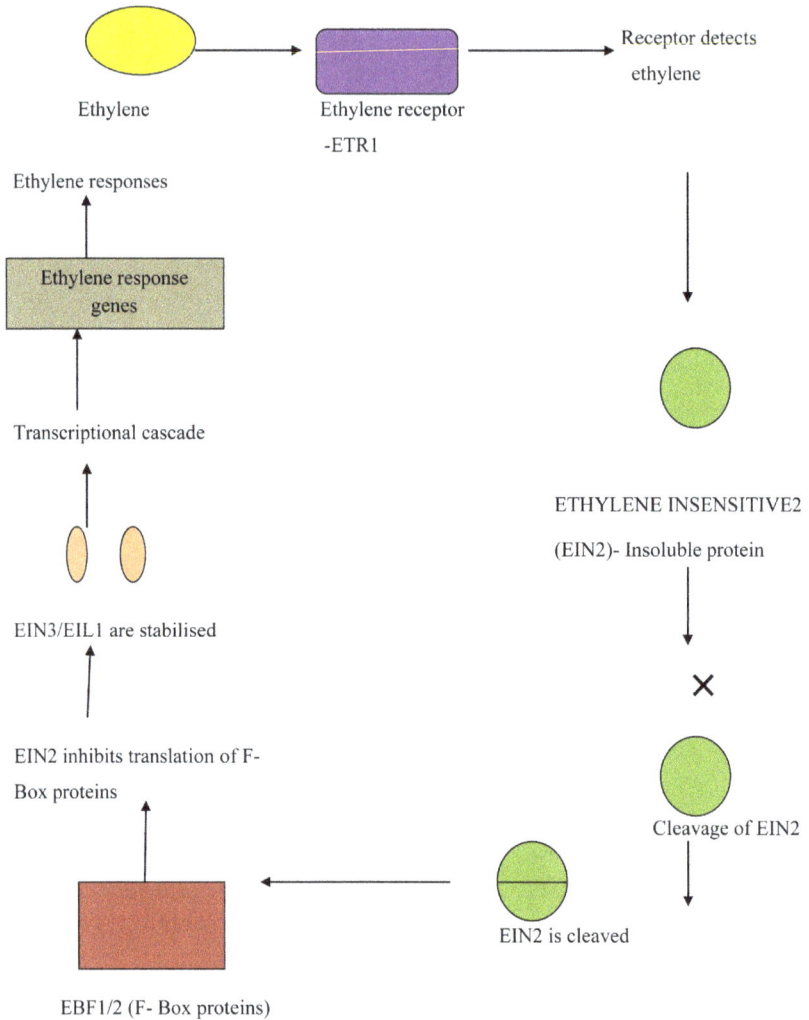

FIGURE 9.1a Ethylene receptor in signaling pathway when ethylene is detected.

An increased ethylene level causes the following:

- Cell death.
- Inhibits the action of plant hormone abscisic acid (Jakubowicz and Nowak, 2010) and gibberellins.
- Ethylene high levels can fasten the aging process in plants and damage the product quality and shelf life (Chang, 2016).

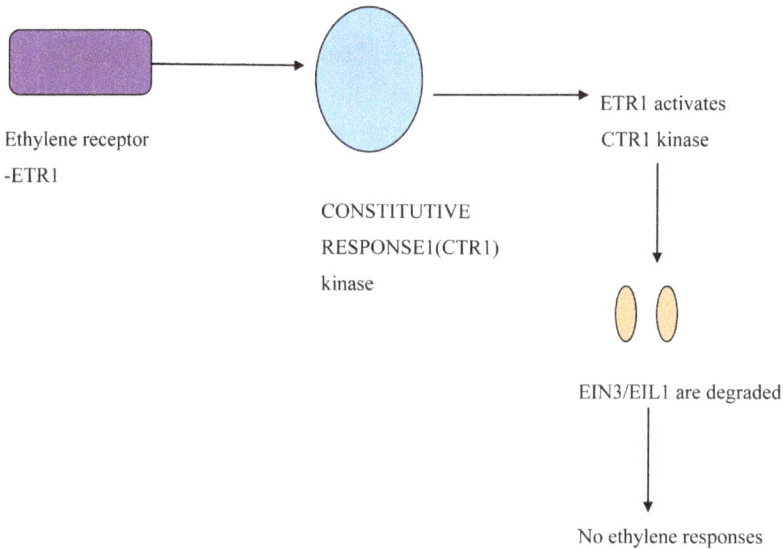

FIGURE 9.1b   Ethylene receptor in signaling pathway when ethylene is not detected.

9.3.4   ETHYLENE VARIATION IN PLANTS

The ethylene transduction cascade is conserved in plants, and helps in regulation under adverse or stress conditions. Plants follow ethylene signaling in response to any adverse conditions. For example, plants in flood areas (e.g., rice and *Rumex palustris*) due to flood-induced acquisition of ethylene, adapt under such conditions (Bailey-Serres and Voesenek, 2008). In contrast to this, few aquatic plants are studied, they adapt to submerged lifestyle and lost many genes, which are essential in ethylene signaling pathway (Street and Schaller, 2016; Voesenel et al., 2015). The increase in the level of ethylene is due to stress of presence of metals, chemicals, high temperature, amount of water, ultraviolet light, phytopathogens, and some injury (Abeles et al., 1992).

9.4   ACC DEAMINASE ENZYME AND ITS BIOSYNTHESIS

9.4.1   ACC DEAMINASE ENZYME

The enzyme ACC deaminase has many polypeptide chains, having molecular weight ranging between 35and 42 kDa. It is a sulfhydral enzyme in which pyridoxal-5-phosphate (PLP), a cofactor binds to each subunit. The enzyme ACC is present in the cytoplasm (Glick et al., 1998). This ACC deaminase is an essential intermediate during the production of ethylene (Adams and Yang, 1979). Ethylene is an important plant hormone, which helps in many developmental processes and also plays a role to regulate adverse conditions (Dubois et al., 2018).

9.4.2 ACC DEAMINASE AND OTHER GENES IN THE REGULATION OF ETHYLENE LEVEL

9.4.2.1 ACC DEAMINASE GENES

ACC deaminase activity is seen in many beneficial microorganisms (Nascimento et al., 2014). The gene *AcdS* responsible for coding ACC deaminase has been studied in many microorganisms (Rashid et al., 2012; Tak et al., 2013). It is also reported in *Rhizobium leguminosarum* (Itoh et al., 1996); *Sinorhizobium meliloti* (Ma et al., 2004) and *Mesorhizobium loti* (Kaneko et al., 2002). *AcdS* gene is similar to the genes, namely, dcyD and yedO (Singh et al., 2015). The sequence of proteins in *AcdS* gene follows as, namely, Lys51, Ser78, Tyr295, Glu296, and Leu322. This sequence of proteins performs an important role in ACC deaminase activity (Nascimento et al., 2014).

9.4.3 ACS AND ACO GENES

The most conserved ACS10 and ACS12 proteins in *Arabidopsis* are Asp, Phe, and Tyr aminotransferases (Yamagami et al., 2003). The specific expression of the ACS gene and regulation of ACS activity at the cellular level will help in the monitoring of ethylene production (Xu and Zhang, 2015). The *ACO* gene family is not fully studied. However, the *ACO* genes are reported as ACO6 to ACO13 based on similar sequences (Clouse and Carraro, 2014). In *Arabidopsis*, ACO is encoded by five genes which include *ACO1, ACO2, ACO3, EFE*, and *ACO5* (Vanderstraeten and Straeten, 2017). ACS and ACO are termed as "controlling step" in ethylene synthesis in plants and many crops (Poel and Straeten, 2014).

9.4.4 BIOSYNTHESIS OF ACC DEAMINASE ENZYME

The biosynthesis of ACC deaminase enzyme is represented in Figure 9.2 (Polko and Kieber, 2019).

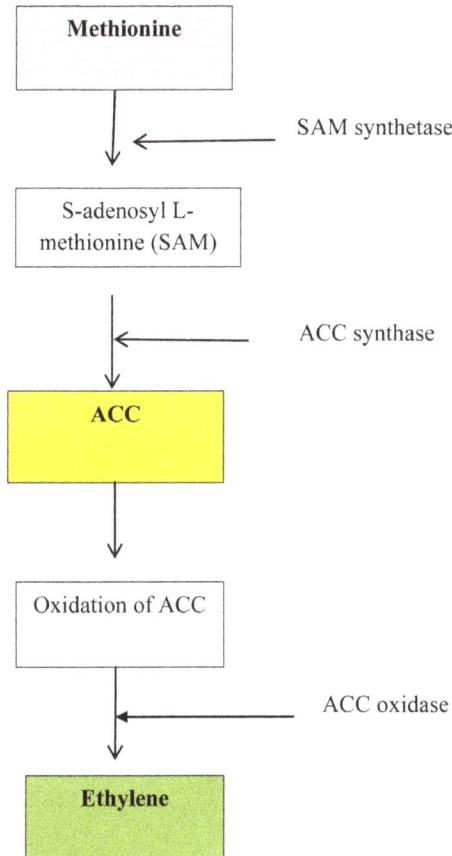

FIGURE 9.2   Biosynthesis of ACC deaminase enzyme.

9.5   REGULATION OF ETHYLENE LEVELS IN THE PLANTS BY ACC DEAMINASE ENZYME

The regulation of ethylene levels in the plants by ACC deaminase enzyme is done by the following ways:

9.5.1   MECHANISM OF ACC DEAMINASE ENZYME

ACC deaminase is important for controlling the effects of ethylene in plants and to lower the level of ACC in plant tissue, thus reducing the production of ethylene. PGPR which produce ACC deaminase convert ACC into α-ketobutyrate and ammonia (Glick, 2005). The PGPR produce IAA, which induces the activity of ACC. The ACC is hydrolyzed to α-ketobutyrate and ammonia by ACC deaminase enzyme (Fig. 9.3). This is known as the "sink effect," where ACC breaks down within the roots or exudates (Saraf et al., 2010). The root-associated bacteria thus reduce the ethylene level near the plant roots (Singh et al., 2015). The ethylene produced in excess feedback inhibits IAA signal transduction and stops IAA to activate ACC synthase transcription (Pierik et al., 2006; Prayitno et al., 2006; Stearns et al., 2012). The PGPR that produces both IAA and ACC deaminase, ethylene levels do not increase. Due to the production of ACC deaminase enzyme, ethylene levels are lowered in the plants. Due to ethylene feedback inhibition of IAA signal transduction, this will increase the plant growth as well as ACC synthase transcription. Thus, there is dual action, IAA will increase the plant growth and ACC deaminase enzyme will lower the ethylene level.

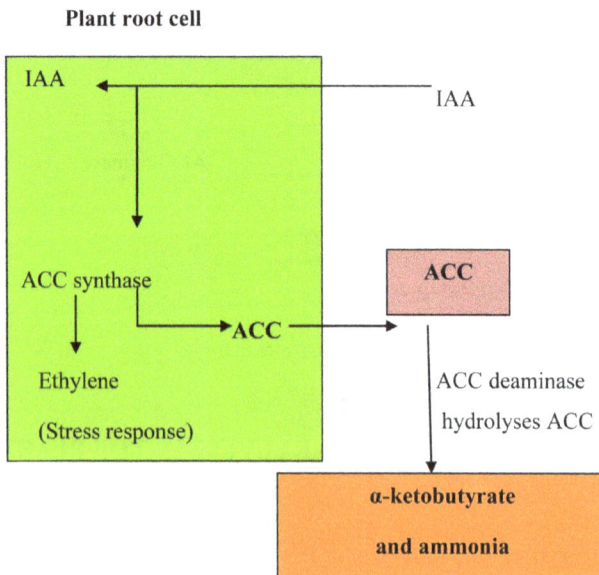

FIGURE 9.3   Mechanism of ACC deaminase enzyme.

9.5.2 INHIBITION OF THE ENZYMES—ACC OXIDASE AND ACC SYNTHASE

Gene silencing approach can reduce ethylene levels (Lin et al., 2009). Ethylene biosynthesis can also be reduced by the addition of inhibitors of enzymes ACC oxidase (ACO) (e.g., cobalt or α-aminoisobutyric acid) or ACC synthase (ACS) (e.g., aminoethoxyvinyl-Gly and aminooxyacetic acid (Lin et al., 2010). There is a report where ACO activity controls ethylene production in *Rumex palustris* during submergence (Vriezemn et al., 1999). Ethylene level is controlled by ACS synthase turnover (Xu and Zhang, 2015). In *Arabidopsis* plant, eight genes encode ACSes, and an additional gene encodes for catalytically inactive enzyme, ACS1 (Tsuchisaka and Theologis, 2004).

The four ways of ethylene regulation in plants are: (Srivastava, 2002)

- Silencing of *ACS* gene expression.
- Silencing of *ACO* gene expression.
- Deamination of ACC to α-ketobutyrate and ammonia.
- Hydrolysis of S-adenosyl L-methionine.

9.6 CONCLUSION

The monitoring and maintaining of ethylene level in the plants is important. The plant enzymes, namely, ACC deaminase, inhibition of ACC oxidase, and ACC synthase will play important role in the regulation of ethylene levels in the plants which is focused in this review. This will not harm the plants from the toxic levels of ethylene, which is essential.

KEYWORDS

- **plant growth promoting rhizobacteria**
- **ethylene**
- **eco-friendly**
- **biological**
- **1-aminocyclopropane1-carboxylate deaminase**

REFERENCES

Abeles, F. B.; Heggestad, H. E. Ethylene: An Urban Air Pollutant. *J. Air Pollut. Cont. Assoc.* **1973**, *23*, 517–521.

Abeles, F. B.; Morgan, P. W.; Saltveit, M. E. *Ethylene in Plant Biology*, 2nd ed.; Academic Press: New York, 1992.

Adams, D. O.; Yang, S. F. Ethylene Biosynthesis—Identification of 1-Aminocyclopropane-1-Carboxylic Acid as an Intermediate in the Conversion of Methionine to Ethylene. *Proc. Natl. Acad. Sci. USA* **1979**, *76*, 170–174.

Bailey-Serres, J.; Voesenek, L. J. Flooding Stress: Acclimations and Genetic Diversity. *Annu Rev Plant Biol* **2008**, *59*, 313–339.

Chang, C. Q & A: How Do Plants Respond to Ethylene and What Is Its Importance? *BMC Biol.* **2016**, *14*, 1–7.

Clouse, R. M.; Carraro, N. A Novel Phylogeny and Morphological Reconstruction of the PIN Genes and First Phylogeny of the ACC Oxidases (ACOs). *Front. Plant Sci.* **2014**, *5*, 1–15.

Dubois, M.; Van den Broeck, L.; Inzé, D. The Pivotal Role of Ethylene in Plant Growth. *Trends Plant Sci.* **2018**, *23*, 1–14.

Egamberdieva, D.; Lugtenberg, B. Use of Plant Growth-Promoting Rhizobacteria to Alleviate Salinity Stress in Plants. In: *Use of Microbes for the Alleviation of Soil Stresses;* Miransari, M., Ed.; Springer: New York, 2014; pp 73–96.

Etienne, P.; Diquelou, S.; Prudent, M.; Salon, C.; Maillard, A.; Ourry, A. Macro and Micronutrient Storage in Plants and Their Remobilization When Facing Scarcity: The Case of Drought. *Agriculture* **2018**, *8*, 1–17.

Freebairn, H. T.; Buddenhagen, I. W. Ethylene Production by *Pseudomonas solanacearum. Nature* **1964**, *202*, 313–314.

Gamalero, E.; Glick, B. R. Bacterial Modulation of Plant Ethylene Levels. *Plant Physiol.* **2015**, *169*, 13–22.

Glick, B. R.; Penrose, D. M.; Li, J. A Model for the Lowering of Plant Ethylene Concentrations by Plant Growth-Promoting Bacteria. *J. Theor. Biol.* **1998**, *190*, 63–68.

Glick, B. R. Modulation of Plant Ethylene Levels by the Bacterial Enzyme ACC Deaminase. *FEMS Microbiol Lett.* **2005**, *251*, 1–7.

Grobelak, A.; Kokot, P.; Hutchison, D.; Grosser, A.; Kacprzak, M. Plant Growth Promoting Rhizobacteria as an Alternative to Mineral Fertilizers in Assisted Bioremediation-Sustainable Land and Waste Management. *J. Environ. Manage.* **2018**, *227*, 1–9.

Itoh, T.; Aiba, H.; Baba, T.; Hayashi, K.; Inada, T.; Isono, K., et al.. A 460-kb DNA Sequence of the *Escherichia coli* K-12 Genome Corresponding to the 40.1–50.0 min Region on the Linkage Map. *DNA Res.* **1996**, *3*, 441–445.

Jakubowicz, M.; Nowak, W. 1-Aminocyclopropane -1-Carboxylate Synthase, an Enzyme of Ethylene Biosynthesis. *Comprehen. Natur. Prod. II Chem. Biol.* **2010**, *5*, 91–120.

Jia, Y. J.; Ito, H.; Matsui, H.; Honma, M. 1-Aminocyclopropane-1-Carboxylate (ACC) Deaminase Induced by ACC Synthesized and Accumulated in *Penicillium citrinum* Intracellular Spaces. *Biosci. Biotechnol. Biochem.* **2006**, *64*, 299–305.

Kaneko, T.; Nakamura, Y.; Sato, S.; Minamisawa, K.; Uchiumi, T.; Sasamoto, S.; Watanabe, A.; Idesawa, K.; Iriguchi, M.; Kawashima, K.; Kohara, M.; Matsumoto, M.; Shimpo, S.; Tsuruoka, H. Complete Genomic Sequence of Nitrogen-fixing Symbiotic Bacterium *Bradyrhizobium japonicum* USDA110. *DNA Res.* **2002**, *9*, 189–197.

Li, G.; Xu, W.; Kronzucker, H.; Shi, W. Ethylene Is Critical to the Maintenance of Primary Root Growth and Fe Homeostasis under Fe Stress in *Arabidopsis. J. Exp. Bot.* **2015,** *66,* 2014–2054.

Lin, Z.; Zhong, S.; Grierson, D. Recent Advances in Ethylene Research. *J. Exp. Bot.* **2009,** *60,* 3311–3336.

Lin, L. C.; Hsu, J. H.; Wang, L. C. Identification of Novel Inhibitors of 1-Aminocyclopropane-1 Carboxylic Acid Synthase by Chemical Screening in *Arabidopsis thaliana. J. Biol. Chem.* **2010,** *285,* 33445–33456.

Ma, W.; Charles, T.; Glick, B. Expression of an Exogenous 1-Aminocyclopropane-1-Carboxylate Deaminase Gene in *Sinorhizobium meliloti* Increases Its Ability to Nodulate *Alfalfa. Appl. Environ. Microbiol.* **2004,** *70,* 5891–5897.

Nagargade, M.; Tyagi, V.; Singh, M. K. Plant Growth Promoting Rhizobacteria: A Biological Approach toward the Production of Sustainable Agriculture. In *Role of Rhizospheric Microbes in Soil: Volume 1: Stress Management and Agricultural Sustainability;* Meena, V. S., Ed.; Springer: Singapore, 2018; pp 205–223.

Nascimento, F. X.; Rossi, M. J.; Soares, C. S.; McConkey, B. J.; Glick, B. R. New Insights into 1-Aminocyclopropane-1-Carboxylate (ACC) Deaminase Phylogeny, Evolution and Ecological Significance. *PLoS ONE* **2014,** *9,* e99168.

Pierik, R.; Tholen, D.; Poorter, H.; Visser, E. J. W.; Voesenek, L. J. The Janus Face of Ethylene: Growth Inhibition and Stimulation. *Trends Plant Sci.* **2006,** *11,* 176–183.

Poel, B. V.; Straeten, D. V. 1-Aminocyclopropane-1-Carboxylic Acid (ACC) in Plants: More Than Just the Precursor of Ethylene. *Front. Plant Sci.* **2014,** *5,* 1–11.

Polko, J. K.; Kieber, J. J. 1-Aminocyclopropane 1-Carboxylic Acid and Its Emerging Role as an Ethylene-Independent Growth Regulator. *Front Plant Sci.* **2019,** *10,* 1–9.

Prayitno, J.; Rolfe, B. G.; Mathesius, U. The Ethylene-Insensitive Sickle Mutant of *Medicago truncatula* Shows Altered Auxin Transport Regulation During Nodulation. *Plant Physiol.* **2006,** *142,* 168–180.

Raghuwanshi, R.; Prasad, J. K. Perspectives of Rhizobacteria with ACC Deaminase Activity in Plant Growth under Abiotic Stress. In *Root Biology;* Giri, B.; Prasad, R.; Varma, A., Eds.; Springer: Cham, 2018; pp 303–321.

Rashid, S.; Charles, T. C.; Glick, B. R. Isolation and Characterization of New Plant Growth-Promoting Bacterial Endophytes. *Agric. Ecosyst. Environ.* **2012,** *61,* 217–224.

Saraf, M.; Jha, C. K.; Patel, D. The Role of ACC Deaminase Producing PGPR in Sustainable Agriculture. In *Plant Growth and Health Promoting Bacteria*; Maheshwari, D. K., Ed.; Microbiology Monographs 18. Springer-Verlag: Berlin Heidelberg, 2010; pp 365–385.

Schaller, G. E. Ethylene and the Regulation of Plant Development. *BMC Biol.* **2012,** *10,* 1–3.

Shameer, S.; Prasad, T. Plant Growth Promoting Rhizobacteria for Sustainable Agricultural Practices with Special Reference to Biotic and Abiotic Stresses. *Plant Growth Regul.* **2018,** *84,* 603–615.

Singh, N.; Kashyap, S. *In-silico* Identification and Characterization of 1-Aminocyclopropane-1-Carboxylate Deaminase from *Phytophthora sojae. J. Mol. Model.* **2012,** *18,* 4101–4111.

Singh, R. P.; Shelke, G. M.; Kumar, A.; Jha, P. N. Biochemistry and Genetics of ACC Deaminase: A Weapon to 'Stress Ethylene' Produced in Plants. *Front. Microbiol.* **2015,** *6,* 1–14.

Srivastava, L. Chapter 11 Ethylene. In *Plant Growth and Development: Hormones and Environment*; Elsevier Publisher, 2002; pp 233–250.

Stearns, J. C.; Woody, O. Z.; McConkey, B. J.; Glick, B. R. Effects of Bacterial ACC Deaminase on *Brassica napus* Gene Expression Measured with an *Arabidopsis thaliana* Microarray. *Mol. Plant-Microb. Interact.* **2012**, *25*, 668–676.

Street, I. H.; Schaller, G. E. Ethylene: A Gaseous Signal in Plants and Bacteria. *Biochem (Lond)* **2016**, *38*, 4–7.

Tak, H. I.; Ahmad, F.; Babalola, O. O. Advances in the Application of Plant Growth Promoting Rhizobacteria in Phytoremediation of Heavy Metals. *Rev. Environ. Contam. Toxicol.* **2013**, *223*, 33–52.

Tsuchisaka, A.; Theologis, A. Unique and Overlapping Expression Patterns among the *Arabidopsis* 1-Aminocyclopropane-1-Carboxylate Synthase Gene Family Members. *Plant Physiol.* **2004**, *136*, 2982–3000.

Vanderstraeten, L.; Straeten, D. Accumulation and Transport of 1-Aminocyclopropane-1-Carboxylic Acid (ACC) in Plants: Current Status, Considerations for Future Research and Agronomic Applications. *Front Plant Sci.* **2017**, *8*, 1–18.

Voesenek, L. J.; Pierik, R.; Sasidharan, R. Plant Life without Ethylene. *Trends Plant Sci.* **2015**, *20*, 783–786.

Vriezen, W.; Hulzink, R.; Mariani, C.; Voesenek, L. 1-Aminocyclopropane-1-Carboxylate Oxidase Activity Limits Ethylene Biosynthesis in *Rumex palustris* during Submergence. *Plant Physiol.* **1999**, *121*, 189–195.

Wada, T.; Yamada, M.; Tabata, S. Complete Genome Sequence of Nitrogen Fixing Symbiotic Bacterium *Bradyrhizobium japonicum* USDA110. *DNA Res.* **2002**, *9*, 189–197.

Xu, J.; Zhang, S. Ethylene Biosynthesis and Regulation in Plants. In *Ethylene in Plants*; Wen, C. K., Ed.; Springer: Berlin, 2015; pp 1–25.

Yamagami, T.; Tsuchisaka, A.; Yamada, K.; Haddon, W. F.; Harden, L. A.; Theologis, A. Biochemical Diversity among the 1-Aminocyclopropane-1-Carboxylate Synthase Isozymes Encoded by the *Arabidopsis* Gene Family. *J. Biol. Chem.* **2003**, *278*, 49102–49112.

CHAPTER 10

Core Microbiome of *Solanum Lycopersicum* for Sustainable Agroecosystems

ANAMIKA CHATTOPADHYAY[1*] and G. THIRIBHUVANAMALA[2]

[1]*Department of Microbiology, Punjab Agricultural University, Ludhiana, Punjab, India*

[2]*Department of Fruit Crops, Tamil Nadu Agricultural University, Coimbatore, Tamil Nadu, India*

Corresponding author. E-mail: anamikachatterjee16@gmail.com

ABSTRACT

Plant microbiota is of enormous importance in determining plant health and productivity. The configuration and construction of plant microbiota differ following the plant tissues and compartments which are specific habitats for microbial colonization. These distinct habitats may include the rhizosphere, endosphere, and phyllosphere of the plant. The plant microbiome is known to benefit the host health in various ways. Studies reveal that species of organisms like *Pseudomonas and Bacillus* isolated from the core microbiome residing in mentioned habitats are being utilized as biocontrol agents. Biocontrol has a great significance in the achievement of sustainable crop production and protection. Sustainable approaches help in the achievement of higher yields without being detrimental to the biodiversity and environment. It includes methods of growing and protecting crops responsibly, avoiding application, and dependence on chemically produced fertilizers and pesticides. The plant host in this study is the tomato which is a high-value crop and an established model system growing in commercial conditions. Tomatoes are a dietary source of the antioxidant lycopene, which possess numerous health benefits such as the low possibility of heart disease and cancer. They are also an abundant source of potassium, vitamin C, and vitamin K. This chapter elaborates the core microbiome present in various habitats and stages of tomato, their interaction, isolation, and utilization in terms of biocontrol in sustainable agroecosystems.

10.1   THE NEED OF SUSTAINABLE AGROECOSYSTEMS

For decades, agriculture has been a crucial sector playing a significant role in the survival of humankind. The livelihood of around 70% of the world's population is reliant on agriculture and its allied sectors. Agriculture is not only a source of livelihood, but also contributes to the national revenue, supplies food and fodder, expands the marketable surplus and also plays a significant role in the international trade. As the demand for food is directly proportional to the increasing in population, intensification, and accretion of agriculture has become necessary and is leading to significant global environmental changes. The emergence of Green Revolution technology is a result of agricultural intensification. It has three principles: people require food to eat, inadequate land resources, and the final process is to enhance the yield through external inputs (Lobell et al., 2014). Hence, agrochemicals, including nitrogenous fertilizers and various pesticides are applied for increasing grain yield (Saikia et al., 2013).

Conversely, consistent usage of these agrochemicals has led to a noticeable change and deterioration in soil quality and diversity of microflora. Besides, it elicited various issues including widening the socio-economic gap, insecurity of food, alarming traditional rural structure, suicide among marginal farmers, and ecological problems like the interruption of the integrity of the ecosystem, its efficiency and stability. The other durable faults of this system of agromanagement are deterioration of soil organic carbon and fertility of the soil, soil erosion, gradual weakening of biodiversity, pesticide pollution, and evolving pest resistance, desertification, climate change, food prices hike, etc. (Lichtfouse et al., 2009). These conditions, altogether, arise the need for a substitute agriculture practice which has an ecologically healthy prospect.

One promising alternate practice of agriculture is sustainability. We need to sustain our natural resources for replenishment and continued support of the growing human population. Agriculture is considered sustainable when the system conserves resources, supports social norms, has high commercial value and fair competition in the market, and is environmentally safe. The health of plants, soil, and humankind relies upon the development of efficient and sustainable agroecosystems (Yadav et al., 2018).

10.2 ROLE OF MICROORGANISMS IN MAINTAINING SUSTAINABILITY

Agroecosystems are known to have an intricate pattern of interaction among macro and microorganisms. There are ongoing studies related to these associations for maximizing their function in biocontrol, nutrient uptake, crop growth enhancement, and yield. This will help in combating pests attack, climate change, and maintaining a sustainable ecosystem with least use of chemicals. However, plant–microbe association optimization is a difficult task to perform because of the intricacy of plant–microbes and microbe–microbe interactions, and their reliance on environmental conditions. Certain microorganisms are assembled or repulsed directly by plant-signaling mechanisms. By this mechanism, the mycorrhizal fungi and nitrogen-fixing bacteria interact and repel different pathogens.

Hence, having a superior insight of multifarious interactions amid environmental conditions, plant genotypes, and microbiome assembly gives a broad idea in breeding programs. Despite the enormous perspective, we are just commencing to discover opportunities in which microbiome data can be assimilated with 'smart farming' practices.

Soil acts as a base for various biological processes where the microbes interact, forming an association with plants and other microbes. These interactions and its processes help in the growth of plants and enhance soil health, which in turn results in maintaining sustainability in the agroecosystems by reducing the use of chemicals. These biological processes include:

- Biological nitrogen fixation
- Plant residue decomposition
- Mineralization/immobilization
- Nutrient cycling
- Denitrification (Yadav et al., 2018).

10.2.1 BIOLOGICAL NITROGEN FIXATION

Biological nitrogen fixation (BNF) is a process carried out by prokaryotes. The prokaryotes reduce molecular atmospheric nitrogen into ammonia. The resultant ammonia afterward assimilates in amino acids with the help of the enzyme glutamine synthetase and provides the Earth's ecosystems with approximately 200 million tons nitrogen/year (Rao et al., 2017).

Though the process includes various complex biochemical reactions, it may be summarized in a relatively less complicated way by the following equation:

$$N_2 + 8H_2 + 16ATP \rightarrow 2\,NH_3 + 2H_2 + 16ADP + 16\,P_i$$

Legumes have a distinctive capability to propagate well in nutrients-exhausted soils and enhance them due to their efficient biological nitrogen-fixing ability (Khaitov et al., 2016). Improving BNF in legumes may upsurge plant-based protein for human intake and the BNF is of great significance as the overuse of nitrogenous fertilizers has caused the severe environmental crisis.

This process of nitrogen fixation takes place in association with Rhizobium. Various aspects like the host plant, its cultivation history, biotic and abiotic factors, soil pH, drought, salinity, soil's organic carbon content, and soil's texture are the known factors to affect the diversity and distribution of Rhizobium. Stress conditions are habitable for some microbes to evolve tolerant traits, including beneficial microbes, to become potential plant growth promoters. However, the symbiosis between Rhizobium–legume is negatively impacted by salinity and other abiotic factors (Abdiev et al., 2019).

10.2.2 PLANT RESIDUE DECOMPOSITION

Efficient methods for crop residue management are the crucial steps for the protection of agricultural soils from wind and water erosion, for the maintenance of soil productivity by recycling plant nutrients, and for the improvement of physical properties of soil. The most effective and sustainable method of degradation of plant residues is with the help of microorganisms present in the ground. The diversity and efficiency of these microbes depend on various factors such as soil organic matter content, the cropping system, the cropping pattern, the type of soil, the source of irrigation, etc. (Arcand et al., 2016).

Agricultural practices have a significant impact on soil microbial populations, which are the chief managers of soil carbon and nutrient cycling methods. Variations in microbial population structure can disturb the fate of carbon and nutrients existing in the soil throughout the process of decomposition and may affect the retention of carbon and the contribution of crop nutrients in agroecosystems. Long-term management encourages alterations in microbial communities and resource availability along with long-term influences on soil carbon accretion and fecundity.

10.2.3 MINERALIZATION AND IMMOBILIZATION

Mineralization is the oxidation of chemical compounds present in organic matter, making the nutrients in those compounds available insoluble inorganic for the uptake of plants. Mineralization benefits in terms of increasing the bioavailability of the nutrients present in the decomposing organic compounds, most notably, because of their quantities, nitrogen, phosphorus, and sulfur.

Immobilization in the soil is the transformation of inorganic compounds to organic compounds by microorganisms or plants, preventing it from being available to plants. Immobilization is the reverse of mineralization where the inorganic nutrients are utilized by soil microbes which make them unavailable for plant uptake. The immobilization process is a biological process dominated by the action of bacteria that consumes biological macromolecules and inorganic nitrogen for amino acids. Immobilization and mineralization occur concurrently, where nitrogen existing in the decomposing system is transformed into organic from the inorganic state by immobilization and vice versa by decay and mineralization.

The primary factor that determines whether nitrogen mineralizes or immobilizes majorly on the carbon–nitrogen ratio of the decomposing organic matter. If the C:N ratio is more than 30:1 then the decomposing microbes tend to absorb nitrogen in mineral forms such as ammonium or nitrate, and this mineral nitrogen will be known as immobilized. On the other hand, if the C:N ratio is less than 25:1, by further decomposition and release of ammonia, mineralization will take place. After the completion of organic matter decomposition, mineralization adds up more nitrogen, increasing total nitrogen content in the soil (Dai et al., 2017).

10.2.4 NUTRIENT CYCLING

Agriculture influences nutrient dynamics, energy fluxes, and hydrologic cycles. Explanations of nutrient cycling and decomposition in agroecosystems are site-specific, and therefore, overviews are of little cogency unless procedures and regulatory mechanisms are taken into account. The nutrient cycle in the unmanaged ecosystem varies with that in the agroecosystem along with the role and the comparative importance of soil organic matter, organic fertilizers, and inorganic fertilizers as sources of nutrient inputs into agroecosystems. The primary nutrient cycle in the agroecosystem includes:

1. Nitrogen cycle
2. Carbon cycle
3. Phosphorous cycle
4. Sulfur cycle

10.2.4.1 NITROGEN CYCLE

Nitrogen includes the majority of the Earth's atmosphere and is the fourth most copious element in cellular biomass. Microbial activities are solely responsible for the exchange between inert nitrogen gas in the existing atmosphere and reactive nitrogen. Before the discovery of the Haber–Bosch process, which is the industrial fixation of N2 into ammonia, NH3, in 1909, microorganisms play a significant role in the production and recycling of almost all the reactive nitrogen in the atmosphere. Though, the Haber–Bosch process extensively enhanced the production of crops, excess use of chemical fertilizers and other human-made sources of fixed nitrogen exceeded natural assistances, leading to unparalleled environmental degradation (Stein and Klotz, 2016) (Table 10.1).

TABLE 10.1 Some Significant Processes in the Nitrogen Cycle Executed by Bacteria.

Process	Microbes involved	Explanation
Biological nitrogen fixation	*Bacillus, Azotobacter, Clostridium,* and *Klebsiella*	In this process, with the help of microbes, the atmospheric molecular nitrogen is converted into nitrogenous compounds.
Ammonification	*Clostridium* sp., *Pseudomonas, Proteus, Bacillus,* and soil actinomycetes	In this process, the proteins and nucleic acid of dead remains of plants and animals and also from excretory products of animals get degraded with the help of microorganisms
Nitrification	Bacteria: *Nitrosomonas, Nitrosococcus, Nitrosolobus, Nitrobacter,* etc.	Ammonia gets oxidized to nitrite and the nitrite formed is further oxidized to nitrate
	Fungus: *Cephalosporium, Aspergillus,* and *Penicillium*	
Denitrification	*Thiobacillus denitrificans, Micrococcus denitrificans, Serratia, Pseudomonas, Bacillus,* and *Paracoccus.*	Nitrates get reduced to nitrogen gas or nitrous oxide with the help of microbes.

10.2.4.2 THE CARBON CYCLE

In biological systems of all living organisms, carbon is considered one of the most crucial elements. The carbon dioxide present in the atmosphere or dissolved in water is the organic carbon compounds in nature. There are two significant steps involved in the carbon cycling process:

1. The transformation of the oxidized form of carbon into a reduced organic way with the help of photosynthetic organisms.
2. Reinstating of the original oxidized form through mineralization of the organic structure with the use of microorganisms.

The transformation of the oxidized form of carbon into a reduced organic way with the help of photosynthetic organisms.

Photosynthetic algae and higher plants reduce CO_2 into organic carbon compounds. The carbon fixation in oceans is done by phytoplankton, which is free-floating, microscopic algae. Annually, over 1.6×10^{10} tons of carbon is fixed by photosynthetic terrestrial plants (Table 10.2).

Two reactions are involved in this step:

1. $CO_2 + 2H_2 \rightarrow (CH_2O)_x + H_2O$
2. $CH_3COCOOH + CO_2 \rightarrow HOOCCH_2.COCOOH$
   (Pyruvic acid) (Oxaloacetic acid)

TABLE 10.2 Description of Microbes Involved in the Decomposition Process.

Type of decomposition	Description	Microbes involved
Cellulose decomposition	The cellulose decomposition involves the following reactions: Cellulose→ cellobiose (Enzyme-Cellulase) Cellobiose→ Glucose (Enzyme-β-glucosidase) Glucose→ CO_2+H_2O (Enzyme system of microbes)	Fungi: *Trichoderma, Aspergillus, Penicillium, Fusarium, Chaetomium, Verticillium, Rhizoctonia, Myrothecium, Pleurotus, Fomes,* etc. Bacteria: *Clostridium, Cellulomonas, Streptomyces, Cytophaga, Bacillus, Pseudomonas, Nocardia,* etc.
Hemicellulose decomposition	Takes place with the help of enzyme hemicellulase	Fungus: *Chaetomium, Aspergillus, Penicillium, Trichoderma,* etc. Bacteria: *Bacillus, Pseudomonas, Cytophaga, Vibrio, Erwinia, Streptomyces,* etc.
Lignin decomposition	It is highly resistant to microbial degradation, however, some microbes are capable of degrading lignin	Fungus: *Aspergillus, Penicillium, Clavaria, Polyporus* Bacteria: *Streptomyces, Nocardia, Flavobacterium,* etc.

We are reinstating of the original oxidized form through mineralization of the organic structure with the help of microorganisms.

There are three different modes of mineralization of organic matter and the release of CO2 in the atmosphere. They are:

i. The respiration process
ii. A forest fire and burning of fuel
iii. Microbial decomposition of organic matter.

For the decomposition of organic matter, there are mainly three types of decay, and there are various microbes involved in this process, as illustrated in the table below:

10.2.4.3   PHOSPHOROUS CYCLE

Phosphorous is one of the essential elements present in organisms. It occurs in both organic and inorganic form. An abundant amount of phosphorous is present in the environment but unavailable form. However, microbes can help in making the unavailable form to an available one.

The phosphorous cycle has two steps involved:

i.   Mineralization: Conversion of organic phosphorous into insoluble inorganic phosphorous.
ii.   Solubilization: Conversion of insoluble inorganic phosphorous into soluble inorganic phosphorous.

Microorganisms involved in the process:
Fungus: *Aspergillus*, *Penicillium,* and *Fusarium.*
Bacteria: *Bacillus, Pseudomonas, Micrococcus, Flavobacterium, etc.*

10.2.4.4   SULFUR CYCLE

Like nitrogen and carbon, sulfur is also an essential component of all living organisms. Plants absorb the sulfur-containing amino acids directly, but that amount of sulfur is not sufficient to meet the sulfur requirements of plants. Therefore, sulfur passes through and a cycle of transformation conciliated by various microorganisms.

Three steps are involved in this cycle:

1.   Organic compounds get degraded and release hydrogen sulfide. Microbes involved: *Desulfotomaculum*
2.   The released hydrogen sulfide gets oxidized into elemental sulfur. Microbes involved: *Chlorobium*, *Chromatium* (sulfur bacteria) and *Rhodospirillum, Rhodopseudomonas,* and *Rhodomicrobium* (non-sulfur bacteria).

10.3   HOST PLANT: TOMATO

Tomato is an edible berry plant with the scientific name-*Solanum Lycopersicum*. The species of tomato originated from Western South America and Central America (Krist, 2020). According to the Food and Agriculture

Organization Corporate Statistical Database (2017), the estimated total world production of tomatoes was 182,301,395 metric tons. There was an increase of 1.6% from 179,508,401 tons in 2016. China was the largest producer, accounting nearly 33% of global production.

Tomatoes and tomato-based products possess various benefits, including the presence of high amount of lycopene content. It is the most abundant carotenoid in tomatoes. This lycopene is full of antioxidant properties. The substantial antioxidant property of this carotenoid is because of its ability to subdue the singlet oxygen and its capacity to trap peroxy radicals. Lycopene has eleven conjugated double bonds which give it a deep red color and are accountable for its antioxidant activity. This is a chief purposeful feature of lycopene. Studies have shown that, by the dietary intake of tomato or tomato-based products, reduces in the risk of chronic diseases such as cancer and cardiovascular diseases (Joshi et al., 2020).

10.3.1   MICROBIAL COMMUNITIES IN DIFFERENT COMPARTMENTS OF TOMATO PLANT

Plants have advanced with an abundance of microorganisms having crucial roles in plant health, growth, and development. Due to the immense potential of the plant's microbiota and current challenges in crop production, different approaches for microbiome improvement have come into action. Agricultural practices or the plant genotype may have an impact on plant microbiota and their functioning. Therefore, the selection of suitable agrarian methods and plant breeding techniques can lead to enhanced plant–microbiome interactions as they are the approach to upsurge the advantage of plant microbiota (Compant et al., 2019). The perception of the microbiome, whether it is the human gut microbiome or plant microbiome, remains a fascinating area in terms of research and studies. The configuration and utility of the plant microbiome are generally managed by several biotic and abiotic factors *viz.*, host's genotype, the health status of plant hosts, the developmental stage of host plants, and different environmental parameters. Based on the engrossment of these factors in plant–microbe interactions, recognizably other microbial communities are detected in three different sections of the plant, namely rhizosphere, phyllosphere, and endosphere (Rossmann et al., 2017).

10.3.2   MICROBIAL POPULATIONS IN RHIZOSPHERE OF TOMATO AND ITS SIGNIFICANCE

The region of the soil surrounding plant roots and tissues colonized by microorganisms is called the rhizosphere. Plant roots deliver exudates to the soil microorganisms, which aid as substrates for their growth and development. These exudates initiate rhizospheric interactions which include plant–microbe interactions and microbe–microbe interactions and affect the soil microbial community. Rhizosphere interactions comprise of plant roots, soil, and microorganisms which alter the physical and chemical properties of the soil, causing alterations in the whole microbial population of the rhizosphere environment. The configuration of the microbial community in the rhizosphere differs from that in non-rhizosphere soil, since, in the rhizospheric zone, the nutrient availability to the soil results from biological interactions between roots and the microbial community present in the soil. Chinakwe et al. (2019) observed the population of rhizospheric microorganisms of two varieties of tomato, that is, cherry and plum. They were cultivated in a greenhouse for 5 weeks, and standard microbiological procedures were applied to this observation. On a study performed by Kazerooni et al. (2017) on conventional farming (CM) and desert farming (DE) systems in Oman on fungal diversity and its configuration, the fungus in rhizosphere of tomato was assessed using: 454-pyrosequencing and culture-based techniques. It has been observed that both methods gave varied results in terms of fungal diversity. Besides, when compared to direct plating, pyrosequencing recovered more taxonomic group of microbes. Both techniques indicated that fungal diversity in the conventional farm was analogous to that in the desert farm.

Conversely, the configuration of fungal classes and taxa in the two farming systems were found to be dissimilar. Pyrosequencing revealed that Microsporidetes and Dothideomycetes are the most common fungal classes in both the farming systems, whereas the culture-based technique revealed that Eurotiomycetes were found to be the most abundant class in both farming systems. Furthermore, some pathogenic fungi like *Pythium* and *Fusarium* were detected in the rhizosphere of tomato. The common fungal species found in the rhizosphere of tomatoes were saprophytes.

The rhizosphere microbiotas have both beneficial and detrimental effects on the development and vigor of the host plant. The most occurring rhizospheric microbes include mycorrhizal fungi and Rhizobia. These microbes deliver vital nutrients like phosphorus and nitrogen to the host plant and play a crucial role in photosynthesis as well. Additionally, siderophore-producing bacteria promote the iron acquisition, and plant growth-promoting rhizobacteria facilitates plant growth and remarkably suppress plant pathogenic diseases. These microbes are frequently detected in the rhizospheric soil. Certain rhizospheric microorganisms are also known to promote plant adaptation in terms of tolerating abiotic stress conditions such as drought, salinity, and temperature as well (Philippot et al., 2013).

It has been revealed that rhizosphere microbiome which augments plant growth and yields is subjective to few environmental factors such as soil type, climate change, plant cultivar, and anthropogenic events. Anthropogenic events mainly include the use of nitrogen-based chemical fertilizers which are linked with environmental destruction and this adverse approach demands for specific eco-friendly measures to improve nitrogen levels in agroecosystems. This can be attained by harnessing nitrogen-fixing endophytic and free-living rhizobacteria. It has been found that microbes such as Rhizobium, Pseudomonas, Azospirillum, and Bacillus have a positive impact on crops by improving both above- and below-ground biomass and in turn, plays optimistic roles in obtaining sustainability in agroecosystems. Therefore, studies on the rhizosphere microbiome are of immense importance, and detailed analyses can be done with refined culture-free techniques like the next generation sequencing, with the outlook of discovering novel bacteria with plant growth-promoting traits (Igiehon and Babalola, 2018).

Rhizosphere microbiomes have been analyzed in mesocosm conditions. It has been observed that the transfer of rhizosphere microbiota from resistant plants repressed disease symptoms in susceptible plants. Rhizosphere metagenomes from resistant and susceptible plants were comparatively investigated, and it was found that detection and association of a flavor-bacterial genome were much more plentiful in the resistant plant rhizosphere microbiome than the susceptible one. According to this study, flavor-bacterium, named TRM1, can suppress *R. solanacearum*. This reveals the role of native microbiota in defending plants from microbial pathogens and its involvement toward sustainability (Kwak et al., 2018).

10.3.3   *MICROBIAL COMMUNITIES IN PHYLLOSPHERE OF TOMATO*

The second compartment for the residence of plant microbiome is the phyllosphere. It has fewer nutrients and a high species richness index compared to rhizosphere. Phyllosphere is the area that surrounds the leaf of plants. As it is exposed to the outside environment and comes in contact with air and various dust particles, phyllosphere is known to possess diverse microflora. And also, surface appendages, cuticles, and waxes help this microflora to adhere on the leaf surface. The survival, multiplication, and death of these microbes depend on the leaf exudates.

According to a study performed using 16S-rRNA gene pyrosequencing by Romero et al. (2014), on endophytic bacterial communities of tomato leaves, it was found that leaf endophytes primarily include five species, out of which *Proteobacteria* is found in profusion (90%), followed by *Actinobacteria* (1.5%), *Planctomycetes* (1.4%), *Verrucomicrobia* (1.1%), and *Acidobacteria* (0.5%). *Gammaproteobacteria* is found to be the amplest class of *Proteobacteria* (84%), while *Alphaproteobacteria* and *Betaproteobacteria* represented 12 and 4% of this species, respectively. Also, Senthilkumar and Krishnamoorthy (2017) isolated and characterized phyllosphere methylotrophic bacteria from tomato leaves via selective ammonium mineral salt medium.

Other than bacterial communities, phyllosphere is also colonized by fungus-like organisms like *Phytophthora* spp. The fungus is present in the phyllosphere as well, and most of them are pathogenic and causes leaf-associated diseases in tomato. These include *Alternaria* sp., *Colletotrichum* sp., *Pseudocercospora fuligena*, *Cercospora fuligena* causing Cercospora leaf mould, *Stemphylium botryosum* f. sp. *lycopersici*, *Stemphylium lycopersici*, *Stemphylium floridanum* and *Stemphylium solani* causing grey leaf spot, *Oidiopsis sicula* and *Leveillula Taurica* causes powdery mildew and *Fulvia fulva* and *Cladosporium fulvum* causing leaf mould. It also includes a genus of Septoria and Verticillium. The phyllosphere comprises enormously diversified and diverse microbial communities. The interactions between these communities result in enhanced plant health and increased productivity of agriculturally important crops. The phyllosphere microbes significantly vary from each other based on their specific ecological niche, different species of host plant, their physicochemical attributes and environmental conditions like water availability, pH, temperature, nutritional accessibility, and geographical locations (Chaudhary et al., 2017) (Fig. 10.1).

Most of the phyllosphere microbiota in tomato plants play a vital role in the bioremediation of pesticides residues and atmospheric hydrocarbon

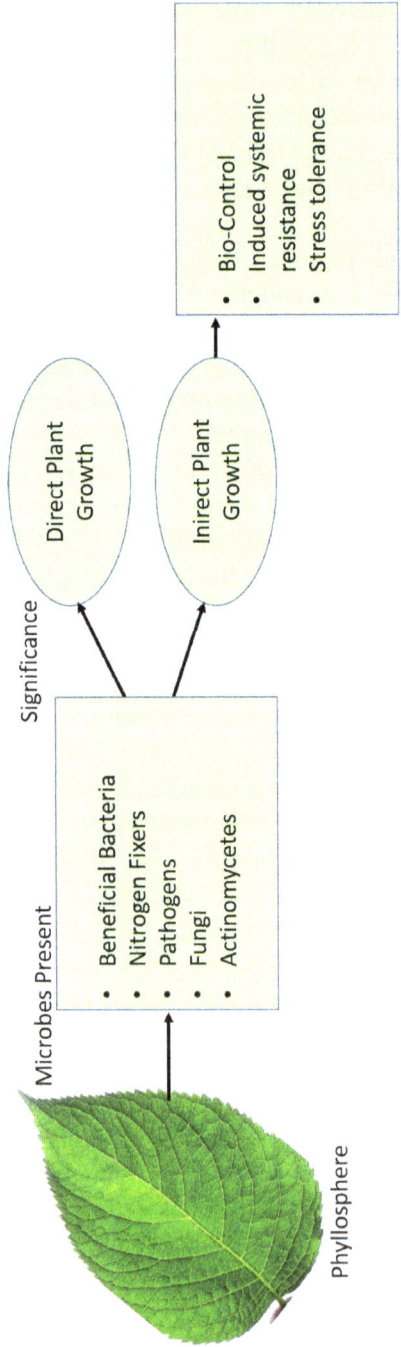

FIGURE 10.1 Phyllosphere of tomato core microbiome (modified sufficiently from Parasuraman et al., 2019).

pollutants. They also help in sustaining the plant's development and health as biofertilizers, phytostimulators, and biopesticides to protect from conquering pathogens (Muller and Ruppel, 2014). The colonization of these microorganisms in phyllosphere strongly depends on the plant species. Secondary metabolites such as 2-phenyl-3-butenyl, alkenyl glucosinolates, and 4-pentenyl have been found to have a positive influence on the growth and survival of phyllosphere bacteria. On the other hand, glucosinolate-2-phenylmethyl harms bacterial diversity (Wagi and Ahmed, 2017). Thus, the performance of phyllosphere microbiota in their interactions with the tomato plant as host enables plant growth promotion, plant protection, and phytoremediation of environmental pollutants, and cultivating stress-tolerance capability in agroecosystems. Leaves of tomato plants are a potential zone accessible for the microbial colonization and growth, on the leaf surface (the phylloplane) or inside the leaf itself as endophytes. Even though the phyllosphere consists of a diversified community of microbes, it can still be a challenging environment for the growth of microbes as it is exposed to the outer environment where there are fluctuations in moisture content, less nutrient availability, and variation in temperature. However, bacteria adapted to phyllosphere may be proficient of altering their environment by expanding the seepage of nutrients from tomato plant and then generating extracellular polysaccharides to survive withering and aid in the interface with the tomato plant (Jackson et al., 2015).

Although phyllosphere microbiota plays a crucial role in plant protection, growth, and development, understanding the mechanisms of interactions between phyllosphere microbiota and their respective plant hosts, that is, tomato, apparently, by recent developments in high-throughput molecular technologies, it has made probable to comprehend the phyllosphere microbiome and its role in accomplishing sustainability in agroecosystems (Rastogi et al., 2013). These molecular approaches not only aid in understanding the dynamics, but also help in recognizing the budding microbes to act as biocontrol agents, and this, in turn, helps in maintaining sustainability. The colonization of microbes in the tomato phyllosphere can have an insightful influence on the health and disease management of its host as well as other crops. This unique aspect advocated a radical improvement in biological sciences aiming plant–microbes interaction studies toward accomplishing a sustainable environment. Along with the discovery of an enormous variation associated with hosts, fruitful research terms the configuration of microbial communities to explicate the central beliefs that direct their assemblage, dynamics, and utilities (Berg et al., 2016).

10.3.4   MICROBIAL COMMUNITIES IN ENDOSPHERE OF TOMATO AND ITS SIGNIFICANCE

Endosphere is the third compartment of plant micro-biome. It involves inner plant tissues, colonized by microorganisms closely interacting with the host plant. This compartment of the plant includes the endophyllo-sphere (internal shoot and leaf tissue), endorhizosphere (inner root tissue), internal plant reproductive tissue, and the inner seed tissue. The endo-sphere allied microbes in tomato play a substantial role in plant growth and development by amending metabolic interactions, improved nutrient uptake and enhance tolerance biotic and abiotic stress tolerance. Besides, the existence of endophytes also includes modifying or prompting the gene expression of plant protection and metabolic pathways (Rossmann et al., 2017).

Most of the bacterial endophytes usually complete their life cycle inside the host plants deprived of being pathogenic to the host plant. Still, their proliferation might be partial by the unique immune system of host plants. Dong et al. (2019) isolated endophytic bacteria from different parts of the tomato plant from both above and below ground parts including roots, stems, leaves, flowers, fruits, tubers, seeds, and ovules. It was found that bacterial genera *Pseudomonas, Acinetobacter,* and *Enterobacter* as endophytes were plentiful in the roots, stems, and leaves of the tomato plant whereas, in fruits, the bacterial endophytes were varied in diverse compartments. *Acinetobacter* was found in abundance in the placenta, *Enterobacter* in the pericarp and seeds, and *Weissella* in the jelly.

Another research by Lee et al. (2019) revealed the presence of endo-phytes in a tomato plant and also their relative abundance. According to this research, it was found that, among the bacterial populations, the most dominant phylum was *Proteobacteria*, with a high close group. *Actinobacteria* was found to be the second most abundant species, in the bulk soil, rhizosphere, and endosphere, respectively. Unlike the bacterial communities, the archaeal communities were also found to be present in abundance in the endosphere. The most dominant archaeal phylum was

Thaumarchaeota, followed by *Euryarchaeota*. The fungal communities were also discovered in the endosphere out of which, *Ascomycota* was the most abundant phylum (Lee et al., 2019). Even though the study of rhizospheric microorganisms is extensive and has always been a fascinating area of research, the detailed analysis of endophytes and their close association with plants, have been only recently practiced. It is now clear that the whole plant microbiome has insightful influences on plant health, its growth and development. Like the other two compartments of microbial habitation in tomato plants, endosphere microbiome is also involved in enhancing the plant heath and contributing to sustainability. These roles include nutrient procurement, nutrient cycling, and interrelating with each other in complex ways. The structure and complexity of tomato endosphere microbiome can vary depending on various factors such as environmental factors, plant genotype, and abiotic or biotic stresses. The microbiome has a fundamental role in the survival of plants, and the endosphere microbes can elucidate the phenotypic and genotypic variation (Doty, 2017). The tomato roots are also internally colonized by a varied range of endophytes. The bacterial endophytes can enter inside the root tissues through root cracks or passive mechanisms or emergence points of lateral roots and also by active processes. The establishment and distribution of endophytes within plants depend on various aspects like the dispersal of plant resources and the capability of the endophyte to colonize plants.

Endophytes in tomatoes can be either beneficial or detrimental and can switch between mutualistic and parasitic life approaches. They benefit the plant in its growth and development by producing growth hormones, antibiotics, and toxicants. They also help in the enhancement of the nutrient uptake and aggregate the plant's tolerance to biotic and abiotic stresses. Moreover, the survival of endophytes can include modifying/encouraging the gene expression of the plant's defense mechanisms and metabolic pathways, and, reliant on the type of interaction, associates of the endosphere microbiome can make both local and systemic alterations in the host. It has been also presented that apart from the advantages above of endophytes; they can also combat pests and diseases and act as biocontrol agents. For example, it has been found that a beneficial endophyte *Fusarium solani* strain K, isolated from the roots of tomato has direct and indirect defense mechanisms against spider mites *Tetranychus urticae* which also a pest of tomato (Pappas et al., 2018). This again helps in reduction in the use of chemicals, in turn playing a significant role in maintaining sustainability in agroecosystems.

10.3.5   UTILIZATION OF TOMATO CORE MICROBIOME AS BIOCONTROL

Biocontrol can be defined as the usage of biological agents or living organisms for suppressing and reducing the invasion of pests such as insects, weeds, and pathogenic microorganisms. This is a sustainable approach for enhancement in plant health without causing a detrimental impact on the environment. Extensive research has been done to assess the potential of endophytes as inoculants to promote plant growth, and as biocontrol agents for maintaining sustainability in agroecosystems. Several analyses and studies have been done to demonstrate the role of tomato core microbiome, its antagonistic properties and role in biocontrol and also several plant growth-promoting endophytic bacteria have been recognized as part of the core tomato microbiome. For example, Tian et al. (2017) identified 49 bacterial endophytic strains belonging to the phyla Firmicutes and Proteobacteria. They isolated those strains from tomato roots on selected media and performed a phylogenetic analysis using 16S rRNA gene sequences. It was found that the isolates clustered into groups of the orders Enterobacteriales, Pseudomonadales, Burkholderiales, Rhizobiales, and Xanthomonadales representing the preponderance of the Proteobacteria groups of tomato root endophytes in addition to phylum Firmicutes. In vitro bioassays indicated that 31 of the stated 49 isolated endophytic strains showed antagonistic activity against examined microbial agents such as *Bacillus subtilis, Escherichia coli, Aspergillus niger, and Staphylococcus aureus. Bacillus* and *Pseudomonas* exhibited maximum antimicrobial activity. Apart from these, many strains of endophytes validated the capability of plant growth promotion. Four Strains belonging to the order Pseudomonadales represented IAA producing potential, 18 strains of *Bacillus* and *Rhizobium* species were found to be nitrogen fixers and certain strains of *Bacillus* and *Pseudomonas* substantiated to contribute in nutrient acquisition and biocontrol, respectively.

10.3.6   MANIPULATION OF THE TOMATO CORE MICROBIOME

To meet the demand for food, the land area utilized by agriculture and the productivity of crops must be increased. As the use of chemicals and anthropogenic hazards are growing, the soil fertility is getting reduced. To avoid that, there is a need for adopting an eco-friendly approach. In this situation, deliberate exploitation of the plant microbiome may be an alternate approach to develop sustainability in agroecosystems. This can be performed by isolation and exploitation of beneficial traits of the core microbiome to decrease the dependency on chemical inputs in agroecosystems. There are specific ways for manipulating the core microbiome, such as promoting good management of soil and crop rotations. This will increase the diversity of microorganisms in the soil, endorsing high adaptability to plant pathogens.

An alternate way of manipulating core microbiome is to stimulate certain microorganisms or introduce inoculants. This method targets to establishing a healthy community that competitively omits plant pathogens. Co-inoculation of beneficial stains will enhance microbial root-colonizing capacity and may also encourage the release of antimicrobial compounds that will enhance the potential to suppress soil-borne pathogens. The inoculation of microorganisms can recover plant nutritional status as well. For example, *Rhizobium* sp. can be inoculated for better nitrogen fixation in tomato plants and phosphorus-solubilizing microorganisms can be inoculated either alone or along with rock phosphate to enhance the phosphorous solubilization in tomato plants. It is essential to contemplate that the inoculants should necessarily be free of metabolites that can be hazardous to humans, animals, plants, and other beneficial local microbes and also, they must be able to survive under differential conditions. As the tomato root exudates may have an antagonistic effect on the inoculants making it difficult for the inoculants to stay and has an impact on their structure, it is crucial to manipulate the microbiome and change the eminence of root exudates through plant breeding methods or biotechnological approaches and genetic modification. Enhancing plant–microbiome interactions by microbiome manipulation can improve crop sustainability in tomatoes, dropping the impacts of traditional agricultural practices. Though, several efforts have been made to comprehend the factors regulating microbiome congregation, manipulating the microbiome is still a challenge to be addressed (Rossmann et al., 2017).

10.4 CONCLUSION

By the increasing demand for food supply, and extensive use of chemicals for the fulfillment of this rapid increase in demand, premature deterioration in the quality of soil and produce has been observed. The chemical residue not only causes the soil less fertile by killing the beneficial microbes, but also gets accumulated as residues in produce creating various health problems in human beings. Because of this extensive use of chemicals, there is also a reduction in the population of beneficial predators which help in biocontrol. Tomato being a nutrient-rich and perishable crop with a high risk of pests and diseases and a lot of chemicals are being used to treat them, it is necessary to find an alternate approach for sustainably maintaining its health and productivity. Promising practices for maintaining sustainability should have the potential to exploit the resources and limit the damages caused by chemicals. Native microbiota may also have the possibility of causing disease outbreaks in agroecosystems, but this can be elucidated by manipulating the core microbiome. The core microbiome extracted from these compartments has enormous potential of maintaining sustainability in agroecosystems. Further enhancements, developments, and identification are to be done for exploiting the tomato core microbiome for maintaining sustainability. Optimizing and genetically manipulating comprehensive configurations of both plant genetic varieties and core microbiomes will be the strategy to manage resource-efficient and pathogen-resistant agroecosystems.

KEYWORDS

- core microbiome
- plant microbiota
- sustainable agroecosystem
- biocontrol

REFERENCES

Abdiev, A.; Khaitov, B.; Toderich, K.; Park, K. W. Growth, Nutrient Uptake and Yield Parameters of Chickpea (Cicer arietinum L.) Enhance by Rhizobium and Azotobacter Inoculations in Saline Soil. *J. Plant Nutr.* **2019,** *42* (20), 2703–2714.

Arcand, M. M.; Helgason, B. L.; Lemke, R. L. Microbial Crop Residue Decomposition Dynamics in Organic and Conventionally Managed Soils. *Appl. Soil Ecol.* **2016,** *107*, 347–359.

Berg, G.; Rybakova, D.; Grube, M.; Koberl, M.; Price, A. The Plant Microbiome Explored: Implications for Experimental Botany. *J. Exp. Bot.* 2016, **67** (4), 995–1002.

Chaudhary, D.; Kumar, R.; Sihag, K.; Rashmi, K. A. Phyllospheric Microflora and Its Impact on Plant Growth: A Review. *Agric. Rev.* 2017, *38* (1), 51–59.

Chinakwe, E. C.; Ibekwe, V. I.; Nwogwugwu, U. N.; Onyemekara, N. N.; Ofoegbu, J.; Mike-Anosike, E.; Emeakaraoha, M.; Adeleye, S.; Chinakwe, P. O. Microbial Population Changes in the Rhizosphere of Tomato Solanum Lycopersicum Varieties during Early Growth in Greenhouse. *Malaysian J. Sustain. Agric.* 2019, *3* (1), 23–27.

Compant, S.; Samad, A.; Faist, H.; Sessitsch, A. A Review on the Plant Microbiome: Ecology, Functions and Emerging Trends in Microbial Application. *J. Adv. Res* 2019.

Dai, X.; Wang, H.; Fu, X. Soil Microbial Community Composition and Its Role in Carbon Mineralization in Long-Term Fertilization Paddy Soils. *Sci. Total Environ.* 2017, *580*, 556–563.

Dong, C.-J.; Wang, L-L.; Li, Q.; Shang, Q-M. Bacterial Communities in the Rhizosphere, Phyllosphere and Endosphere of Tomato Plants. *PLoS ONE* 2019, *14* (11), e0223847. https://doi.org/10.1371/journal.pone.0223847

Doty, S. L. Functional Importance of the Plant Endophytic Microbiome: Implications for Agriculture, Forestry, and Bioenergy. In *Functional Importance of the Plant Microbiome*; Doty S., Eds.; Springer: Cham, 2017.

Igiehon, N. O.; Babalola, O. O. Rhizosphere Microbiome Modulators: Contributions of Nitrogen Fixing Bacteria towards Sustainable Agriculture. *Int. J. Environ. Res. Public Health* 2018, *15*, 574.

Joshi, B.; Kar, S. K.; Yadav, P. K.; Yadav, S.; Shrestha, L. Bera, T. K. Therapeutic and Medicinal Uses of Lycopene: A Systematic Review. *Int. J. Res. Med. Sci.* 2020, *8* (3), 1195.

Kazerooni, E. A.; Maharachchikumbura, S. S. N.; Rethinasamy, V.; Al-Mahrouqi, H.; Al-Sadi, A. M. Fungal Diversity in Tomato Rhizosphere Soil under Conventional and Desert Farming Systems. *Front. Microbiol.* 2017, *8*, 1462. doi: 10.3389/fmicb.2017.01462

Khaitov, B.; Kurbonov, A.; Abdiev, A.; Adilov, M. Effect of Chickpea in Association with Rhizobium to Crop Productivity and Soil Fertility. *Eurasian J. Soil Sci.* 2016, *5* (2), 105.

Krist, S. Tomato Seed Oil. In *Vegetable Fats and Oils*; Springer: Cham, 2020; pp 761–765.

Kwak, M.; Kong, H.; Choi, K., et al. Rhizosphere Microbiome Structure Alters to Enable Wilt Resistance in Tomato. *Nat. Biotechnol.* 2018, *36*, 1100–1109. https://doi.org/10.1038/nbt.4232

Lee, S. A.; Kim, Y.; Kim, J. M., et al. A Preliminary Examination of Bacterial, Archaeal, and Fungal Communities Inhabiting Different Rhizocompartments of Tomato Plants under Real-World Environments. *Sci. Rep.* 2019, *9*, 9300. https://doi.org/10.1038/s41598-019-45660-8

Lichtfouse, E.; Navarrete, M.; Debaeke, P.; Souchère, V.; Alberola, C.; Ménassieu, J. Agronomy for Sustainable Agriculture: A Review. *Agron. Sustain. Dev.* 2009, *29*, 1–6.

Lobell, D. B.; Baldos, U. L. C.; Hertel, T. W. Climate Adaptation as Mitigation: The Case of Agricultural Investments. *Environ. Res. Lett.* 2014, *8* , 1–12.

Muller, T.; Ruppel, S. Progress in Cultivation-Independent Phyllosphere microbiology. *FEMS Microbiol. Ecol.* 2014, *87*, 2–17.

Ottesen, A. R.; González Peña, A.; White, J. R. et al. Baseline Survey of the Anatomical Microbial Ecology of an Important Food Plant: *Solanum lycopersicum* (Tomato). *BMC Microbiol.* 2013, *13*, 114. https://doi.org/10.1186/1471-2180-13-114

Pappas, M. L.; Liapoura, M.; Papantoniou, D.; Avramidou, M.; Kavroulakis, N.; Weinhold, A.; Papadopoulou, K. K. The Beneficial Endophytic Fungus Fusariumsolani Strain K

Alters Tomato Responses against Spider Mites to the Benefit of the Plant. *Front. Plant Sci.* **2018**, *9*, 1603.

Philippot, L.; Raiijmakers, J. M.; Lemanceau, P.; Puttem, W.H. Going Back to the Roots: The Microbial Ecology of the Rhizosphere. *Nat. Rev. Microbiol.* 2013, *11*, 789–799.

Rao, A. S.; Jha, P.; Meena, B. P.; Biswas, A. K.; Lakaria, B. L.; Patra, A. K. Nitrogen Processes in Agroecosystems of India. In *The Indian Nitrogen Assessment* . Elsevier, 2017; pp 59–76.

Rastogi, G.; Coaker, G. L.; Leveau, J. H. J. New Insights into the Structure and Function of Phyllosphere Microbiota through High-throughput Molecular Approaches. *FEMS Microbiol. Lett.*2013, *348*, 1–10.

Romero, F. M.; Marina, M.; Pieckenstain, F. L. The Communities of Tomato (*Solanum lycopersicum* L.) Leaf Endophytic Bacteria, Analyzed by 16S-ribosomal RNA Gene Pyrosequencing. *FEMS Microbiol. Lett. 2014 Feb*, *351* (2), 187–194. https://doi.org/10.1111/1574-6968.12377

Rossmann, M.;, Sarango-Flores, S. W.; Chiaramonte, J. B.; Kmit, M. C. P.; Mendes R. Plant Microbiome: Composition and Functions in Plant Compartments. In *The Brazilian Microbiome*; Pylro, V.; Roesch, L., Eds.; SpringerL Cham, 2017.

Saikia, S. P.; Bora, D.; Goswami, A.; Mudoi, K. D.; Gogoi, A. A Review on the Role of Azospirillum in the Yield Improvement of Non Leguminous Crops. *Afr. J. Microbiol. Res.* 2013, *6*, 1085–1102.

Senthilkumar, M.; Krishnamoorthy, R. Isolation and Characterization of Tomato Leaf Phyllosphere Methylobacterium and Their Effect on Plant Growth. *Int. J. Curr. Microbiol. App. Sci* 2017, *6* (11), 2121–2136.

Srivastavaa P.; Singh, R.; Tripathi, Sc.; RaghubanshiA, S. An Urgent Need for Sustainable Thinking in Agriculture–An Indian Scenario. *Ecol. Indicat.* **2016** Aug; *67*, 611–622.

Stein, L. Y.; Klotz, M. G. The Nitrogen Cycle. *Curr. Biol.* **2016**, *26* (3), R94–R98. doi:10.1016/j.cub.2015.12.021

Tian, B.; Zhang, C.; Ye, Y.; Wen, J.; Wu, Y.; Wang, H.; Lei, S. Beneficial Traits of Bacterial Endophytes Belonging to the Core Communities of the Tomato Root Microbiome. *Agric. Ecosyst. Environ.* **2017**, *247*, 149–156.

Wagi, S.; Ahmed, A. Phyllospheric Plant Growth Promoting Bacteria. *J. Bacteriol. Mycol.* 2017, *5* (1), 00119.

Yadav, S. K.; Soni, R.; Rajput, A. S. Role of Microbes in Organic Farming for Sustainable Agro-Ecosystem. In *Microorganisms for Green Revolution. Microorganisms for Sustainability*; Panpatte, D.; Jhala, Y.; Shelat, H.; Vyas, R., Eds., Vol 7; Springer: Singapore, 2018.

Index

For Product Safety Concerns and Information please contact our EU
representative GPSR@taylorandfrancis.com
Taylor & Francis Verlag GmbH, Kaufingerstraße 24, 80331 München, Germany